Impact of Functional Ingredients on Technological, Sensory and Health Properties of Bakery Products

Impact of Functional Ingredients on Technological, Sensory and Health Properties of Bakery Products

Guest Editors

Barbara Simonato
Roberta Tolve

MDPI

Basel • Beijing • Wuhan • Barcelona • Belgrade • Novi Sad • Cluj • Manchester

Guest Editors

Barbara Simonato
Department of Biotechnology
University of Verona
Verona
Italy

Roberta Tolve
Department of Biotechnology
University of Verona
Verona
Italy

Editorial Office
MDPI AG
Grosspeteranlage 5
4052 Basel, Switzerland

This is a reprint of the Special Issue, published open access by the journal *Foods* (ISSN 2304-8158), freely accessible at: https://www.mdpi.com/journal/foods/special_issues/JV4SL35BG4.

For citation purposes, cite each article independently as indicated on the article page online and as indicated below:

Lastname, A.A.; Lastname, B.B. Article Title. *Journal Name* **Year**, *Volume Number*, Page Range.

ISBN 978-3-7258-6852-0 (Hbk)
ISBN 978-3-7258-6853-7 (PDF)
https://doi.org/10.3390/books978-3-7258-6853-7

Contents

About the Editors

Barbara Simonato

Barbara Simonato is an associate professor of Food Science and Technology at the Department of Biotechnology, University of Verona, Italy. Her research focuses mainly on the recovery and valorization of agro-industrial byproducts for the development of functional food ingredients and products, particularly bakery goods and pasta. Her main research interests include the optimization of food formulations from a nutritional and technological perspective, the study of bioactive compounds and their stability during processing and digestion, the evaluation of starch and protein digestibility, and the assessment of predicted glycemic index. She also has extensive expertise in food processing technologies, analytical characterization of food matrices, and the evaluation of nutritional, technological, and sensory properties of fortified foods.

Roberta Tolve

Roberta Tolve is a temporary assistant professor of Food Science and Technology at the Department of Biotechnology, University of Verona, Italy. Her research focuses on food technologies aimed at the development of innovative and functional food products with potential health benefits. Initially centered on microencapsulation techniques for the preservation of bioactive compounds, her research has expanded toward sustainable approaches based on circular economy principles, including the recovery and valorization of food industry byproducts. Her main interests include green extraction processes, microencapsulation of nutraceutical compounds, and the evaluation of technological, nutritional, and sensory properties of fortified foods. She has expertise in food processing, sensory analysis, functional food design, and the application of sustainable technologies in the food sector.

Preface

Improving human well-being, promoting sustainable lifestyles, and enhancing waste management are key goals of the United Nations' 2030 Agenda for Sustainable Development. Within this framework, the valorization of agro-industrial byproducts has gained increasing attention, as they represent a source of nutrients and bioactive compounds that can be converted into functional ingredients for the development of new food formulations. The valorization of agro-industrial byproducts also represents a valuable strategy to manage environmental challenges while creating economic value.

At the same time, consumer demand has driven the development of functional foods that offer health benefits beyond basic nutrition. Functional ingredients from agro-industrial byproducts, such as dietary fiber, proteins, vitamins, minerals, and antioxidant compounds, positively influence physiological functions and support disease prevention and can be easily added to staple foods, such as bakery products, which are ideal vehicles for their incorporation.

This Reprint presents innovative approaches to recovering and reusing agro-industrial byproducts, highlighting their transformation into value-added ingredients for bakery product formulations. The studies demonstrate how the functional ingredients can improve food's nutritional quality, technological performance, and sensory properties while reducing the environmental impact of food production.

It aims to inspire researchers, food scientists, and industry stakeholders by showcasing the potential of agro-industrial byproducts to drive sustainable food innovations aligned with environmental and economic goals.

Barbara Simonato and Roberta Tolve
Guest Editors

foods MDPI

Editorial

Impact of Functional Ingredients on the Technological, Sensory, and Health Properties of Bakery Products

Roberta Tolve and Barbara Simonato *

Department of Biotechnology, University of Verona, Strada Le Grazie 15, 37134 Verona, Italy; roberta.tolve@univr.it
* Correspondence: barbara.simonato@univr.it

Improving well-being, allowing for sustainable lifestyles, and enhancing waste control are aims of the United Nations in the 2030 Agenda for Sustainable Development [1]. These objectives have driven researchers to exploit agro-industrial byproducts as a resource for creating innovative ingredients and fortified foods with enhanced nutritional properties. While agro-food industry waste disposal is a critical environmental issue, recent attainments suggest that the recovery of such byproducts has remarkable economic advantages [2,3].

The emerging popularity of healthy and sustainable diets and the growing consumer demand for functional foods represent recent challenges for investigators and producers. Functional ingredients are represented by different compounds with health-promoting properties that benefit human health [4].

This Special Issue focuses on the various strategies proposed by different authors to recover and re-utilize agro-industrial byproducts that are rich in healthy molecules, such as vitamins, minerals, proteins, and other bioactive compounds involved in modulating humans' biological functions. These are used to formulate functional bakery goods, creating staple foods that are consumed worldwide.

In the framework of the circular economy and the interest in the recovery and reutilization of agri-food industry byproducts, Pasqualoni et al. developed a low-glycemic bread fortified with red chicory products. There is growing interest in bread formulations with a lower glycemic index (GI). From this perspective, the authors replaced wheat with chickpea flour, various types of resistant starch, and powder from red chicory byproducts (RCP), valuable polyphenols, and dietary fiber sources. Combining RCP with chemically modified tapioca (RS IV) and retrograded starch from tapioca (RS III) allows for the production of bread with a low predicted glycemic index of 52 and 55, respectively, where 55 is the upper value for classifying food as low-pGI [5]. From a technological point of view, the experimental bread had a greater baking loss and increased hardness and chewiness, with decreased cohesiveness, while the volume and specific volume decreased. Incorporating RCP, chickpea flour, and resistant starch from tapioca (RS III and IV) into bread dough is an encouraging approach for developing high-fiber, low-glycemic-index bread, making it an attractive option for health-conscious consumers.

Prasad et al. investigated the inclusion of spinach as agronomic waste to enhance bread's nutritional value. It has been shown that incorporating spinach produced bread of analogous height but with an altered color and texture compared to the control. As the spinach content increased, total carbohydrates (including starch) decreased, while protein, fiber, and essential molecules such as some vitamins and minerals increased. Additionally, the inclusion of spinach increased bread porosity, while an analysis of the texture showed decreased hardness, chewiness, gumminess, and firmness, and improved cohesiveness. The sensory analysis showed that the taste and odor were less preferred, and general liking reduced with increasing spinach levels, while texture was shown to be acceptable. Nevertheless, the 20% spinach fortification was accepted similarly to the control. Also, in this case, adding a vegetable byproduct to a staple food, such as bread, could be a favorable approach to increasing daily vegetable intake by humans.

Koriyama et al. studied the effects of fortification with the nutrient-rich Moringa oleifera leaf powder (MLP) and roasted MLP (RMLP) on bread characteristics at variable temperatures and times. The authors reported an increase in bread volume with the addition of RMLP treated at 130 °C for ≥20 min. Furthermore, RMLP allowed for regular dough fermentation, while MLP interfered with carbon dioxide production. Therefore, bread fortified with MLP bread generally shows low fermentation and crumb expansion, but these effects are decreased by heat treatment. These results emphasize the effect of heat treatment in diminishing the influences of MLP on bread fermentation and swelling.

Guan et al. evaluated the effect of tannic acid (TA) as a dough additive on enhancing bread quality. TA supplementation at 0.1% and 0.3% concentrations significantly improved bread's antioxidant activity and enhanced its texture by decreasing crumb hardness, gumminess, and chewiness, and increasing the specific volume. Moreover, adding TA did not modify the crust color but increased the bread's porosity. Additionally, the study demonstrated no negative impact on sensory properties, suggesting TA's commercial viability as a dough improver.

As reported by Cerdá-Bernad et al., saffron (*Crocus sativus* L.) has many byproducts in the production of the most expensive spice in the world. To obtain 1 kg of the spice, about 350 kg of tepals are discarded. Responding to this challenge, some authors aimed to study the nutritional, physicochemical, and sensory properties of wheat and spelt bread supplemented with saffron floral byproducts in a 0, 2.5%, 5%, and 10% (*w*/*w*) ratio. The antioxidant compounds during *in vitro* digestion were also evaluated. The results demonstrated improved dietary fiber and mineral content in the 10% saffron floral byproduct- fortified bread.

Moreover, the increments in phenolic content and antioxidant ability were stable throughout the *in vitro* digestion processes. Finally, the addition of saffron flowers altered the organoleptic properties of bread. Thus, saffron floral byproducts represent sustainable ingredients that can be used to design new functional foods.

Ozón et al., focusing on sustainability, developed another approach, fortifying bread with chia seed byproducts: bioactive peptides from industrial oilseeds. Wheat bread was supplemented with chia expeller hydrolysate flour. ThisClick or tap here to enter text. fortified bread achieved good overall consumer liking according to an affective test. The peptides obtained from the protein hydrolysis of the chia expeller demonstrated technological properties and chemical compositions that are suitable for developing bakery products.

This method provides a novel, eco-friendly option for recovering nutrients from agro-industrial byproducts, obtaining bread with antioxidant properties due to the functional peptides derived from the chia expeller.

Bianchi et al. developed an alternative approach, reducing muffin fat content by using wine lees (WL), a cheap byproduct from wine production, as a fat substitute. The results showed that by substituting sunflower oil with WL, the batter consistency of the fat-replaced muffins was lower than the control, while the hardness and chewiness were higher. Muffins with 25% and 50% fat substitutions had a better volume, increased fiber content, and decreased calorie content.

The food industry has recently explored the use of other types of fats to substitute traditional solid ones [6]. Oleogels containing hydroxypropyl methylcellulose (HPMC), xanthan gum (XG), and olive oil replaced shortening in white pan bread preparation to reduce the saturated fat and trans-fat. The study conducted by Kim et al. displayed that a 50% shortening replacement had no impact on pan bread volume compared to the control and reduced bread retrogradation.

This provides a practical solution to decrease harmful fats in bread through using oleogels as healthier alternative fats.

The rise in gluten-related disorders has driven the demand for gluten-free (GF) products by consumers. Koh et al. evaluated the effect of adding different kinds of spent tea leaf powder (STLP) from green, oolong, and black teas at 8% *w*/*w* in GF shortbread cookies. The addition of STLP to cookies increased moisture, total dietary fiber (soluble and insoluble), and ash contents, but decreased carbohydrate and energy value compared to control cookies. Adding STLP improved the cookies' antioxidant properties and the samples' shelf-life quality. The moisture content, water activity, texture, and microbiology were retained. Thus, STLP represents a valuable byproduct and could be a promising ingredient in the formulation of gluten-free goods with improved nutritional, technological, and sensory properties.

Baek et al. assessed the use of O-free wheat, a lower-allergy grain in ω-gliadins with a reduced amount of glutenins and gliadins, to formulate cookies.

The authors showed that the experimental cookies were like the ones prepared with flour with low gluten content, which is suitable for preparing these baked goods. This research suggests adopting O-free flour in developing allergy-friendly cookies.

Conclusions

In conclusion, this Special Issue showcases the diverse approaches and innovations in the utilization of functional ingredients from agro-industrial byproducts to improve bakery products' nutritional, sensory, and technological properties. The research highlights not only the potential to enhance health benefits, such as lowering the glycemic index and increasing fiber content and antioxidant capacities, but also the ability to repurpose waste into valuable components of food products in line with sustainability goals. From fortifying bread with resistant starch, spinach, and Moringa to incorporating chia peptides and wine lees in muffins, each case presents a compelling argument for eco-friendly solutions to food production. The studies also underscore the importance of consumer acceptance in developing these functional foods, as sensory attributes play a crucial role in their success. Ultimately, these studies align with global efforts towards developing sustainable food systems, reducing waste, and promoting healthier diets, contributing significantly to the objectives of the United Nations 2030 Agenda for Sustainable Development.

Funding: This research received no external funding.

Acknowledgments: As Guest Editors of the Special Issue "Impact of Functional Ingredients on Technological, Sensory, and Health Properties of Bakery Products", we would like to express our deep appreciation to all authors whose valuable work was published under this issue and thus contributed to the success of the edition.

Conflicts of Interest: The authors declare no conflicts of interest.

List of Contributions:

1. Pasqualoni, I.; Tolve, R.; Simonato, B.; Bianchi, F. The Impact of Selected Ingredients on the Predicted Glycemic Index and Technological Properties of Bread. *Foods* **2024**, *13*, 2488.
2. Prasad, R.V.; Dhital, S.; Williamson, G.; Barber, E. Nutrient Composition, Physical Characteristics and Sensory Quality of Spinach-Enriched Wheat Bread. *Foods* **2024**, *13*, 2401.
3. Koriyama, T.; Saikawa, M.; Kurosu, Y.; Kumagai, M.; Hosoya, T. Effects of Roasting on the Quality of Moringa oleifera Leaf Powder and Loaf Volume of Moringa oleifera-Supplemented Bread. *Foods* **2023**, *12*, 3760.
4. Guan, Y.; Yang, X.; Pan, C.; Kong, J.; Wu, R.; Liu, X.; Wang, Y.; Chen, M.; Li, M.; Wang, Q.; et al. Comprehensive Analyses of Breads Supplemented with Tannic Acids. *Foods* **2023**, *12*, 3756.
5. Cerdá-Bernad, D.; Frutos, M.J. Saffron Floral By-Products as Novel Sustainable Vegan Ingredients for the Functional and Nutritional Improvement of Traditional Wheat and Spelt Breads. *Foods* **2023**, *12*, 2380.
6. Ozón, B.; Cotabarren, J.; Geier, F.R.; Kise, M.P.; García-Pardo, J.; Parisi, M.G.; Obregón, W.D. Development of Fortified Breads Enriched with Plant-Based Bioactive Peptides Derived from the Chia (*Salvia hispanica* L.) Expeller. *Foods* **2023**, *12*, 3382.
7. Bianchi, F.; Cervini, M.; Giuberti, G.; Simonato, B. The Potential of Wine Lees as a Fat Substitute for Muffin Formulations. *Foods* **2023**, *12*, 2584.
8. Kim, S.H.; Jo, Y.J.; Lee, S.H.; Park, S.H. Development of Oleogel-Based Fat Replacer and Its Application in Pan Bread Making. *Foods* **2024**, *13*, 1678.

9. Koh, W.Y.; Lim, X.X.; Tan, T.C.; Mamat, H.; Kobun, R.; Rasti, B. Utilising Spent Tea Leaves Powder as Functional Ingredient to Enhance the Quality of Non-Gluten Shortbread Cookies. *Foods* **2023**, *12*, 1557.
10. Baek, N.; Moon, Y.; Kim, J.; Kweon, M. Effect of Gluten Composition in Low-Allergy O-Free Wheat Flour on Cookie-Making Performance Compared with Flours with Different Gluten Strengths. *Foods* **2023**, *12*, 3843.

References

1. United Nations. Resolution Adopted by the General Assembly on 25 September 2015, 526 A/RES/70/1, Transforming Our World: The 2030 Agenda for Sustainable Development, 527, United Nations. 2015. Available online: https://undocs.org/en/A/RES/70/1 (accessed on 17 September 2024).
2. Soceanu, A.; Dobrinas, S.; Sirbu, A.; Manea, N.; Popescu, V. Economic aspects of waste recovery in the wine industry. A multidisciplinary approach. *Sci. Total Environ.* **2021**, *759*, 143543. [CrossRef] [PubMed]
3. Rohner, F.; Wirth, J.P.; Zeng, W.; Petry, N.; Donkor, W.; Neufeld, L.M.; Mkambula, P.; Groll, S.; Mbuya, M.N.; Friesen, V.M. Global coverage of mandatory large-scale food fortification programs: A systematic review and meta-analysis. *Adv. Nutr.* **2023**, *14*, 1197–1210. [CrossRef] [PubMed]
4. Biasini, B.; Rosi, A.; Giopp, F.; Turgut, R.; Scazzina, F.; Menozzi, D. Understanding, promoting and predicting sustainable diets: A systematic review. *Trends Food Sci. Technol.* **2021**, *111*, 191–207. [CrossRef]
5. Atkinson, F.S.; Brand-Miller, J.C.; Foster-Powell, K.; Buyken, A.E.; Goletzke, J. International Tables of Glycemic Index and Glycemic Load Values 2021: A Systematic Review. *Am. J. Clin. Nutr.* **2021**, *114*, 1625–1632. [CrossRef] [PubMed]
6. Silva, R.C.D.; Ferdaus, M.J.; Foguel, A.; da Silva, T.L.T. Oleogels as a fat substitute in food: A current review. *Gels* **2023**, *9*, 180. [CrossRef] [PubMed]

foods

MDPI

Article

Utilising Spent Tea Leaves Powder as Functional Ingredient to Enhance the Quality of Non-Gluten Shortbread Cookies

Wee Yin Koh [1,*], Xiao Xian Lim [2], Thuan Chew Tan [2,3], Hasmadi Mamat [1], Rovina Kobun [1] and Babak Rasti [4]

1 Functional Foods Research Group, Faculty of Food Science and Nutrition, Universiti Malaysia Sabah, Kota Kinabalu 88400, Malaysia
2 Food Technology Division, School of Industrial Technology, Universiti Sains Malaysia, Minden 11800, Malaysia
3 Renewable Biomass Transformation Cluster, School of Industrial Technology, Universiti Sains Malaysia, Minden 11800, Malaysia
4 Australasian Nanoscience and Nanotechnology Initiative, 8054 Monash University, Clayton, VIC 3168, Australia
* Correspondence: weeyin@ums.edu.my

Abstract: The increasing prevalence of gluten-related disorders has led to higher consumer demand for convenient, gluten-free bakery products with health-promoting properties. In this study, non-gluten shortbread cookies were incorporated with various kinds of spent (green, oolong, and black) tea leaves powder (STLP) at 8% *w*/*w*. Cookies with STLP had significantly higher ($p < 0.05$) moisture (2.18–2.35%), crude fibre (14.5–14.9%), total dietary fibre (22.38–22.59%), insoluble dietary fibre (15.32–15.83%), soluble dietary fibre (7.06–7.66%), and ash (1.9–2.0%) contents, but were significantly lower ($p < 0.05$) in carbohydrate (53.2–53.9%) and energy value (426.4–428.2 kcal) compared to control cookies (1.62%; 1.43%; 6.82%; 4.15%; 2.67%; 7.70%; 62.2%; and 457.8 kcal, respectively). The addition of STLP significantly enhanced ($p < 0.05$) the antioxidant properties of the cookies. Non-gluten shortbread cookies with spent green tea leaves powder (GTC) received the highest ($p < 0.05$) score for all sensory attributes, including overall acceptability. In addition, the shelf-life quality of the formulated cookie samples in terms of the moisture content, water activity, colour, texture, microbiology, and sensory properties was maintained ($p > 0.05$) for at least 22 days at 25 °C. STLP, which would have been previously thrown away, could be utilized as a potential functional ingredient to produce non-gluten shortbread cookies with enhanced nutritional, physicochemical, microbiological, sensory, and antioxidative properties.

Keywords: green tea; oolong tea; black tea; polyphenols; dietary fibre

1. Introduction

Cookies (or biscuits) are small, flat-baked products formulated with three main ingredients: flour, fat, and sugar. While bread is usually consumed as a staple food, cookies are widely consumed as snacks by all age groups all over the globe. Cookies have long been found to have good consumer acceptance owing to their nutrient availability, organoleptic properties, palatability, convenience (ready-to-eat), low cost, and comparatively longer shelf-life [1]. In particular, shortbread cookies are soft-type biscuits characterised by their relatively higher fat content (20–60%), and signature short and crumbly texture.

Traditionally, wheat flour is the main ingredient in cookies. However, the high gluten content in wheat flour could aggravate adverse health effects, including Celiac disease, wheat allergies, and non-Celiac gluten sensitivities, in certain consumers [1]. Today, gluten-related disorders have become a global health problem, whereby the global prevalence is estimated at ~5% [1,2]. Hence, the development of gluten-free cookies is necessary for those with gluten-related disorders. According to the Codex standard for gluten-free foods, a food designated as gluten-free can be made using one or more ingredients that do not contain wheat, rye, barley, and oats [3]. Rice flour is the most widely used flour in developing gluten-free bakery products. Nevertheless, rice flour does not supply an adequate amount of dietary fibre and nutritional value. Hence, numerous attempts have been conducted to produce non-gluten bakery products by utilising agro-industrial waste that is rich in nutrients and bioactive compounds, e.g., coffee extract residues [4], goji berry by-product [5], grape pomace [6], and hog plum bagasse [7].

Tea (*Camellia sinensis* L.) is the second-most drunk beverage in the world, next to drinking water. Depending upon the fermentation degree, tea can be categorised into green (unfermented), oolong (semi-fermented, 30–60 % oxidation), and black (fully fermented, 80–90 % oxidation) [8]. Tea leaves are rich in polyphenols and have been reported to possess health-promoting effects, including antioxidant, antiviral, antiallergic, anti-inflammatory, anticancer, cardiovascular, and neuroprotective properties [9,10]. Studies have shown that the polyphenols found in both black and green tea possess the ability to impede the reproduction of the adenovirus [11] and herpes simplex virus [12]. After one to two weeks of consumption, oolong tea (10 g/1000 mL per day, taken after three daily meals) was found to significantly ($p < 0.05$) reduce allergic reactions in patients with severe atopic dermatitis [13]. In addition, the consumption of green tea extracts (1000 mg, two capsules/day) for 12 weeks significantly reduced the disease activity in patients with systemic lupus erythematosus, an inflammatory autoimmune disease [14]. Green tea extract (EGCG, epigallocatechin-3-O-gallate, 44.9 mg/day for 4 weeks) treatment reduced the proliferation rate of tumour cells [15]. The visual recognition memory of adults (aged 16–34 years) with Down's syndrome was also seen to improve with a 12-month treatment of green tea extract (EGCG, 9 mg/kg/day) and cognitive training [16]. Furthermore, high content of lignin, which functions as dietary fibre, has also been found in the leaves of tea [17].

According to the latest data from Statista, in the year 2020, the global production and consumption of tea reached 7.02 billion metric tons [18] and 6.3 billion kilograms [19], respectively. In response to the increasing tea consumption, the amount of spent tea leaves has also increased and has subsequently caused environmental pollution. Because spent tea leaves are said to contain similar components with amounts similar to new tea leaves, the disposal of spent tea leaves also represents a loss of valuable resources [20]. Studies have reported on the application of spent tea leaves in the area of animal feed [21] and agricultural compost [22]. However, there is a scarcity of research on the usage of spent tea leaves in the development of new food products [17,23]. Thus, this work illustrates a novel approach utilising STLP as an ingredient for non-gluten shortbread cookie preparation.

In the present study, various (green, oolong, and black teas) kinds of STLP were incorporated into non-gluten shortbread cookies with the objective to investigate the effect of STLP incorporation on the proximate composition, antioxidative, microbiological, and sensory characteristics, as well as the shelf-life stability, of non-gluten shortbread cookies.

2. Materials and Methods

2.1. Ingredients

Green, oolong and black tea leaves produced from the same cultivar of tea leaves (*Camellia sinensis* var. *assamica*) were bought from Kong Wooi Fong (KWF) Food Industries Sdn Bhd, Kuala Lumpur, Malaysia. Rice flour, maize flour, sugar, unsalted butter, eggs, and almond powder were bought from a local supermarket in Pulau Pinang, Malaysia.

2.2. Preparation of STLP

Tea leaves (100 g) were steeped in boiling water (1 L, 100 °C) and swirled with a magnetic stirrer on a hot plate (StableTemp, Cole-Parmer, Vernon Hills, IL, USA) at a rate of 5 min for 100× *g*. Subsequently, the tea was filtered through a gauze cloth to separate the spent tea leaves. After being dried in an oven (Memmert, UF-260, Schwabach, Germany) at 40 °C for 48 h, the dried spent tea leaves were ground (Orimas FFC 23, Agrowindo, Malang,

Indonesia), sieved (125 μm, Endecotts Ltd., London, UK), sealed in aluminium-laminated film bags, and stored at 4 °C.

2.3. Preparation of Non-Gluten Shortbread Cookies with STLP

The non-gluten shortbread cookies were produced in accordance with the recipe specified in Table 1. The level of replacement of rice flour with STLP was selected based on a preliminary sensory acceptability test (result not shown) [24]. Among the replacement levels (2, 4, 6, 8, and 10 % *w*/*w*), the highest possible replacement of rice flour with STLP was 8 %.

Table 1. Formulation of non-gluten shortbread cookies with various spent tea leaf powders.

Ingredients (% *w*/*w*)	Cookies			
	CCO	GTC	OTC	BTC
Rice flour	32	24	24	24
Maize flour	15	15	15	15
Spent tea leaves powder	0	8	8	8
Unsalted butter	2.5	2.5	2.5	2.5
Sugar	24	24	24	24
Egg yolk	11.5	11.5	11.5	11.5
Almond powder	15	15	15	15

CCO = control; GTC = non-gluten shortbread cookies with spent green tea leaves powder; OTC = non-gluten shortbread cookies with spent oolong tea leaves powder; BTC = non-gluten shortbread cookies with spent black tea leaves powder.

The rice flour, maize flour, and STLP were mixed in a dough mixer (MK-HKM200, Panasonic, Osaka, Japan). The sugar, unsalted butter, precisely weighed egg yolk, and almond powder were added later into the dough and mixed thoroughly. The dough was then wrapped and covered with a plastic sheet. After being stored for 1 h in the refrigerator (4 °C, MC-4L1005, Midea Co., Ltd., Hefei, Anhui, China), the dough was rolled to 5 mm thickness, cut into squares (10 g, 3.5 × 3.5 cm), baked at 170 °C for 18 min in an oven (MBE-201SE-Z, Murni Bakery Equipment, Selangor, Malaysia), and left to cool at ambient temperature (25 °C) for 1 h. All the cookies (Figure 1) were sealed in airtight polyethylene bags, kept under ambient temperature (25 °C), and shielded against light.

(**a**) CCO (**b**) GTC (**c**) OTC (**d**) BTC

Figure 1. Non-gluten shortbread cookies with various spent tea leaf powders: (**a**) control non-gluten shortbread cookies without spent tea leaves powder (CCO), (**b**) non-gluten shortbread cookies with spent green tea leaves powder (GTC), (**c**) non-gluten shortbread cookies with spent oolong tea leaves powder (OTC), and (**d**) non-gluten shortbread cookies with spent black tea leaves powder (BTC).

2.4. Quality Properties of Non-Gluten Shortbread Cookies with STLP

2.4.1. Proximate Composition

The proximate composition of the non-gluten shortbread cookies was conducted according to the Association of Official Analytical Chemists (AOAC) method [25]. The moisture, crude fat, total ash, crude protein, and crude fibre content of the cookies were determined through the oven drying method (AOAC 934.01; laboratory scale oven (UN-55, Memmert GmbH Co. KG, Schwabach, Germany), 105 °C), Soxhlet extraction (AOAC 930.09; solvent = diethyl ether (Sigma-Aldrich, St. Louis, MO, USA)), incineration method (AOAC 940.26; muffle furnace (Carbolite 1400, Carbolite Ltd., London, UK), 600 °C), Kjeldahl method (AOAC 979.09; protein-nitrogen conversion factor = 6.25), and fritted glass crucible method (AOAC 978.10). The insoluble (IDF) and soluble dietary fibre content (SDF) of the non-gluten shortbread cookies were determined by enzymatic-gravimetric methods (AOAC 991.42 and AOAC 993.19), whereas the total dietary fibre content (TDF) was calculated as the sum of IDF and SDF. The total carbohydrate and energy values were obtained based on percentage differences (Equation (1)) and conversion factors (Equation (2)).

$$\text{Total carbohydrate } (\%) = 100\% - (\text{moisture} + \text{crude fat} + \text{ash} + \text{crude protein} + \text{crude fibre})\% \tag{1}$$

$$\text{Energy value (kcal/100 g)} = [\text{total carbohydrate (g/100 g)} \times 4\ \text{kcal/g}] + [\text{crude protein (g/100 g)} \times 4\ \text{kcal/g}] + [\text{crude fat (g/100 g)} \times 9\ \text{kcal/g}] \tag{2}$$

2.4.2. Determination of Tannin

The tannin content of the non-gluten shortbread cookies was determined according to the AOAC (AOAC 952.03) method [25]. First, cookies were crushed and dispersed in distilled water (1:10, *w*/*v*). After shaking for 30 min, the mixture was centrifuged and filtered to obtain the extract. Subsequently, 2 mL of cookie extract was mixed with 35 mL of distilled water, 1 mL of Folin–Denis reagent (Sigma-Aldrich, St. Louis, MO, USA), and 2.5 mL of sodium carbonate (35% *w*/*v*, Merck, Darmstadt, Germany). After diluting to 50 mL with distilled water, the mixture solution was incubated at room temperature (25 °C) for 90 min. The absorbance was then read at 760 nm using a spectrophotometer (Shimadzu, UV-160A PC, Kyoto, Japan).

2.4.3. Characteristics of Cookie Doughs (Density and pH Value)

The density of the cookie dough was determined in accordance with the water displacement method described in [4]. About 3 g of the dough was weighed into a graduated cylinder partially filled with water. The volume of water displaced was recorded as the volume reading. The dough density was calculated as the ratio of the weight to the volume (g/mL). For the pH determination, 10 g of the dough was homogenised in 40 mL of distilled water. After being subjected to centrifugation (1000× *g*, 15 min), the pH value of the supernatant was measured.

2.4.4. Characteristics of Cookies (Spread Factor, Loss Rate, Hardness, Colour, and Water Activity)

The spread factor (diameter/thickness) of the non-gluten shortbread cookies was measured and calculated according to the American Association of Cereal Chemistry (AACC) (10–50.05) [26]. The loss rate was determined by comparing the weight of the cookies before and after baking. The hardness of the cookies was measured by using a texture analyser (TA.XT2i Texture Analyzer, Stable Micro System, Goldaming, Surrey, UK) with a P5 cylindrical probe under the following operating conditions: compression (return to start), test speed of 1.5 mm/s, strain deformation of 80%, 20 mm limit, and trigger force and speed of 10 gf and 10 mm/min, respectively [4]. The colour values (CIE L*; lightness, CIE a*; redness, and CIE b*; yellowness) of the cookies were measured by using a reflectance spectrophotometer (CM-3500d, Konica Minolta Sensing, Inc., Osaka, Japan). The water activity of the cookies was measured by using a water activity meter (Aqualab, Pullman, WA, USA).

2.5. Antioxidative Properties of Non-Gluten Shortbread Cookies with STLP

The total phenolic (TPC) and flavonoid (TFC) contents, ferric-reducing antioxidant power (FRAP), and 2,2-diphenylpicryl hydrazine (DPPH) scavenging activity of the non-gluten shortbread cookies were determined according to the methods proposed by Rhee et al. [27], the National Food Research Institute [28], Blois [29], and Oyaizu [30], respectively, with slight modifications.

2.5.1. Sample Preparations

The cookie samples were crushed and mixed with distilled water in a ratio of 1:10, agitated (60 min), centrifuged, and filtered to obtain the extracts. The resultant extract was concentrated to 1 mg/mL and kept in the refrigerator at −20 °C until used [4,31].

2.5.2. Total Phenolic (TPC) and Flavonoid (TFC) Contents

To determine the TPC, 0.1 mL of cookie extract was mixed with 10 mL of 2% sodium bicarbonate (Merck, Darmstadt, Germany) and agitated for 2 min. After the addition of 0.1 mL of 50% Folin–Ciocalteu's reagent (Sigma-Aldrich, St. Louis, MO, USA), the mixture was left in the dark for 30 min, and then the absorbance was taken at 750 nm using a spectrophotometer (Shimadzu, UV-160A PC, Kyoto, Japan). For the determination of TFC, 0.1 mL of cookie extract was incubated in a mixture of diethylene glycol (1 mL, Sigma-Aldrich, St. Louis, MO, USA) and 1 N sodium hydroxide (0.1 mL, Emsure, Merck KGaA, Darmstadt, Germany) for 60 min at 37 °C. The absorbance was read at 420 nm.

2.5.3. Antioxidative Activities

The DPPH radical scavenging activity and FRAP assay were used to assess the antioxidant activities of the non-gluten shortbread cookies. To determine the DPPH radical scavenging activity, cookie extract (0.5 mL) was combined with 5 mL of a DPPH methanol solution (0.4 mM, HPLC grade, JT Baker, Phillipsburg, NJ, USA). After being left in the dark for 30 min, the absorbance of the mixture was measured at 517 nm. To perform the FRAP assay, 0.1 mL of cookie extract was dispersed in a mixture of 1% potassium ferricyanide (0.25 mL, Sigma-Aldrich, St. Louis, MO, USA) and 0.2 M phosphate buffer (0.25 mL, pH 6.6, Merck, Darmstadt, Germany), followed by incubation for 20 min at 50 °C. After the addition of 0.25 mL of 10% trichloroacetic acid (Bendosen Laboratory Chemicals, Bendosen, Norway), the mixture was centrifuged at 2500 rpm for 10 min. Subsequently, the supernatant (0.1 mL) was mixed with 0.1 mL of distilled water and 20 μL of 0.1% ferric chloride (Sigma-Aldrich, St. Louis, MO, USA), and the absorbance of the mixture was measured at 700 nm.

2.6. Microbiological Safety Analysis

The microbiological safety analysis of the non-gluten shortbread cookies was conducted prior to the hedonic sensory test. The cookie samples (25 g) were crushed under aseptic conditions and homogenised in sterile saline solution (225 mL, 0.85% *w/v*, HmbG Chemicals, Hamburg, Germany) for 30 min [32]. Subsequently, ten-fold serial dilutions were prepared. The dilutions of cookie samples were then plated on agar plates for enumeration. The total aerobic mesophilic count was estimated on Plate Count Agar (Merck, Frankfurt, Germany) using the pour plate technique (incubation: 30 °C, 72 h) [33]. The yeast and mould count were conducted on the Dichloran-Rose Bengal-Chloramphenicol agar (Merck, Frankfurt, Germany) using the spread plate technique (incubation: 25 °C, yeasts = 72 h, moulds = 120 h) [34,35]. The count of *Staphylococcus* spp. was carried out using the Baird Parker plates supplemented with Rabbit Plasma Fibrinogen (Sigma-Aldrich, St. Louis, MO, USA) with spread plate technique (incubation: 37 °C, 48 h) [36]. The detection of *Salmonella* spp. and *Escherichia coli* was performed on modified semi-solid Rappaport-Vassiliadis agar (Sigma-Aldrich, St. Louis, MO, USA) using the spread plate technique (incubation: 41.5 °C, 48 h) [37] and Tryptone Bile-glucuronide agar (Sigma-Aldrich, St. Louis, MO, USA) using the pour plate technique (incubation: 44 °C, 24 h) [38], respectively. Coliform bacteria were enumerated through pour plating on Violet Red Bile Agar (Sigma-Aldrich, St. Louis, MO, USA) (incubation: 30 °C, 48 h) [39]. For *Bacillus cereus*, the dilution was spread on a plate of Mannitol Egg Yolk Agar supplemented with polymyxin (Sigma-Aldrich, St. Louis, MO, USA) (incubation: 30 °C, 48 h) [40].

2.7. Sensory Analysis

The hedonic sensory test was conducted in a sensory evaluation lab according to the ISO 11136 standard [24]. An untrained group of 50 panellists (25 males, 25 females, between the ages of 20 and 60) was recruited to evaluate the sensory attributes (colour, smell, taste, texture (crumbly), and overall acceptability) of the non-gluten shortbread cookies using a seven-point hedonic scale (from "extremely like" to "extremely dislike"). Natural mineral water (Spritzer, Spritzer Bhd., Taiping, Malaysia) was served to rinse their mouth during the sensory test.

2.8. Storage Study

The storage stability study of the non-gluten shortbread cookies was carried out by sealing them in airtight polyethylene bags, storing them at ambient temperature (25 °C), and analysing them at an interval of 7 days for moisture content, water activity, colour, hardness, microbial load, and sensory attributes up to 22 days [41]. The moisture content of the cookies was determined using the oven drying method (AOAC 934.01) [25]. The water activity and colour of the cookies were measured by using a water activity meter (Aqualab, Pullman, WA, USA) and reflectance spectrophotometer (CM-3500d, Konica Minolta Sensing, Inc., Osaka, Japan), respectively. The hardness value was determined following the method described in Section 2.4.4 [4]. The microbial load (total aerobic mesophilic [33], yeasts and moulds [34,35], *Staphylococcus* spp. [36], *Salmonella* spp. [37], *Escherichia coli* [38], coliforms [39], and *Bacillus cereus* [40]) of the cookies during storage was determined according to methods proposed by the International Standard Organization (ISO). Sensory analysis was carried out according to the method mentioned in Section 2.7 [24].

2.9. Statistical Analysis

The significance of data was evaluated by a one-way analysis of variance and Duncan's multiple range test at $p < 0.05$ using SPSS Statistics version 27 (IBM, Chicago, IL, USA) software. Results were shown as the mean ± SD from three independent parallel experiments. The Pearson test was used to study the correlation between TPC, TFC, and antioxidant activities (coefficients are significant at the significance level of 0.05).

3. Results and Discussion

3.1. Quality Properties of Non-Gluten Shortbread Cookies with STLP

3.1.1. Proximate Composition

The nutritional composition of non-gluten shortbread cookies with and without the incorporation of STLP is presented in Table 2. In general, all the cookies formulated with STLP exhibited a significantly ($p < 0.05$) higher moisture, crude ash, and dietary fibre content, but significantly ($p < 0.05$) lower available carbohydrate content compared to the control cookies (CCO). There was no significant difference in the crude fat and protein content among the cookies with and without STLP.

Moisture content plays a vital role in the production of bakery goods, in which a lower moisture content could provide a crunchy structure and a longer shelf life to the cookies [7]. The moisture content of all the cookies formulated in this study ranged from 1.62 to 2.35%, which is in line with the findings that reported 1–5% moisture in cookies [19]. All the cookies with STLP had a significantly ($p < 0.05$) higher moisture content (2.18–2.35%) than that of the control (1.62%), indicating that the incorporation of STLP had significantly ($p < 0.05$) raised the moisture content of the cookies. The increase in the moisture content of GTC, OTC, and BTC could be due to the higher fibre content of STLP compared to rice flour. The cellulose in tea fibre readily interacts with water molecules through hydrogen bonding owing to their preponderance of hydroxyl functional groups. This allows the cookies to bind and keep more water, thus resulting in a higher moisture content [17]. From a food safety standpoint, the maximum allowed value of moisture for cookies recommended by the Food and Agriculture Organization/World Health Organization (FAO/WHO) is 4.5% [42]. The moisture content of all the cookies formulated in this study was lower than 4.5%, hence, they were within the acceptable moisture level and expected to be higher in microbiological durability and more stable against moisture-dependent biochemical and enzymatic deterioration.

Table 2. Proximate composition, energy value, and tannin content of non-gluten shortbread cookies with various spent tea leaf powders.

Component	Cookies			
	CCO	GTC	OTC	BTC
Moisture (%)	1.62 ± 0.24 [c]	2.18 ± 0.03 [b]	2.35 ± 0.01 [a]	2.19 ± 0.01 [b]
Crude fat (%)	20.17 ± 0.03 [a]	20.60 ± 0.55 [a]	20.80 ± 0.38 [a]	20.39 ± 0.46 [a]
Total ash (%)	1.43 ± 0.04 [c]	2.04 ± 0.04 [a]	1.91 ± 0.06 [b]	1.96 ± 0.04 [b]
Crude protein (%)	6.86 ± 0.10 [a]	6.88 ± 0.04 [a]	6.89 ± 0.10 [a]	6.84 ± 0.05 [a]
Crude fibre (%)	7.70 ± 0.04 [b]	14.47 ± 0.45 [a]	14.89 ± 0.09 [a]	14.75 ± 0.06 [a]
Total dietary fibre (%)	6.82 ± 1.35 [b]	22.38 ± 1.07 [a]	22.59 ± 1.28 [a]	22.48 ± 1.16 [a]
Insoluble dietary fibre (%)	4.15 ± 0.62 [b]	15.32 ± 0.78 [a]	15.44 ± 0.99 [a]	15.83 ± 1.63 [a]
Soluble dietary fibre (%)	2.67 ± 0.79 [b]	7.06 ± 0.38 [a]	7.15 ± 0.35 [a]	7.66 ± 0.50 [a]
Total carbohydrate (%)	62.21 ± 0.68 [a]	53.83 ± 0.80 [b]	53.16 ± 0.48 [b]	53.87 ± 0.48 [b]
Energy (kcal/100 g)	457.84 ± 0.71 [a]	428.21 ± 2.68 [b]	427.36 ± 1.80 [b]	426.37 ± 2.44 [b]
Tannin (mg/100 g)	0.28 ± 0.31 [d]	2.33 ± 0.27 [c]	4.22 ± 0.79 [b]	6.10 ± 0.26 [a]

The mean ± SD ($n = 3$). Different letters indicate significant differences between the values (Duncan's multiple range test, $p < 0.05$, [a–d] cookie samples). CCO = control; GTC = non-gluten shortbread cookies with spent green tea leaves powder; OTC = non-gluten shortbread cookies with spent oolong tea leaves powder; and BTC = non-gluten shortbread cookies with spent black tea leaves powder.

The total ash content is a measure of the total amount of minerals present in a food. The total ash content of all the cookies was in the range of 1.43–2.04%. Significant ($p < 0.05$) increments were observed when STLP was added to the cookie formulation. The increase in the total ash content suggested that a significant amount of minerals was present in the STLP. Tea contains a rich amount of minerals, such as manganese, copper, magnesium, zinc, sodium, calcium, and iron [9].

The results of the crude fibre, total dietary fibre (TDF), insoluble dietary fibre (IDF), and soluble dietary fibre (SDF) analyses of the cookies are shown in Table 2. Significant differences ($p < 0.05$) were observed between the control (CCO; 7.7%) and cookies with STLP in terms of crude fibre content (14.5–14.9%). The cookies formulated with STLP also had a significantly higher TDF (22.38–22.59%), IDF (15.32–15.83%), and SDF (7.06–7.66%) compared to the control cookies (TDF: 6.82%; IDF: 4.15%; and SDF: 2.67%). The increase in fibre content was expected as spent tea leaves contain a high level of cellulose, lignin, hemicellulose, and crude fibre components [17]. According to the European Regulation (EC) No 1924/2006, a food product can be classified as "high in dietary fibre" with at least 6 g/100 g dietary fibre [43]. In the present study, all the cookies formulated can be considered food products with high dietary fibre. A good balance of IDF and SDF (ratio: 1.0 to 2.3 [44]) is essential for providing adequate nutritional effects and fibre functionality in the human diet. The IDF/SDF ratios of the control cookies and cookies with STLP were 1.55 and 2.07–2.17, respectively. Therefore, the dietary fibre proportion of all the cookie samples in the study was considered well-balanced.

Based on the results, the available carbohydrate content (53.2–53.9%) and energy value (426.4–428.2 kcal/100 g) of all the cookies with STLP were significantly ($p < 0.05$) lower than that of control (62.2% and 457.8 kcal/100 g, respectively). These findings suggested that the incorporation of STLP decreased the available carbohydrate content and energy value of cookies. The reduction in the available carbohydrate content and energy value makes STLP-incorporated cookies a potential candidate for health-promoting food when compared to the control cookies [10].

3.1.2. Tannin Content

Tannin is a bioactive component associated with several health benefits but exhibits antinutritional properties when its value exceeds the acceptable safe level (10 mg/100 g) [45]. The tannin content of the non-gluten shortbread cookies with and without the incorporation of STLP is shown in Table 2. All the cookies formulated with STLP had a significantly ($p < 0.05$) higher tannin content than the control cookies (CCO). This might be attributed to the high tannin content of STLP. For instance, the tannin content in STLP is generally 3170–12,600 mg/100 g [46]. The tannin contents in cookies were in the order of BTC (6.10 mg/100 g) > OTC (4.21 mg/100 g) > GTC (2.33 mg/100 g) > CCO (0.28 mg/100 g). The tannin content of spent black tea powder was higher than that of spent oolong and green tea powders owing to the oxidation step involved in their production, which breaks down organic matter to produce tannins [47]. In the present study, the tannin contents of the cookie samples formulated in this study were considerably low. This could be attributed to the thermal degradation of tannin during the baking process [48]. Based on the results, the tannin content of all the cookies was within the acceptable range of its antinutritional effect on human consumption.

3.1.3. Characteristics of Cookie Doughs (Density and pH Value)

The density of cookie dough is often measured to evaluate the quality of the final baked product. In general, dough with a lower density produces harder cookies with greater crispiness whereas dough with a higher density results in cookies with a weaker texture [4]. The dough density of all the cookie samples in this study ranged from 1.05 to 1.27 g/mL (Table 3), with no significant difference ($p > 0.05$) observed among the cookies. The incorporation of STLP neither affected the dough density nor the quality of the cookies.

Table 3. The density and pH of the dough of non-gluten shortbread cookies with various spent tea leaf powders.

Cookies	Density (g/mL)	pH
CCO	1.27 ± 0.05^{a}	7.75 ± 0.03^{a}
GTC	1.08 ± 0.08^{a}	7.15 ± 0.01^{c}
OTC	1.07 ± 0.06^{a}	7.16 ± 0.01^{c}
BTC	1.05 ± 0.09^{a}	7.38 ± 0.02^{b}

The mean ± SD ($n = 3$). Different letters indicate significant differences between the values (Duncan's multiple range test, $p < 0.05$, $^{a-c}$ cookie samples). CCO = control; GTC = non-gluten shortbread cookies with spent green tea leaves powder; OTC = non-gluten shortbread cookies with spent oolong tea leaves powder; and BTC = non-gluten shortbread cookies with spent black tea leaves powder.

The pH value of the cookie doughs varied from 7.15 to 7.75 (Table 3) and decreased ($p < 0.05$) upon the incorporation of STLP. This could be attributed to the slightly acidic nature of STLP [49]. Naturally, tea leaves contain a variety of acidic compounds such as caffeine and tannins.

3.1.4. Characteristics of Cookies (Spread Factor, Loss Rate, Hardness, Colour, and Water Activity)

The effect of the partial replacement of rice flour with STLP on the spread factor, loss rate, and hardness of the cookies are shown in Table 4. The spread factor (width/thickness ratio) is a physical attribute that has long been used to evaluate the quality of cookies. In most cases, cookies with higher spread factors are known to have a better quality and higher consumer acceptability. Control cookies (CCO) were found to exhibit the lowest spread factor (4.58), followed by GTC (4.98), BTC (5.01), and OTC (5.36). The incorporation of STLP increased ($p < 0.05$) the spread factor of cookies, which could be due to the higher water-retaining capacity of the STLP-cookie dough (Table 2: moisture content = cookies with STLP > cookies without STLP). The stronger the ability of the dough to retain water, the lower the viscosity of the cookies. A lower dough viscosity could lead to a faster spreading rate during the baking, thus contributing to a higher spread factor for the cookies [50].

Table 4. Spread factor, loss rate, and hardness of non-gluten shortbread cookies with various spent tea leaf powders.

Cookies	Texture		
	Spread Factor	**Loss Rate (%)**	**Hardness (g)**
CCO	4.58 ± 0.02 [c]	13.76 ± 0.05 [b]	3242.32 ± 384.18 [a]
GTC	4.98 ± 0.01 [b]	12.01 ± 0.04 [c]	3391.29 ± 359.72 [a]
OTC	5.36 ± 0.02 [a]	13.65 ± 0.07 [b]	3398.64 ± 142.54 [a]
BTC	5.01 ± 0.03 [b]	15.08 ± 0.04 [a]	3374.28 ± 531.18 [a]

The mean ± SD ($n = 3$). Different letters indicate significant differences between the values (Duncan's multiple range test, $p < 0.05$, [a–c] cookie samples). CCO = control; GTC = non-gluten shortbread cookies with spent green tea leaves powder; OTC = non-gluten shortbread cookies with spent oolong tea leaves powder; and BTC = non-gluten shortbread cookies with spent black tea leaves powder.

The weight loss of cookies is primarily due to the moisture loss from the dough during the baking process. According to the results, the loss rate of the cookies with STLP differed significantly ($p < 0.05$), with GTC (12.01%) demonstrating the lowest loss rate, followed by OTC (13.65%), and BTC (15.08%), which had the highest loss rate. The partial replacement of rice flour with spent green tea leaves powder significantly ($p < 0.05$) reduced the loss rate of cookies. However, an increase in loss rate was observed with spent black tea leaves powder.

The textural characteristic (hardness) is one of the most important criteria considered by consumers to evaluate the quality of a cookie. The cookie's hardness corresponds to the force required to compress a cookie between the molars in the first bite until the cookie is cracked into two pieces. No significant difference ($p > 0.05$) in the hardness was found between the control cookies (3242.32 g) and cookies with STLP (3374.28–3398.64 g) (Table 4). According to Han and Lee [4], the hardness of cookies is greatly dependent on the protein content and the density of the cookie dough. In this study, the differences in protein content (Table 2) and dough density (Table 3) between the control cookies and cookies with STLP were not significant.

Colour is another important quality factor that affects the consumers' overall acceptability of cookies. The appearance and colour of the cookies are presented in Figure 1 and Table 5, respectively. Compared to the control cookies, cookies with STLP had significantly ($p < 0.05$) lower CIE L*, b*, and a* values, indicating that the addition of STLP imparted darker colour to the cookies. These could be attributed to the original colour of STLP. STLP rich in polyphenols is darker in colour compared to white rice flour. During the production of tea, tea leaves undergo an enzymatic oxidation process catalysed by polyphenol oxidase. The oxidation process converts the colourless catechins in fresh tea leaves into theaflavins and thearubigins with a reddish-brown coloration, which later transforms the tea leaves to a dark brown colour [51]. Our observations are in agreement with Arı Akın et al. [17] who also found an increase in the darkness of bread upon the incorporation of tea fibre into the bread formulation.

Table 5. Colours of non-gluten shortbread cookies with various spent tea leaves powder.

Cookies	Colour Parameters		
	L*	a*	b*
CCO	77.78 ± 0.65 [a]	6.35 ± 0.48 [a]	40.71 ± 0.92 [a]
GTC	60.43 ± 1.12 [b]	2.81 ± 0.15 [b]	27.08 ± 1.36 [b]
OTC	59.03 ± 0.98 [b]	2.65 ± 0.29 [c]	26.98 ± 0.22 [b]
BTC	58.41 ± 1.03 [b]	2.89 ± 0.21 [b]	26.84 ± 0.74 [b]

The mean ± SD ($n = 3$). Different letters indicate significant differences between the values (Duncan's multiple range test, $p < 0.05$, [a–c] cookie samples). CCO = control; GTC = non-gluten shortbread cookies with spent green tea leaves powder; OTC = non-gluten shortbread cookies with spent oolong tea leaves powder; BTC = non-gluten shortbread cookies with spent black tea leaves powder; and L* lightness, a* redness, and b* yellowness.

Water activity is the measurement of the unbound water molecules in food and is one of the most important factors influencing the storage stability of dried foods. The United States Food and Drug Administration has established a 0.60 water activity threshold for preventing or limiting microbial growth in food [52]. Table 6 shows the water activities of the non-gluten shortbread cookies with various spent tea leaf powders. According to the results, cookies with STLP (0.22–0.24) had a significantly ($p < 0.05$) higher water activity compared to the control cookies (0.15). The increase in water activity after the addition of STLP was associated with the high moisture content of the STLP [17]. The water activity of the cookie samples in this study ranged from 0.15 to 0.24, which is significantly less than 0.6, thereby indicating the potential high storage stability of the cookies.

Table 6. The water activity of non-gluten shortbread cookies with various spent tea leaf powders.

Cookies	Water Activity
CCO	0.15 ± 0.01 [b]
GTC	0.23 ± 0.02 [a]
OTC	0.24 ± 0.02 [a]
BTC	0.22 ± 0.02 [a]

The mean ± SD ($n = 3$). Different letters indicate significant differences between the values (Duncan's multiple range test, $p < 0.05$, [a,b] cookie samples). CCO = control; GTC = non-gluten shortbread cookies with spent green tea leaves powder; OTC = non-gluten shortbread cookies with spent oolong tea leaves powder; and BTC = non-gluten shortbread cookies with spent black tea leaves powder.

3.2. *Antioxidative Properties of Non-Gluten Shortbread Cookies with STLP*

3.2.1. Total Phenolic (TPC) and Flavonoid (TFC) Contents

In this study, the TPC and TFC of the cookies with and without STLP were measured by using water extraction and are presented in Figure 2a,b. The results showed that both the TPC and TFC of the control cookies (purely cereal-based) were significantly ($p < 0.05$) lower than those containing STLP. Tea is known to have rich phenolic compounds, especially catechin, tannin, theaflavin, and thearubigin [10]. When comparing the type of STLP used, cookies prepared from spent green tea leaves powder contained the highest TPC. The variation in the TPC and TFC of the cookies could be attributed to the extent of tea fermentation. In contrast to spent black and oolong tea leaf powders, spent green tea leaves powder retained more polyphenolic compounds as the fermentation (oxidation) step is bypassed in green tea production [53,54].

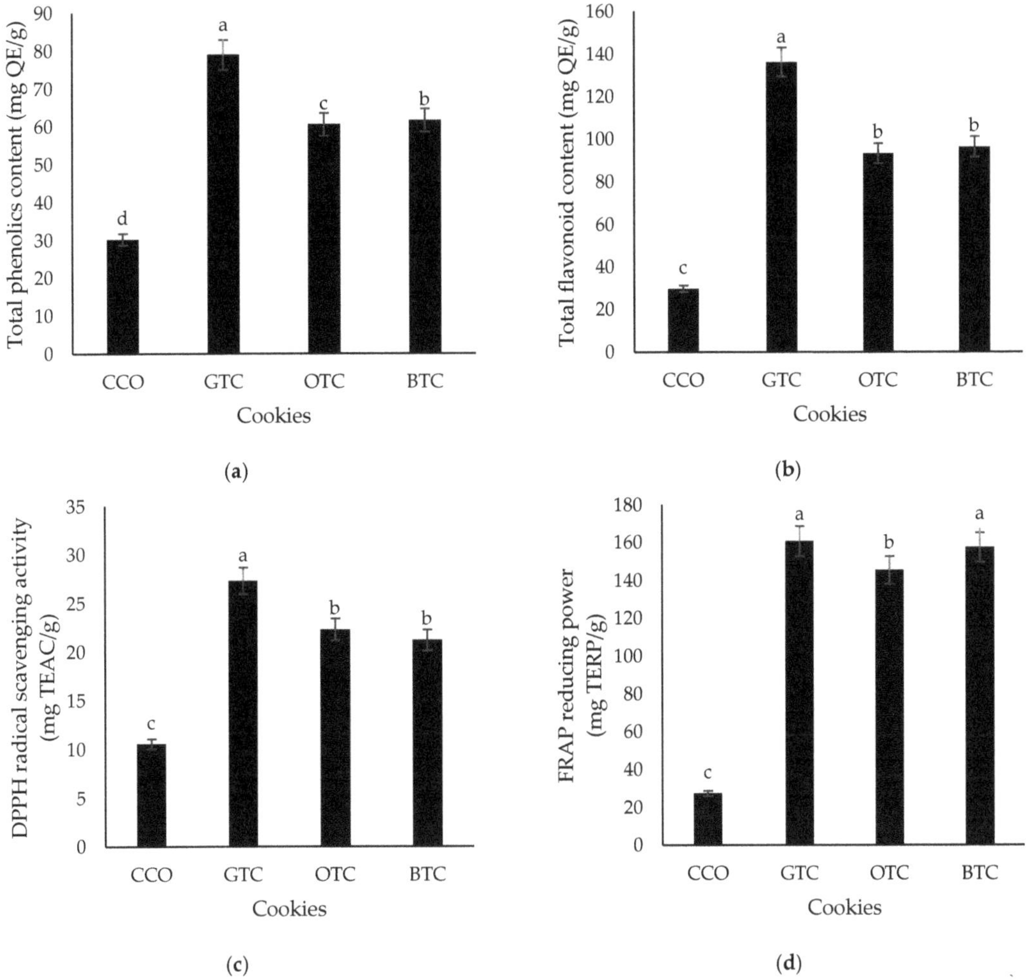

Figure 2. The antioxidative properties (**a**) total phenolics content, (**b**) the total flavonoid content, (**c**) DPPH radical scavenging activity, and (**d**) ferric reducing antioxidant power (FRAP) of non-gluten shortbread cookies with various spent tea leaf powders. The mean ± SD (n = 3). Different letters indicate significant differences between the values (Duncan's multiple range test, $p < 0.05$, $^{a–d}$ cookie samples). CCO = control; GTC = non-gluten shortbread cookie with spent green tea leaves powder; OTC = non-gluten shortbread cookies with spent oolong tea leaves powder; and BTC = non-gluten shortbread cookies with spent black tea leaves powder.

3.2.2. Antioxidative Activities

The DPPH and FRAP values of all the cookies are shown in Figure 2c,d, respectively. The DPPH value of the control cookies (CCO) was significantly ($p < 0.05$) lower compared to the cookies containing STLP, implying that the incorporation of STLP had increased the antioxidant activities of cookies. STLP is rich in polyphenols with a great free radical scavenging ability. This result is in line with the TPC finding. Among the cookies with STLP, GTC had a significantly ($p < 0.05$) higher DPPH value than that of OTC and BTC. This indicated that spent green tea leaves powder was more effective in scavenging radicals and demonstrated a higher antioxidant activity compared to spent oolong and black tea leaf powders. The FRAP values of the cookies followed the same trend as TPC: GTC > BTC > OTC > CCO.

3.2.3. The Relationship between TPC, TFC, and Antioxidative Activities

A high positive correlation was found between TPC and TFC (r = 0.999, $p < 0.01$). While TPC comprises both flavonoid and non-flavonoid phenolic compounds, the strong correlation indicated that flavonoids were the main phenolic compounds in cookies. The TPC and TFC of cookies showed high, positive correlations with DPPH (TPC: r = 0.991, $p < 0.01$; TFC: r = 0.853, $p < 0.01$) and FRAP (TPC: r = 0.986, $p < 0.01$; TFC: r = 0.872, $p < 0.05$). The results suggested that the phenolic compounds in the cookies were the predominant free radical scavenger and ferric reducer. FRAP was highly correlated ($p < 0.01$) with DPPH, with a correlation coefficient of 0.888. The high correlation of FRAP with DPPH suggested that the phenolic compounds present in the cookies could scavenge DPPH free radicals and reduce ferric ions.

3.3. *Microbiological Quality*

Microbiological testing was carried out to ensure the safety of cookie samples for human consumption. According to the maximum acceptable limits established by the FAO/WHO [42], at the time of consumption, the total count of aerobic mesophilic bacteria, yeasts and moulds, *Staphylococcus* spp., coliforms, and *Bacillus cereus* in baked products must not exceed 1×10^4 CFU/g, 1×10^2 CFU/g, 1×10 CFU/g, 1×10 CFU/g, and 1×10 CFU/g, respectively, with the absence of *Salmonella* spp. (0 CFU/25 g) and *Escherichia coli* (0 CFU/10 g). Based on the results obtained (Table 7), no growth of bacteria was observed in any of the cookie samples. The good microbial qualities of the cookies were attributed to the baking process. The temperature attained during baking was high enough to kill all viable microorganisms. Similar to our studies, no microbial growth was observed in the freshly baked edible insect powder-enriched cookies developed by Sriprablom et al. [55].

Table 7. Microbiological quality of non-gluten shortbread cookies with various spent tea leaf powders.

Microbial Load (CFU/g)	Reference Value (Maximum Level) *	Cookies			
		CCO	GTC	OTC	BTC
Total aerobic mesophilic	1×10^4 CFU/g	<10	<10	<10	<10
Yeasts and moulds	1×10^2 CFU/g	<10	<10	<10	<10
Staphylococcus spp.	1×10 CFU/g	<10	<10	<10	<10
Salmonella spp.	Absent in 25 g	Absent	Absent	Absent	Absent
Total coliform	1×10 CFU/g	<10	<10	<10	<10
Escherichia coli	Absent in 10 g	Absent	Absent	Absent	Absent
Bacillus cereus	1×10 CFU/g	<10	<10	<10	<10

The mean ± SD ($n = 3$). CCO = control; GTC = non-gluten shortbread cookies with spent green tea leaves powder; OTC = non-gluten shortbread cookies with spent oolong tea leaves powder; BTC = non-gluten shortbread cookies with spent black tea leaves powder; and < 10 = absence of test microorganisms in 1 g of sample. * = reference value from FAO/WHO [42].

3.4. *Sensory Analysis*

Sensory evaluation is the simplest and most direct method of testing the consumers' acceptability of a food product. The sensory evaluation results of the non-gluten shortbread cookies are presented in a radar plot (Figure 3).

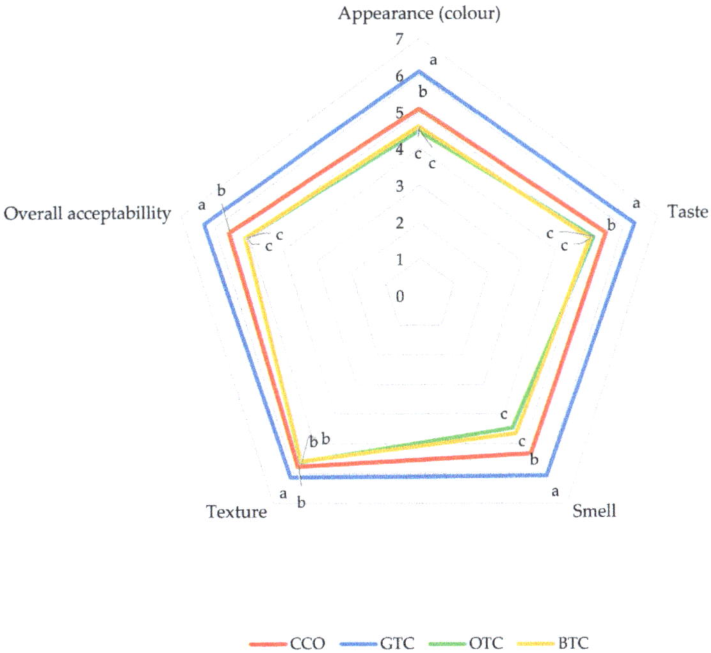

Figure 3. Sensory properties of non-gluten shortbread cookies with various spent tea leaf powders. The mean ± SD ($n = 3$). Different letters indicate significant differences between the values (Duncan's multiple range test, $p < 0.05$, $^{a–c}$ cookie samples). CCO = control; GTC = non-gluten shortbread cookie with spent green tea leaves powder; OTC = non-gluten shortbread cookies with spent oolong tea leaves powder; BTC = non-gluten shortbread cookies with spent black tea leaves powder; 1 = extremely dislike; 2 = moderately dislike; 3 = slightly dislike; 4 = neither like nor dislike; 5 = slightly like; 6 = moderately like; and 7 = extremely like.

Based on the sensory results, GTC (non-gluten shortbread cookies with spent green tea leaves powder) were the best rated ($p < 0.05$) for all of the attributes. Among the four cookie samples, GTC scored the highest for its appearance. Spent green tea leaves powder gave GTC a light greenish colour that was desirable for some panellists. OTC and BTC were less acceptable due to their unappealing darker colour (Figure 1). Panellists preferred the texture, taste, and smell of GTC the most among the cookie samples. The partial replacement of rice flour with spent black and oolong tea leaves powder caused a significant negative effect on the texture, taste, and smell of the cookies. Most tea products possess a bitter aftertaste owing to the presence of catechins [56]. Panellists found no significant astringent aftertaste in GTC.

Overall acceptability is an important parameter in organoleptic estimation as it includes all the sensory attributes. Among all the cookies, panellists preferred GTC the most, followed by CCO, OTC, and BTC. Considering that GTC received the highest overall acceptance score, the partial replacement of rice flour with spent green tea leaves powder in the non-gluten shortbread cookie formulation clearly provided a better eating quality than the spent oolong and black tea leaf powders.

3.5. Storage Study

The moisture content and water activity are critical factors in determining the quality and microbiological stability of dried food products such as cookies [7]. The changes in the moisture content and water activity in the cookie samples during storage are shown in Figures 4 and 5, respectively. The moisture content and water activity of all the cookie samples significantly ($p < 0.05$) increased during the storage. The increase in moisture content in the cookie samples could be attributed to environmental moisture absorption, which corresponds to the hygroscopic nature of cookies. The increased water activity of the cookies was related to the increased moisture. Although the moisture content and water activity of all the cookie samples increased upon storage, the values were still lower than their respective maximum limiting values (moisture content = 4.5% [42]; water activity = 0.60 [52]) at the end of the storage period. The hardness of the cookies decreased significantly ($p < 0.05$) after being stored for 22 days (Figure 6). The decrease in the hardness of cookies during storage could be attributed to the increase in moisture content (Figure 4) and water activity (Figure 5). The ingress moisture softened the cookies [57] and the colour of the cookies was almost stable during 22 days of storage (Figure 7).

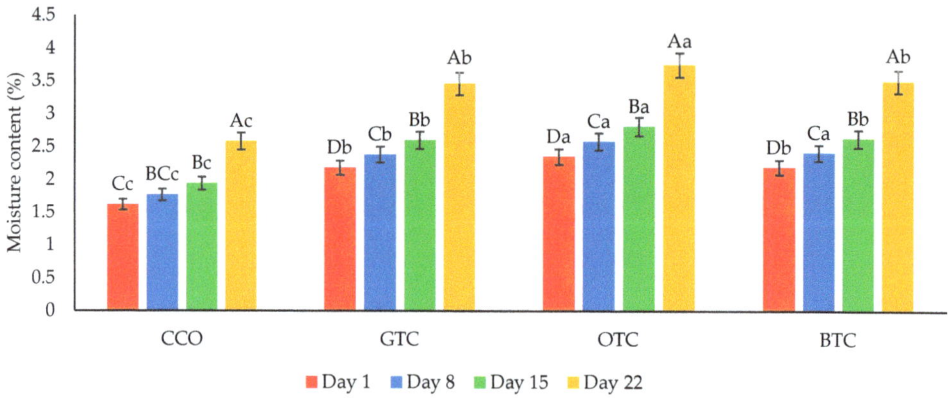

Figure 4. The moisture content of non-gluten shortbread cookies with various spent tea leaf powders during different storage periods (day 1, 8, 15, and 22) at 25 °C. The mean ± SD (n = 3). Different letters indicate significant differences between the values (Duncan's multiple range test, $p < 0.05$, [a–c] cookie samples, [A–D] storage). CCO = control; GTC = non-gluten shortbread cookies with spent green tea leaves powder; OTC = non-gluten shortbread cookies with spent oolong tea leaves powder; and BTC = non-gluten shortbread cookies with spent black tea leaves powder.

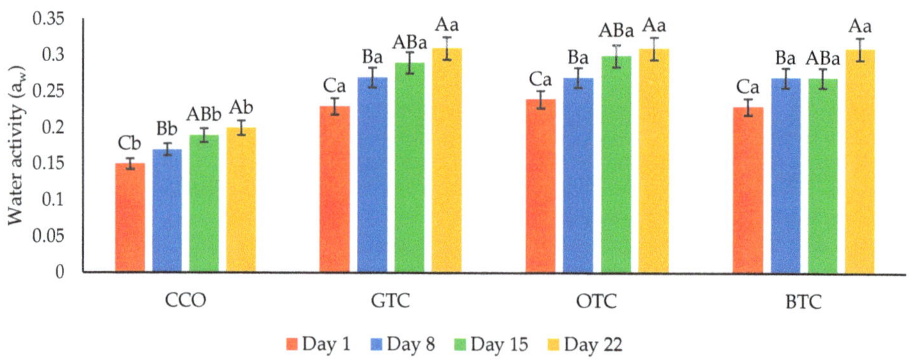

Figure 5. Water activity (a_w) of non-gluten shortbread cookies with various spent tea leaf powders during different storage periods (day 1, 8, 15, and 22) at 25 °C. The mean ± SD (n = 3). Different letters indicate significant differences between the values (Duncan's multiple range test, $p < 0.05$, [a,b] cookie samples, [A–C] storage). CCO = control; GTC = non-gluten shortbread cookies with spent green tea leaves powder; OTC = non-gluten shortbread cookies with spent oolong tea leaves powder; and BTC = non-gluten shortbread cookies with spent black tea leaves powder.

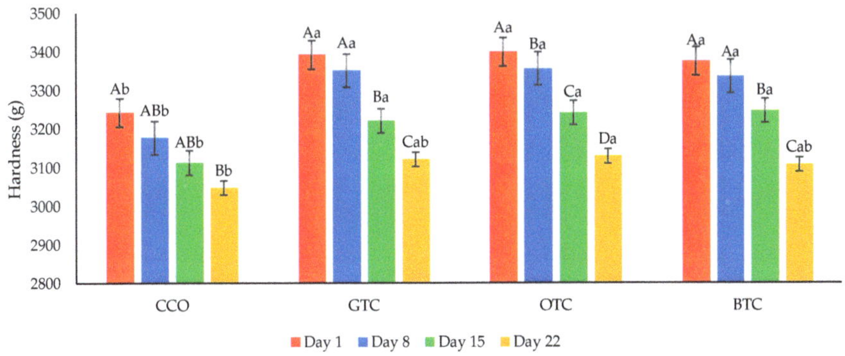

Figure 6. The hardness of non-gluten shortbread cookies with various spent tea leaf powders during different storage periods (day 1, 8, 15, and 22) at 25 °C. The mean ± SD (n = 3). Different letters indicate significant differences between the values (Duncan's multiple range test, $p < 0.05$, ^{a,b} cookie samples, $^{A–D}$ storage). CCO = control; GTC = non-gluten shortbread cookies with spent green tea leaves powder; OTC = non-gluten shortbread cookies with spent oolong tea leaves powder; and BTC = non-gluten shortbread cookies with spent black tea leaves powder.

The growth of microorganisms is a critical factor in determining the shelf life of food products as microbiological deterioration remains the primary cause of food spoilage. The microbiological quality assessment of the non-gluten shortbread cookies carried out during the storage period is presented in Table 8. As per the results acquired, none of the cookie samples showed any visible signs of microbial growth until the 15th day of storage. On the 22nd day of storage, aerobic mesophilic bacteria were detected in all the cookies, with bacterial counts ranging from 2.23×10^2 CFU/g to 2.68×10^2 CFU/g, which is much lower than the permissible limit of 1×10^4 CFU/g [42]. Based on the results, yeasts and moulds, *Staphylococcus* spp., *Salmonella* spp., coliforms, *Escherichia coli*, and *Bacillus cereus* were absent in all the cookie samples throughout the 22-day storage period. Based on the findings of this microbiological study, all of the cookie samples were deemed safe for consumption for a period of 22 days at 25 °C.

Table 8. The microbiological quality of non-gluten shortbread cookies with various spent tea leaf powders during different storage periods (day 1, 8, 15, and 22) at 25 °C.

Microbial Load (CFU/g)	Reference Value (Maximum Level) *	Days	Cookies			
			CCO	GTC	OTC	BTC
Total aerobic mesophilic	1×10^4 CFU/g	1	<10	<10	<10	<10
		8	<10	<10	<10	<10
		15	<10	<10	<10	<10
		22	2.41×10^2	2.68×10^2	2.23×10^2	2.55×10^2
Yeasts and moulds	1×10^2 CFU/g	1	<10	<10	<10	<10
		8	<10	<10	<10	<10
		15	<10	<10	<10	<10
		22	<10	<10	<10	<10
Staphylococcus spp.	1 × 10 CFU/g	1	<10	<10	<10	<10
		8	<10	<10	<10	<10
		15	<10	<10	<10	<10
		22	<10	<10	<10	<10
Salmonella	Absent in 25 g	1	Absent	Absent	Absent	Absent
		8	Absent	Absent	Absent	Absent
		15	Absent	Absent	Absent	Absent
		22	Absent	Absent	Absent	Absent
Total coliform	1 × 10 CFU/g	1	<10	<10	<10	<10
		8	<10	<10	<10	<10
		15	<10	<10	<10	<10
		22	<10	<10	<10	<10

Table 8. *Cont.*

Microbial Load (CFU/g)	Reference Value (Maximum Level) *	Days	Cookies			
			CCO	GTC	OTC	BTC
Escherichia coli	Absent in 10 g	1	Absent	Absent	Absent	Absent
		8	Absent	Absent	Absent	Absent
		15	Absent	Absent	Absent	Absent
		22	Absent	Absent	Absent	Absent
Bacillus cereus	1 × 10 CFU/g	1	<10	<10	<10	<10
		8	<10	<10	<10	<10
		15	<10	<10	<10	<10
		22	<10	<10	<10	<10

The mean ± SD (n = 3). CCO = control; GTC = non-gluten shortbread cookies with spent green tea leaves powder; OTC = non-gluten shortbread cookies with spent oolong tea leaves powder; BTC = non-gluten shortbread cookies with spent black tea leaves powder; and < 10 = absence of test microorganisms in 1 g of sample. * = reference value from FAO/WHO [42].

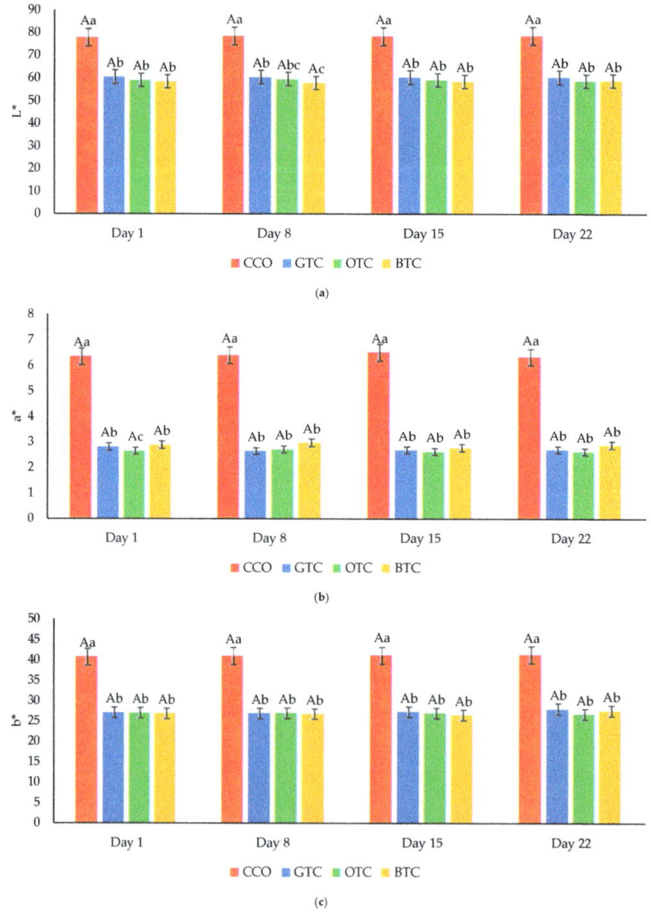

Figure 7. The colour (**a**) lightness, (**b**) redness, and (**c**) yellowness of non-gluten shortbread cookies with various spent tea leaf powders during different storage periods (day 1, 8, 15, and 22) at 25 °C. The mean ± SD (n = 3). Different letters indicate significant differences between the values (Duncan's multiple range test, $p < 0.05$, ^{a,b} cookie samples, ^{A} storage). CCO = control; GTC = non-gluten shortbread cookies with spent green tea leaves powder; OTC = non-gluten shortbread cookies with spent oolong tea leaves powder; BTC = non-gluten shortbread cookies with spent black tea leaves powder; and L* lightness, a* redness, and b* yellowness.

The sensory evaluation scores of the freshly baked and stored cookie samples are shown in Figure 8. The sensory scores of all the cookies were stable throughout the storage period of 22 days. The sensory panelists were unable to discern any differences between the freshly baked cookies and those that were stored for 22 days relative to all sensory attributes ($p > 0.05$). Results indicated that all the cookie samples retained good quality characteristics in terms of organoleptic properties until the end of the storage period.

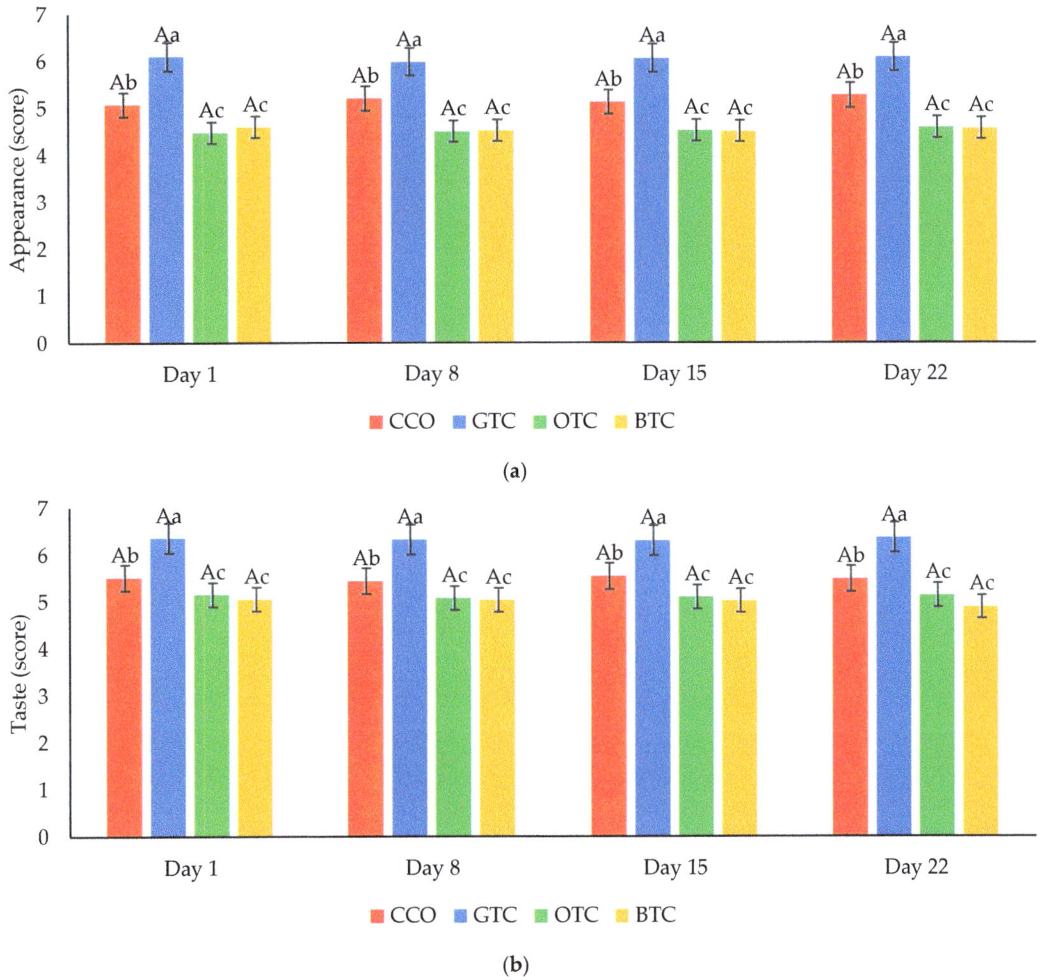

Figure 8. *Cont.*

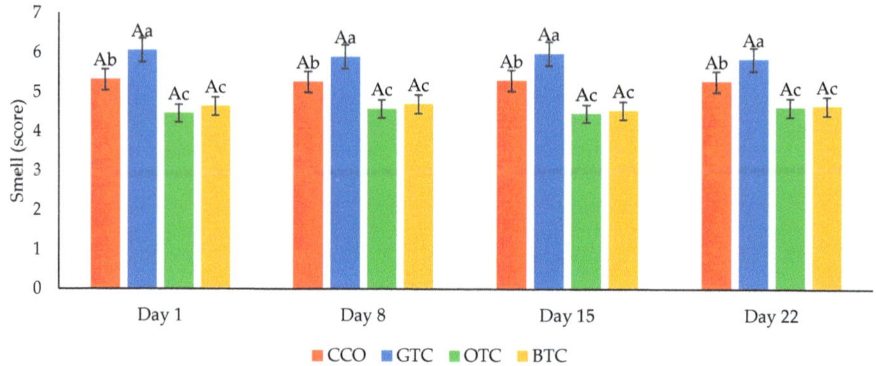

(c)

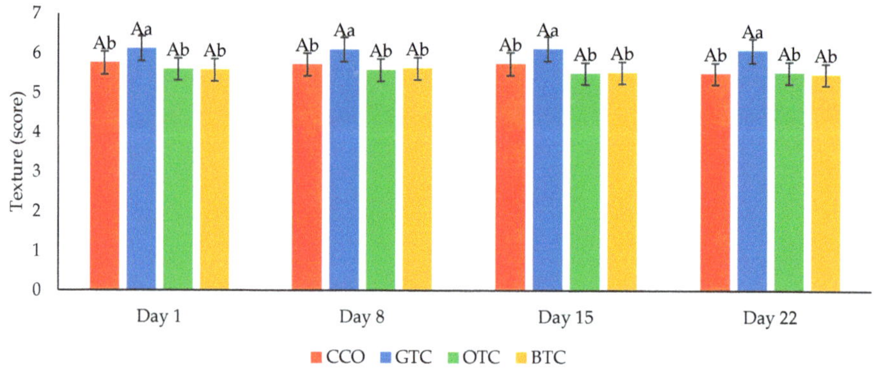

(d)

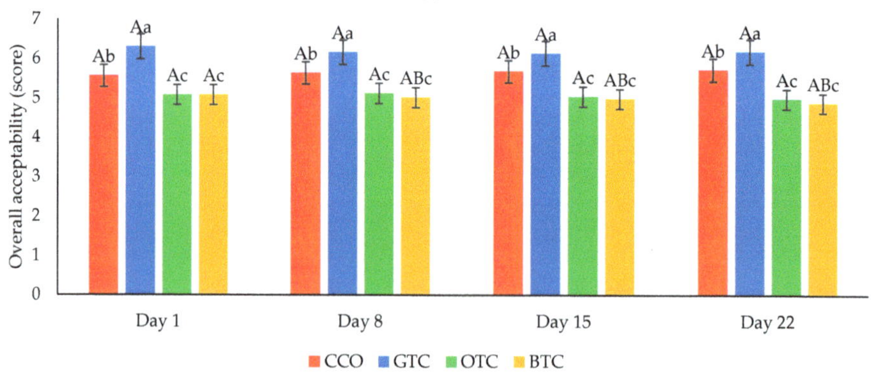

(e)

Figure 8. The sensory properties (**a**) appearance, (**b**) taste, (**c**) smell, (**d**) texture, and (**e**) overall acceptability of non-gluten shortbread cookies with various spent tea leaf powders during different storage periods (day 1, 8, 15, and 22) at 25 °C. The mean ± SD (n = 3). Different letters indicate significant differences between the values (Duncan's multiple range test, $p < 0.05$, $^{a-c}$ cookie samples, ^{A,B} storage). CCO = control; GTC = non-gluten shortbread cookie with spent green tea leaves powder; OTC = non-gluten shortbread cookies with spent oolong tea leaves powder; BTC = non-gluten shortbread cookies with spent black tea leaves powder; 1 = extremely dislike; 2 = moderately dislike; 3 = slightly dislike; 4 = neither like nor dislike; 5 = slightly like; 6 = moderately like; and 7 = extremely like.

4. Conclusions

Non-gluten shortbread cookies with spent green (GTC), oolong (OTC), and black (BTC) tea leaf powders were formulated and their nutritional composition, dough pH and density, physical, antioxidative, microbiological, and sensory characteristics, as well as the shelf-life stability, were investigated and compared with control cookies (CCO). The addition of STLP did not degrade the quality of the cookies and, in turn, enhanced their nutritional and antioxidant properties. Therefore, STLP, which would have been discarded in the past, is a potential, functional ingredient to be used in the production of non-gluten shortbread cookies. STLP could be used as a baking ingredient to develop gluten-free cookies with enhanced functional properties and achieve a sustainable food supply chain.

Author Contributions: Project administration, resources, funding acquisition, supervision, investigation, methodology, and visualisation, W.Y.K.; data curation, writing—original draft, and formal analysis, W.Y.K. and X.X.L.; writing—review and editing, W.Y.K., X.X.L., T.C.T., H.M., R.K. and B.R. All authors have read and agreed to the published version of the manuscript.

Funding: This research was supported by an SPLB grant (SLB2228) provided by Universiti Malaysia. Sabah (UMS): Malaysia.

Institutional Review Board Statement: Not applicable.

Informed Consent Statement: All sensory panellists were informed of the procedure and their consent for participation was taken before the start of the sensory test.

Data Availability Statement: Data are contained within the article.

Conflicts of Interest: The authors declare no conflict of interest.

References

1. Yildiz, E.; Gocmen, D. Use of almond flour and stevia in rice-based gluten-free cookie production. *JFST* **2021**, *58*, 940–951. [CrossRef]
2. Taraghikhah, N.; Ashtari, S.; Asri, N.; Shahbazkhani, B.; Al-Dulaimi, D.; Rostami-Nejad, M.; Rezaei-Tavirani, M.; Razzaghi, M.R.; Zali, M.R. An updated overview of spectrum of gluten-related disorders: Clinical and diagnostic aspects. *BMC Gastroenterol.* **2020**, *20*, 258. [CrossRef]
3. Codex Alimentarius, FAO/WHO. *Standard for Foods for Special Dietary Use for Persons Intolerant to Gluten*; CODEX STAN 118–1979; Codex Alimentarius Commission: Rome, Italy, 2015.
4. Han, I.; Lee, C.-S. Quality properties and bioactivities of American cookies with coffee extract residues. *LWT* **2021**, *151*, 112173. [CrossRef]
5. Bora, P.; Ragaee, S.; Abdel-Aal, E.-S.M. Effect of incorporation of goji berry by-product on biochemical, physical and sensory properties of selected bakery products. *LWT* **2019**, *112*, 108225. [CrossRef]
6. Theagarajan, R.; Malur Narayanaswamy, L.; Dutta, S.; Moses, J.A.; Chinnaswamy, A. Valorisation of grape pomace (cv. *Muscat*) for development of functional cookies. *Int. J. Food Sci. Technol.* **2019**, *54*, 1299–1305. [CrossRef]
7. Oladunjoye, A.O.; Eziama, S.C.; Aderibigbe, O.R. Proximate composition, physical, sensory and microbial properties of wheat-hog plum bagasse composite cookies. *LWT* **2021**, *141*, 111038. [CrossRef]
8. Wang, S.; Zeng, T.; Zhao, S.; Zhu, Y.; Feng, C.; Zhan, J.; Li, S.; Ho, C.-T.; Gosslau, A. Multifunctional health-promoting effects of oolong tea and its products. *Food Sci. Hum. Wellness* **2022**, *11*, 512–523. [CrossRef]
9. Klepacka, J.; Tońska, E.; Rafałowski, R.; Czarnowska-Kujawska, M.; Opara, B. Tea as a source of biologically active compounds in the human diet. *Molecules* **2021**, *26*, 1487. [CrossRef]
10. Zhang, L.; Ho, C.T.; Zhou, J.; Santos, J.S.; Armstrong, L.; Granato, D. Chemistry and biological activities of processed *Camellia sinensis* teas: A comprehensive review. *CRFSFS* **2019**, *18*, 1474–1495. [CrossRef]
11. Karimi, A.; Moradi, M.-T.; Alidadi, S.; Hashemi, L. Anti-adenovirus activity, antioxidant potential, and phenolic content of black tea (*Camellia sinensis* Kuntze) extract. *JCIM* **2016**, *13*, 357–363. [CrossRef] [PubMed]
12. Isaacs, C.E.; Wen, G.Y.; Xu, W.; Jia, J.H.; Rohan, L.; Corbo, C.; Di Maggio, V.; Jenkins Jr, E.C.; Hillier, S. Epigallocatechin gallate inactivates clinical isolates of herpes simplex virus. *Antimicrob. Agents Chemother.* **2008**, *52*, 962–970. [CrossRef] [PubMed]
13. Uehara, M.; Sugiura, H.; Sakurai, K. A trial of oolong tea in the management of recalcitrant atopic dermatitis. *Arch. Dermatol.* **2001**, *137*, 42–43. [CrossRef]
14. Shamekhi, Z.; Amani, R.; Habibagahi, Z.; Namjoyan, F.; Ghadiri, A.; Saki Malehi, A. A randomized, double-blind, placebo-controlled clinical trial examining the effects of green tea extract on systemic lupus erythematosus disease activity and quality of life. *Phytother. Res.* **2017**, *31*, 1063–1071. [CrossRef]

15. Lazzeroni, M.; Guerrieri-Gonzaga, A.; Gandini, S.; Johansson, H.; Serrano, D.; Cazzaniga, M.; Aristarco, V.; Macis, D.; Mora, S.; Caldarella, P. A presurgical study of lecithin formulation of green tea extract in women with early breast cancer. *J. Cancer Prev.* **2017**, *10*, 363–370. [CrossRef]
16. de la Torre, R.; de Sola, S.; Hernandez, G.; Farré, M.; Pujol, J.; Rodriguez, J.; Espadaler, J.M.; Langohr, K.; Cuenca-Royo, A.; Principe, A. Safety and efficacy of cognitive training plus epigallocatechin-3-gallate in young adults with Down's syndrome (TESDAD): A double-blind, randomised, placebo-controlled, phase 2 trial. *Lancet Neurol.* **2016**, *15*, 801–810. [CrossRef]
17. Arı Akın, P.; Tayfun, K.E.; Tamer, U.; Boyacı, İ.H. Use of tea fibers as a source of dietary fiber in wheat flour and bread. *Cereal Chem.* **2021**, *98*, 1049–1058. [CrossRef]
18. Statista. Global Production and Exports of Tea from 2004 to 2020. Available online: https://www.statista.com/statistics/264183/global-production-and-exports-of-tea-since-2004/ (accessed on 13 January 2023).
19. Statista. Volume of Tea Consumption Worldwide from 2012 to 2025. Available online: https://www.statista.com/statistics/940102/global-tea-consumption/ (accessed on 13 January 2023).
20. Guo, S.; Awasthi, M.K.; Wang, Y.; Xu, P. Current understanding in conversion and application of tea waste biomass: A review. *Bioresour. Technol.* **2021**, *338*, 125530. [CrossRef] [PubMed]
21. Kondo, M.; Hidaka, M.; Hirano, Y.; Kita, K.; Jayanegara, A.; Yokota, H.O. Nutrient digestibility, fecal output of fractionated proteins, and ruminal fermentation parameters of goats fed a diet supplemented with spent green tea and black tea leaf silage. *J. Anim. Sci.* **2022**, *93*, e13681. [CrossRef] [PubMed]
22. Gammoudi, N.; Nagaz, K.; Ferchichi, A. Potential use of spent coffee grounds and spent tea leaves extracts in priming treatment to promote in vitro early growth of salt-and drought-stressed seedlings of *Capsicum annuum* L. *Waste Biomass Valorization* **2021**, *12*, 3341–3353. [CrossRef]
23. Nguyen, N.D.T.; Phan, T.T.H.; Tran, T.T.T.; Ton, N.M.N.; Vo, D.L.T.; Man Le, V.V. Enzymatic treatment of spent green tea leaves and their use in high-fibre cookie production. *Food Technol. Biotechnol.* **2022**, *60*, 396–405. [CrossRef]
24. *ISO 11136*; Sensory Analysis-Methodology-General Guidance for Conducting Hedonic Tests with Consumers in a Controlled Area. ISO: Geneva, Switzerland, 2014.
25. AOAC. *Official Methods of Analysis*, 17th ed.; The Association of Official Analytical Chemists: Gaithersburg, MD, USA, 2000.
26. AACC International. *Approved Methods of Analysis*, 10th ed.; American Association of Cereal Chemists: St. Paul, MN, USA, 2009.
27. Rhee, K.S.; Ziprin, Y.A.; Rhee, K.C. Antioxidant activity of methanolic extracts of various oilseed protein ingredients. *J. Food Sci.* **1981**, *46*, 75–77. [CrossRef]
28. National Food Research Institute. *Manuals of Quality Characteristic Analysis for Food Quality Evaluation (2)*; National Food Research Institute: Skuba, Japan, 1990.
29. Blois, M.S. Antioxidant determinations by the use of a stable free radical. *Nature* **1958**, *181*, 1199–1200. [CrossRef]
30. Oyaizu, M. Studies on products of browning reaction antioxidative activities of products of browning reaction prepared from glucosamine. *Jpn. J. Nutr. Diet.* **1986**, *44*, 307–315. [CrossRef]
31. Niccolai, A.; Venturi, M.; Galli, V.; Pini, N.; Rodolfi, L.; Biondi, N.; D'Ottavio, M.; Batista, A.P.; Raymundo, A.; Granchi, L. Development of new microalgae-based sourdough "crostini": Functional effects of *Arthrospira platensis* (spirulina) addition. *Sci. Rep.* **2019**, *9*, 19433. [CrossRef] [PubMed]
32. *ISO 6887-2:2017*; Microbiology of the Food Chain—Preparation of Test Samples, Initial Suspension and Decimal Dilutions for Microbiological Examination—Part 2: Specific Rules for the Preparation of Meat and Meat Products. ISO: Geneva, Switzerland, 2017.
33. *ISO 4833-1:2013*; Microbiology of the Food Chain—Horizontal Method for the Enumeration of Microorganisms—Part 1: Colony Count at 30 Degrees C by the Pour Plate Technique. ISO: Geneva, Switzerland, 2013.
34. *ISO 21527-1:2008*; Microbiology of Food and Animal Feeding Stuffs. Horizontal Method for the Enumeration of Yeasts and Moulds—Part 1: Colony Count Technique in Products with Water Activity Greater than 0.95. ISO: Geneva, Switzerland, 2008.
35. *ISO 21527-2:2008*; Microbiology of Food and Animal Feeding Stuffs. Horizontal Method for the Enumeration of Yeasts and Moulds—Part 2: Colony Count Technique in Products with Water Activity less than or Equal to 0.95. ISO: Geneva, Switzerland, 2008.
36. *ISO 6888-2:2021*; Microbiology of the Food Chain—Horizontal Method for the Enumeration of Coagulase-Positive Staphylococci (Staphylococcus aureus and Other Species)—Part 2: Method Using Rabbit Plasma Fibrinogen Agar Medium. ISO: Geneva, Switzerland, 2021.
37. *ISO 6579-1:2017*; Microbiology of the Food Chain—Horizontal Method for the Detection, Enumeration and Serotyping of Salmonella—Part 1: Detection of Salmonella spp. ISO: Geneva, Switzerland, 2017.
38. *ISO 16649-2:2001*; Microbiology of Food and Animal Feeding Stuffs—Horizontal Method for the Enumeration of Beta-Glucuronidase-Positive Escherichia Coli—Part 2: Colony-Count Technique at 44 Degrees C Using 5-Bromo-4-Chloro-3-Indolyl Beta-D-Glucuronide. ISO: Geneva, Switzerland, 2001.
39. *ISO 4832:2006*; Microbiology of Food and Animal Feeding Stuffs—Horizontal Method for the Enumeration of Coliforms—Colony-Count Technique. ISO: Geneva, Switzerland, 2006.
40. *ISO 7932:2004*; Microbiology of Food and Animal Feeding Stuffs—Horizontal Method for the Enumeration of Presumptive Bacillus Cereus—Colony-Count Technique at 30 Degrees C. ISO: Geneva, Switzerland, 2004.

41. Zbikowska, A.; Kowalska, M.; Zbikowska, K.; Onacik-Gür, S.; Łempicka, U.; Turek, P. Study on the incorporation of oat and yeast β-glucan into shortbread biscuits as a basis for designing healthier and high quality food products. *Molecules* **2022**, *27*, 1393. [CrossRef]
42. FAO/WHO. Codex Alimentarius Commission Food additives and contaminants. Joint FAO/WHO Food Standards Programme. In *Codex Alimentarius: General Requirements (Food Hygiene) ALINORM*; 01/12A; World Health Organization: Geneva, Switzerland, 2001; pp. 1–289.
43. EC (European Commission). Commission Regulation (EC) No 1924/2006 of 20 December 2006 on Nutrition and Health Claims Made on Foods. Available online: http://data.europa.eu/eli/reg/2006/1924/oj (accessed on 13 January 2023).
44. Spiller, G. Suggestions for a basis on which to determine a desirable intake of dietary fiber. In *Handbook of Dietary Fiber in Human Nutrition*; CRC Press: Boca Raton, FL, USA, 1993; pp. 281–283.
45. Sharma, K.; Kumar, V.; Kaur, J.; Tanwar, B.; Goyal, A.; Sharma, R.; Gat, Y.; Kumar, A. Health effects, sources, utilization and safety of tannins: A critical review. *Toxin Rev.* **2021**, *40*, 432–444. [CrossRef]
46. Negi, T.; Kumar, Y.; Sirohi, R.; Singh, S.; Tarafdar, A.; Pareek, S.; Awasthi, M.K.; Sagar, N.A. Advances in bioconversion of spent tea leaves to value-added products. *Bioresour. Technol.* **2021**, *346*, 126409. [CrossRef] [PubMed]
47. Dhillon, B.; Awasthi, T.; Sodhi, N.S.; Sogi, D.S.; Jaiswal, S. A comparative study to investigate the effects of addition of milk and sugar on total polyphenol, flavonoid, catechin and tannin contents of green and black teas consumed in India. *J. Food Meas. Charact.* **2021**, *15*, 4652–4658. [CrossRef]
48. Sokač, T.; Gunjević, V.; Pušek, A.; Tušek, A.J.; Dujmić, F.; Brnčić, M.; Ganić, K.K.; Jakovljević, T.; Uher, D.; Mitrić, G. Comparison of drying methods and their effect on the stability of graševina grape pomace biologically active compounds. *Foods* **2022**, *11*, 112. [CrossRef] [PubMed]
49. Jakubczyk, K.; Kałduńska, J.; Kochman, J.; Janda, K. Chemical profile and antioxidant activity of the kombucha beverage derived from white, green, black and red tea. *Antioxidants* **2020**, *9*, 447. [CrossRef]
50. Ramashia, S.E.; Mamadisa, F.M.; Mashau, M.E. Effect of *Parinari curatellifolia* peel flour on the nutritional, physical and antioxidant properties of biscuits. *Processes* **2021**, *9*, 1262. [CrossRef]
51. Das, A.K.; Ghosh, A.; Majumder, S.; Saha, S.; Acharyya, S.; Sarkar, S.; Chakraborty, S.; Mukherjee, M.; Bhattacharya, M. Characterization of tea and tea infusion: A study of marketed black tea samples from some tea-growing regions of India. *J. Pharmacogn. Phytochem.* **2020**, *9*, 1532–1540. [CrossRef]
52. Food and Drug Administration. *Water Activity (Aw) in Foods | FDA*; U.S. Food and Drug Administration: Silver Spring, MD, USA, 2014.
53. He, Y.; Lin, Y.; Li, Q.; Gu, Y. The contribution ratio of various characteristic tea compounds in antioxidant capacity by DPPH assay. *J. Food Biochem.* **2020**, *44*, e13270. [CrossRef] [PubMed]
54. Islam, S.N.; Farooq, S.; Sehgal, A. Effect of consecutive steeping on antioxidant potential of green, oolong and black tea. *Int. J. Food Sci. Technol.* **2018**, *53*, 182–187. [CrossRef]
55. Sriprablom, J.; Kitthawee, S.; Suphantharika, M. Functional and physicochemical properties of cookies enriched with edible insect (*Tenebrio molitor* and *Zophobas atratus*) powders. *J. Food Meas. Charact.* **2022**, *16*, 2181–2190. [CrossRef]
56. Zhang, L.; Cao, Q.-Q.; Granato, D.; Xu, Y.-Q.; Ho, C.-T. Association between chemistry and taste of tea: A review. *Trends Food Sci. Technol.* **2020**, *101*, 139–149. [CrossRef]
57. Bhat, N.A.; Wani, I.A.; Hamdani, A.M. Tomato powder and crude lycopene as a source of natural antioxidants in whole wheat flour cookies. *Heliyon* **2020**, *6*, e03042. [CrossRef]

foods

MDPI

Article

Saffron Floral By-Products as Novel Sustainable Vegan Ingredients for the Functional and Nutritional Improvement of Traditional Wheat and Spelt Breads

Débora Cerdá-Bernad and María José Frutos *

Research Group on Quality and Safety, Agro-Food Technology Department, CIAGRO-UMH, Centro de Investigación e Innovación Agroalimentaria y Agroambiental, Miguel Hernández University, 03312 Orihuela, Spain; dcerda@umh.es

* Correspondence: mj.frutos@umh.es; Tel.: +34-96-674-97-44

Abstract: Saffron (*Crocus sativus* L.) is a traditional Mediterranean plant whose stigmas are used to obtain the most expensive spice in the world. Nevertheless, there is a lack of sustainability in its production, since, to produce 1 kg of saffron, about 350 kg of tepals are discarded. Therefore, this study aimed to develop wheat and spelt breads enriched with saffron floral by-products at a ratio of 0, 2.5, 5, and 10% (*w*/*w*), respectively, and to evaluate their nutritional, physicochemical, functional, and sensory properties, as well as the stability of antioxidant compounds during the in vitro digestion. The results revealed that the addition of saffron floral by-products, especially at 10%, increased the dietary fiber content by 25–30% of traditional wheat and spelt breads; improved their mineral content (270–290 mg/100 g for K, 90–95 mg/100 g for Ca, 40–50 mg/100 g for Mg, and 15–18 mg/100 g for Fe); changed their textural properties; and significantly enhanced the phenolic content and antioxidant ability (at 5 and 10%), which remained stable throughout the in vitro oral and gastrointestinal digestion processes. From a sensory point of view, the addition of saffron flowers modified the organoleptic properties of breads. Thus, these novel vegan enriched breads could exert beneficial effects on human health after their intake, making saffron floral by-products suitable and sustainable ingredients to develop new functional foods such as healthier alternative vegan bakery products.

Keywords: saffron flowers; polyphenols; antioxidant capacity; bioactive compounds; in vitro digestion; vegan bakery foods; sensory analysis; food waste; upcycling

1. Introduction

In recent years, consumers are more aware of the relationship between food and health, considering veganism as a suitable option to change their lifestyle and diet. Therefore, worldwide, vegetarian and vegan diets have become widespread in the population due to greater consumer attention to nutritional, environmental, and ethical aspects [1].

Current trends open up great opportunities for new alternative markets, so food industries are actively seeking the development of novel innovative and sustainable vegan products to meet the needs of consumers. Meat alternatives are dominated by soy derivatives since soy protein is abundant, cheap, and, after hydration, has a meat-like texture, but due to allergies issues, new vegetal sources must be sought [2]. Thus, the recovery and valorization of vegetable by-products from plants and vegetables have become of utmost interest to the food industry. These food by-products are considered high-added value ingredients whose function within the food matrix is to provide health-promoting bioactive compounds.

The production of saffron (*Crocus sativus* L.) spice generates large amounts of waste, since only flower stigmas are employed while the rest of the plant is discarded, generating around 350 kg of tepals to obtain 1 kg of the commercial product (Figure 1). However, several studies have reported the potential of these saffron floral by-products at nutritional and biological levels as a source of interesting compounds, such as polyphenols, especially flavonoids (anthocyanins, flavonols), with high antioxidant activity. Their nutritional composition includes carbohydrates (~70%), proteins (~8%), lipids (~4%), fiber (~27%), macro and microminerals, soluble sugars, and vitamin C, among others [3–5]. Therefore, the valorization of these saffron floral by-products through their use as novel sustainable vegan ingredients, with high-added value for the production of new functional food products, would lead to the minimization of their environmental impact, in addition to ensuring sustainability and profitability of the saffron industrial sector, developing, at the same time, an alternative vegetal source that would represent a new income for saffron farmers.

Figure 1. *Crocus sativus* L. flower.

The selection of food matrices as carriers of these plant ingredients is an essential factor to ensure the stability of their functional properties in the final product. Bread, a natural and balanced staple food, which contains a large amount of essential nutrients—carbohydrates, proteins, vitamins A, B1, B2, niacin, folic acid, and minerals—could be considered as a promising carrier matrix of functional substances [6]. Bread has been considered the food par excellence by many civilizations since ancient times, being the result of baking a mixture of flour, water, yeast, and salt. Furthermore, the high consumption and availability of this traditional bakery product, especially wheat and spelt breads, together with its quick and easy production, makes it a suitable matrix for its enrichment with functional ingredients providing bioactive compounds in the diet with positive health effects through the prevention or reduction of the risk of certain diseases [7]. It should be noted that spelt flour was used for its nutritional quality, presenting a high content in protein, and is an excellent source of minerals and vitamin E and B-complex vitamins.

The main aim of this research was to develop different improved traditional formulations of wheat and spelt breads enriched with saffron floral by-products as a sustainable vegan ingredient to produce breads with improved nutritional and functional quality compared to their traditional counterparts. Their physicochemical properties—namely, texture, content of minerals, organic acids, and soluble sugars—and the stability of the antioxidant capacity by ABTS and FRAP assays and total phenolic compounds during the oral and gastrointestinal in vitro digestion, as well as their sensory evaluation, were evaluated. Therefore, this research contributes to the valorization of a currently unexploited biomass through its use as a high-added value ingredient to develop novel functional food products, providing new information regarding the suitability of saffron floral by-products as sustainable alternatives to obtain healthier vegan bakery products.

2. Materials and Methods

2.1. Raw Materials

Saffron floral by-products, mainly composed of tepals, were obtained from the Castilla-La Mancha region (Spain) after harvesting of saffron flowers and the manual removal of the stigmas during the 2021 harvest season. The fresh flowers were frozen in liquid nitrogen and kept at −80 °C until dehydration in a vacuum oven (VACIOTEM, JP SELECTA®, Barcelona, Spain) at 50 ± 3 °C, 28 mbar, for 36 h, and then crushed, sieved (500 µm mesh), and kept at −20 °C until further use.

For the elaboration of different formulations of bread, two types of flour were used: wheat flour (*Triticum aestivum*; W: 360, P: 100, L: 100, 14.5% of protein) or organic spelt flour (*Triticum spelta*; W:95, P:36, L:105, 13.9% of protein) (El Amasadero, Malaga, Spain). Baker's active dry yeast (Mauripan, Barcelona, Spain), common commercial salt, and water were also employed.

2.2. Bread Baking Procedure

The bread formulations were prepared according to a traditional recipe with wheat flour or spelt flour (100%) and adding—based on the weight of flour and following the procedure of Świeca, Gawlik-Dziki, Dziki and Baraniak [8] with slight modifications—water (65–70%), yeast (1%), salt (2%), and dried saffron floral by-products in different concentrations (0%, 2.5%, 5%, and 10%). Eight different formulations were made, four using wheat flour at 0 (WB0%), 2.5 (WB2.5%), 5 (WB5%), and 10% (WB10%) of dried saffron floral by-products; and four formulations using spelt flour at 0 (SB0%), 2.5 (SB2.5%), 5 (SB5%), and 10% (SB10%) of dried saffron floral by-products.

The ingredients were mixed and kneaded for 10 min. A first short fermentation (30 min) was carried out at 30 ± 2 °C and humidity of 40 ± 5%. Loaves were manually round-shaped and left for a second short fermentation (30 min) under the same controlled temperature and humidity conditions. Finally, different bread loaves were baked at 210 °C for 20 min in a domestic oven, the final weight being around 40 g per piece. All formulations were prepared in triplicate.

Once cooled, a group of breads was stored at −20 °C until further physicochemical analysis, texture, and in vitro digestion tests. Another group was freeze-dried (Christ Alpha 2–4, B. Braun Biotech International, Melsungen, Germany) for 24 h (−25 ± 2 °C, 0.220 mbar). Then, they were crushed, sieved (500 µm mesh), and stored in polyethylene bags at −20 °C until further analysis (minerals, organic acids, and sugar composition).

2.3. Physicochemical Characterization

2.3.1. Moisture and Ash

The moisture (%) of fresh samples was determined by keeping bread formulations in an oven at 105 ± 5 °C until they reached constant weight, according to the AOAC [9]. The ash content (%) was carried out by incineration of samples obtained after moisture in a muffle oven (Habersal PR 1300 PAD, Barcelona, Spain) at 550 ± 25 °C, for one hour [9].

2.3.2. pH, Acidity, TSS, aw, Colur

The pH, acidity, TSS, aw, and color were determined according to Cerdá-Bernad, Valero-Cases, Pastor, Frutos and Pérez-Llamas [10], with some modifications. The pH and titratable acidity (expressed as % citric acid) were measured using an automatic titrator (TitroMatic Crison pH-Matic 23, Barcelona, Spain). The determination of total soluble solids (TSS) was carried out with a digital refractometer (Hanna® HI 96801, Bedfordshire, UK) and expressed as °Brix. Fresh samples were previously mixed with distilled water (1:10, *w*/*v*) with an ULTRA-TURRAX® (IKA T18, Werke GmbH & Co, Staufen, Germany) at 5000 rpm for 10 s.

The water activity (aw) of the different fresh samples was determined using a water activity meter (Novasina AW Sprint TH 500, Pfäffikon, Switzerland) at room temperature.

The color was measured with a Minolta CR-300 Chroma Meter (Japan) colorimeter, using the L*, a*, b* scale (CIELAB system). The results were expressed as luminosity L*, a* (greenness/redness), and b* (blueness/yellowness).

2.4. Texture

Texture profile analysis was performed on bread slices (2.5 cm width) using a TA-XTPlus Texture Analyser (Stable Micro Systems Ltd., Godalming, UK) according to García-Segovia, Igual and Martínez-Monzó [11], a cylindrical aluminum probe (SMS P100, 10 cm in diameter), and a 50 kg load cell. The parameters of the assay were defined as crosshead speed 1.70 mm/s and 40% deformation of the original length. The textural parameters' hardness (N), cohesiveness, springiness (mm), gumminess (N), and chewiness (N) were determined.

2.5. Minerals

The minerals' composition was determined according to Serrano-Díaz, Sánchez, Martínez-Tomé, Winterhalter and Alonso [4], with slight modifications. Freeze-dried bread samples weighing 0.5 g were digested with 10 mL of 65% HNO_3 (*v*/*v*) using a microwave reactor digestor (CEM Mars one, Matthews, NC, USA) for 30 min with a temperature ramp whose final temperature was 200 °C. All samples were filtered (Whatman qualitative filter paper 90 mm), diluted with ultrapure deionized water 1:50 (*v*/*v*), and stored at 4 °C. Total concentrations of macronutrients (Ca, Mg, Na, and K) and micronutrients (Zn, Cu, Mn, and Fe) in the previously mineralized samples were quantified with an Inductively Coupled Plasma Mass Spectrometer (ICPMS-2030, Shimadzu, Kyoto, Japan). Internal standards included calcium (44Ca), magnesium (26Mg), sodium (23Na), potassium (39K), zinc (66Zn), copper (65Cu), manganese (55Mn), and iron (56Fe), and the calibration curves used for quantification showed good linearity ($R^2 \geq 0.998$). The results were expressed as mg/100 g of dw (dry weight).

2.6. Organic Acids, Soluble Sugars, and Inulin Content

The extractions were prepared using ultrapure water and a freeze-dried sample/water ratio 1:20 (*w*/*v*). The extracts were shaken for 5 min at 300 rpm on a magnetic stirrer at room temperature (Ovan, mod. MultiMix Heat D-MMH30E, Barcelona, Spain), sonicated for 15 min, and centrifuged at 11,200× *g* for 10 min at 4 °C. The supernatants were filtered (0.45 μm PTFE filter, Millipore, Spain) and stored at −20 °C.

The identification and quantification of sugars, inulin, and organic acids were carried out by high-performance liquid chromatography using Hewlett-Packard HPLC series 1100 equipment equipped with a Supelcogel C-610H column (30 cm × 7.8 mm) and a Supelcoguard C-610H pre-column (5 cm × 4.6 mm) (Sigma Aldrich, St. Louis, MO, USA). The organic acids were measured at 210 nm in UV-Vis with a diode array detector (DAD G1315A). For sugars, a refractive index detector (G1362A RID) was used. As a mobile phase, 0.1% orthophosphoric acid was used with an injection volume of 20 μL and the flow rate of 0.5 mL/min under isocratic conditions, following the methodology described by Cerdá-Bernad, Valero-Cases, Pastor, Frutos and Pérez-Llamas [10]. The results were expressed as mg/100 g of dw.

2.7. In Vitro Digestion

The oral phase of in vitro digestion was performed according to Gawlik-Dziki, Dziki, Baraniak and Lin [12] with slight modifications. Fresh bread samples (2 g) were homogenized in the presence of 30 mL of PBS (phosphate buffered saline) in a stomacher for 30 s to simulate mastication. The solution was previously adjusted to pH 6.75 and alpha-amylase (E.C. 3.2.1.1.) was added to obtain 100 U per mL of enzymatic activity.

Gastrointestinal digestion was carried out following the methodology described by Cerdá-Bernad, Valero-Cases, Pastor, Frutos and Pérez-Llamas [10] for 180 min at 37 °C under continuous stirring. Aliquots were taken after 60 min of gastric digestion (pH = 3), and at 60 min and 120 min of intestinal digestion (pH = 7). All samples were filtered (0.45 μm; Millipore, Spain) and stored at −20 °C until further analysis (antioxidant capacity and total phenolic content).

2.8. Antioxidant Properties and Total Phenolic Content (TPC)

The antioxidant capacity determined by using the Ferric Reducing Antioxidant Power (FRAP) method and the 2,2′-Azinobis (3-ethylbenzothiazoline-6-sulfonic acid) (ABTS) radical scavenging method was carried out as described by Cerdá-Bernad, Clemente-Villalba, Valero-Cases, Pastor and Frutos [3]. Briefly, the FRAP reagent was prepared by mixing 300 mM acetate buffer (pH 3.6), 10 mM 2,4,6-tris(2-pyridyl)-s-triazine (TPTZ) solution in 40 mM HCl, and 20 mM $FeCl_3 \cdot 6H_2O$ solution in a volume ratio of 10:1:1, respectively. The absorbance was measured at 593 nm and Trolox (10 mM) was used as the standard solution (0.01–5.00 mM). ABTS radical was prepared mixing ABTS (7 mM) with potassium persulfate (2.45 mM), reacting for 16 h in the dark at room temperature and diluting the solution with ultrapure water until its absorbance was adjusted to 0.70 ± 0.02 at 734 nm. Trolox (10 mM) was used as a reference standard (0.20–3.00 mM). The results were expressed as mmol Trolox/100 g of bread.

The TPC was determined using the Folin–Ciocalteu methodology as described by Cerdá-Bernad, Clemente-Villalba, Valero-Cases, Pastor and Frutos [3]. Briefly, 100 μL of the different digested samples were mixed with 400 μL of PBS (50 mM) at pH 7.8 and 2.5 mL of Folin–Ciocalteu reagent previously mixed with ultrapure water 1:10 (v/v). After 2 min, 2 mL of Na_2CO_3 (75 g/L) were added and kept at 50 °C for 10 min. The absorbance was measured at 760 nm and the results were expressed as mg of gallic acid equivalents/100 g of bread.

2.9. Sensory Analysis

A quantitative descriptive analysis (QDA) of selected bread formulations was performed by a sensory panel composed of 12 trained judges (five males and seven females) aged between 21 and 60 years, at the Miguel Hernández University (Spain), following the procedure of Sicari, Romeo, Mincione, Santacaterina, Tundis and Loizzo [13] with some modifications. All the panel members had neither allergies nor food intolerances and were regular consumers of bakery products. The samples were labeled with an alphanumeric code and distributed in a random order. A total number of 17 sensory descriptors were considered regarding the appearance, flavor, and texture, using a 10-point scale (Table S1).

2.10. Statistical Analysis

Results were expressed as mean ± standard deviation. All experiments were carried out in triplicate. The mean comparisons were undertaken using an analysis of variance (ANOVA) and by the Tukey multiple range test, using the SPSS version 21.0 software package (SPSS Inc., Chicago, IL, USA). Principal Component Analysis (PCA) was conducted using XLSTAT (Microsoft Corp., Washington, DC, USA). The significant differences were established as ($p \leq 0.05$).

3. Results and Discussion

3.1. Characterization of Bread Formulations Enriched with Saffron Floral By-Products: Physicochemical, Nutritional, and Technological Properties

3.1.1. Physicochemical Parameters and Minerals' Content

Physicochemical and technological properties of the novel bread formulations were evaluated in order to study the influence of adding different amounts of dried saffron floral by-products compared to the control sample.

Table 1 shows the results of the physicochemical parameters (moisture, ash, pH, acidity, aw, and TSS) of the experimental wheat and spelt bread formulations with saffron floral by-products. Moisture is a relevant factor in the quality of bread since it is related to the shelf life but also influences the texture of the final product. Moisture content regulates the firming rate of bread since its decrease accelerates the process of cross-linking of proteins and starch, which further accelerates the process of firming [14].

Table 1. Physicochemical parameters of the different wheat and spelt bread formulations enriched with saffron floral by-products (moisture, ash, pH, acidity, TSS, aw) [1].

	Moisture (%)	Ash (%)	pH	TA (% Citric Acid)	Aw	TSS (°Brix)
WB0%	31.12 ± 2.43 ^{ab}	2.75 ± 0.07 ^{c}	5.62 ± 0.01 ^{a}	0.09 ± 0.01 ^{d}	0.85 ± 0.03	0.77 ± 0.01 ^{d}
WB2.5%	25.47 ± 2.49 ^{b}	2.95 ± 0.17 ^{c}	5.50 ± 0.01 ^{b}	0.12 ± 0.01 ^{c}	0.84 ± 0.02	0.93 ± 0.02 ^{c}
WB5%	31.12 ± 2.40 ^{ab}	3.44 ± 0.10 ^{b}	5.31 ± 0.02 ^{c}	0.23 ± 0.02 ^{b}	0.84 ± 0.04	1.07 ± 0.02 ^{b}
WB10%	32.44 ± 2.44 ^{a}	3.79 ± 0.12 ^{a}	5.20 ± 0.01 ^{d}	0.28 ± 0.02 ^{a}	0.84 ± 0.02	1.30 ± 0.01 ^{a}
SB0%	32.71 ± 2.61	3.05 ± 0.15 ^{c}	5.54 ± 0.01 ^{a}	0.10 ± 0.02 ^{b}	0.86 ± 0.01	0.97 ± 0.01 ^{d}
SB2.5%	29.51 ± 2.46	3.37 ± 0.11 ^{b}	5.23 ± 0.00 ^{b}	0.21 ± 0.01 ^{a}	0.85 ± 0.00	1.03 ± 0.03 ^{c}
SB5%	31.89 ± 2.36	3.39 ± 0.08 ^{b}	5.21 ± 0.00 ^{c}	0.23 ± 0.02 ^{a}	0.85 ± 0.01	1.07 ± 0.02 ^{b}
SB10%	31.50 ± 2.33	3.95 ± 0.16 ^{a}	5.18 ± 0.01 ^{d}	0.24 ± 0.01 ^{a}	0.85 ± 0.03	1.30 ± 0.01 ^{a}

[1] Means ± standard deviation in the same column followed by different lowercase letters indicate statistically significant differences at ($p \leq 0.05$) for each wheat or spelt sample (n = 3); TA: titratable acidity; TSS: total soluble sugars; WB0%, WB2.5%, WB5%, and WB10%: wheat bread formulations with the addition of 0%, 2.5%, 5%, and 10% of dried saffron floral by-products, respectively; SB0%, SB2.5%, SB5%, and SB10%: spelt bread formulations with the addition of 0%, 2.5%, 5%, and 10% of dried saffron floral by-products, respectively.

Regarding the results, in the spelt bread formulations (SB), no statistically significant differences were found between the spelt samples with different concentrations of saffron floral by-products formulations, showing values around 30% of moisture. In the wheat bread samples (WB), moisture values were between 25 and 32%, indicating statistically significant differences between WB2.5% (25.47 ± 2.49%) and WB10% (32.44 ± 2.44%). Moreover, water activity (aw) is also related to the firming process in the bakery products. The addition of different concentrations of dried saffron flowers' powder did not affect aw compared with the control bread in both wheat and spelt bread formulations since no statistically significant differences were found. These results were similar to those reported in other studies, in which breads enriched with 5 and 10% of insect-based proteins (*Alphitobius diaperinus*) were developed, showing aw values between 0.877 and 0.894 [11]. Therefore, the addition of a new sustainable vegan ingredient to improve the functional properties of spelt and wheat bread did not change the moisture and water activity compared to the control samples.

As expected, the addition of saffron floral by-products in wheat and spelt bread increased the ash values which are related to the average level of mineral components, since saffron flowers are good sources of minerals [4]. For wheat formulations, the addition of 5% of saffron flower ingredient (3.44 ± 0.10%) significantly increased the ash content with respect to the control sample, WB0% (2.75 ± 0.07%). However, in spelt breads, the ash content (3.37 ± 0.11%) in the formulations with 2.5% of the saffron flower ingredient was significantly higher than the control, SB0% (3.05 ± 0.15%). Furthermore, as shown in Table 2, the macromineral and micromineral concentrations revealed that the enrichment process of wheat and spelt breads with saffron floral by-products had a nutritional positive effect due to an increase in the content of beneficial minerals.

Table 2. Composition of minerals (mg/100 g dw) of the different wheat and spelt bread formulations enriched with saffron floral by-products [1].

		WB0%	WB2.5%	WB5%	WB10%	SB0%	SB2.5%	SB5%	SB10%
Macrominerals	Ca	39.22 ± 1.24^{b}	56.18 ± 5.15^{b}	67.92 ± 8.02^{a}	91.63 ± 10.02^{a}	33.47 ± 6.00^{c}	52.74 ± 8.10^{bc}	65.98 ± 1.33^{b}	94.56 ± 3.14^{a}
	K	162 ± 3^{d}	196 ± 9^{c}	226 ± 8^{b}	277 ± 14^{a}	153 ± 10^{c}	178 ± 11^{b}	209 ± 16^{b}	289 ± 18^{a}
	Mg	31.21 ± 1.54^{c}	34.72 ± 1.82^{bc}	36.97 ± 1.43^{b}	41.61 ± 1.93^{a}	33.95 ± 3.01^{b}	34.46 ± 3.11^{b}	40.81 ± 1.04^{a}	48.46 ± 8.03^{a}
	Na	641 ± 18^{ab}	713 ± 32^{a}	682 ± 23^{ab}	605 ± 85^{b}	810 ± 55	713 ± 55	743 ± 13	739 ± 12
Microminerals	Fe	2.01 ± 0.10^{c}	6.01 ± 1.15^{b}	9.15 ± 1.96^{b}	15.87 ± 1.25^{a}	3.27 ± 1.83^{c}	6.27 ± 0.35^{c}	10.63 ± 0.39^{b}	17.85 ± 0.40^{a}
	Mn	1.03 ± 0.19^{b}	1.22 ± 0.03^{ab}	1.25 ± 0.05^{ab}	1.42 ± 0.21^{a}	0.74 ± 0.09^{b}	0.91 ± 0.10^{ab}	1.04 ± 0.01^{ab}	1.37 ± 0.20^{a}
	Zn	0.77 ± 0.11	0.97 ± 0.00	0.97 ± 0.01	1.03 ± 0.06	0.71 ± 0.02	0.78 ± 0.06	0.81 ± 0.10	0.96 ± 0.20

[1] Means ± standard deviation in the same row followed by different lowercase letters indicate statistically significant differences at ($p \leq 0.05$) for each wheat or spelt sample (n = 3); WB0%, WB2.5%, WB5%, and WB10%: wheat bread formulations with the addition of 0%, 2.5%, 5%, and 10% of dried saffron floral by-products, respectively; SB0%, SB2.5%, SB5%, and SB10%: spelt bread formulations with the addition of 0%, 2.5%, 5%, and 10% of dried saffron floral by-products, respectively.

The most abundant mineral that presented in all wheat and spelt breads was Na due to the salt used in the formulation (600–800 mg/100 g), and no significant differences were shown between bread formulations with different concentrations of saffron floral by-products. Potassium was the macromineral also found in high levels, with increasing concentrations as the amount of saffron floral by-products added was higher, showing that WB10% and SB10% had the highest K content (277 ± 14 and 289 ± 18 mg/100 g, respectively). These values represent 7.91% and 8.26%, approximately, of the Recommended Daily Intake (RDI) of potassium in adults (3500 mg per day) with a consumption of 100 g of WB10% or SB10%, respectively [15]. Moreover, calcium and magnesium were also present in significantly higher amounts than the control bread in those with 5 or 10% of saffron floral by-products added, representing a vegan ingredient that improves the nutritional value of traditional spelt and wheat breads because of its rich mineral composition. The WB10% and SB10% samples showed a concentration of Ca of 91.63 ± 10.02 and 94.56 ± 3.14 mg/100 g, respectively, representing 9.65% and 9.95%, approximately, of the RDI of calcium in adults (950 mg per day) with a consumption of 100 g of wheat or spelt bread with saffron floral by-products ingredient at 10% [15]. Regarding Mg content, WB10% and SB10% had 41.61 ± 1.93 and 48.46 ± 8.03 mg/100 g, respectively, indicating around 13.87% and 16.15% of the RDI of magnesium in adults (300 mg per day), respectively, with a consumption of 100 g wheat or spelt bread with 10% of saffron flowers [15]. These results were in accordance with Fahim, Sadat, Janati and Feizy [16] who reported a high content of K and Ca in saffron petals.

In addition to the macrominerals content, which is relevant for human physiological functions, the microminerals iron, zinc, and manganese were also present in the different bread formulations although in a lower amount compared to Ca, Mg, Na, and K (Table 2). The most abundant micromineral was Fe in all samples, followed by Mn and Zn. The WB10% and SB10% samples showed significantly higher levels of Fe (15.87 ± 1.25 and 17.85 ± 0.40 mg/100 g, respectively) with respect to bread formulations enriched with 2.5 or 5% of saffron floral by-products (WB5%, WB10%, SB5%, SB10%), and with respect to the control samples (WB0%, SB0%). The consumption of 100 g of WB10% and SB10% provides the RDI in adults (11 mg per day) [15]. The amount of Mn in the control samples (1.03 ± 0.19 mg/100 g for WB0% and 0.74 ± 0.09 mg/100 g for SB0%) was statistically significantly lower than WB10% and SB10%, at 1.42 ± 0.21 and 1.37 ± 0.20 mg/100 g, respectively. With respect to Zn content, no statistically significant differences were found in both the wheat and spelt bread formulations, showing concentrations ranging from 0.77 to 1.03 for WB and from 0.71 to 0.96 for SB.

Minerals are essential micronutrients for human health and wheat flour is an important source of these components, such as Ca (100–200 mg/100 g), Mg (100–200 mg/100 g), Fe (1–5 mg per 100 g), Zn (1–5 mg/100 g), and Cu (0.1–1 mg/100 g) [17]. In addition, saffron floral by-products can provide mineral enrichment through their use in these novel vegan bread formulations, especially at 10%, showing values around 270–290 mg/100 g for K, 90–95 mg/100g for Ca, 40–50 mg/100 g for Mg, and 15–18 mg/100 g for Fe.

The pH value and acidity are also relevant factors to determine the quality of breads and may influence the sensory properties since both could have an impact on the taste of the final product as well as on the texture, with the solubility of gluten constituent proteins being greater at acidic pH values [18]. Table 1 shows pH values between 5.20 and 5.62 for wheat breads and 5.18 and 5.54 for spelt breads, being significantly lower in breads with a higher concentration of saffron floral by-products with respect to the control samples. At the same time, acidity was higher in bread formulations with higher amount of flowers with respect to the control samples, which is related to the concentration of organic acids in the saffron floral ingredient. Acidity values ranged from 0.09 to 0.24% citric acid in all wheat and spelt bread formulations.

3.1.2. Organic Acids, Soluble Sugars, and Inulin Content

The study of the composition of organic acids is an important quality parameter since they have a considerable effect on the technological characteristics, relevant sensory properties, and shelf life of breads. Data on the organic acid composition of bread formulations are shown in Table 3. Important organic acids were found in the different bread formulations, not only due to their presence in wheat and spelt flour as well as in saffron floral by-products but also as a result of the fermentation process during bread making. In the course of the fermentation period, the *Saccharomyces cerevisiae* yeast used in the formulation, together with the microbiota (bacteria and yeasts) naturally present in wheat and spelt flour, played an important role in terms of bread characteristics [19]. The biochemical changes due to the effects of fermentation on the degradation of carbohydrates and proteins mainly by the yeast and lactic acid bacteria led to the production of organic acids and other metabolites. Lactic acid was one of the main organic acids present in all bread formulations, being significantly higher in breads with 10% of saffron floral by-products (595 ± 20 and 676 ± 11 mg/100 g for WB10% and SB10%, respectively) compared to the other samples. Malic and citric acids were also found in wheat and spelt formulations in high amounts, and previous studies have reported their presence in saffron tepals [4] and in wheat flour [20,21]. Statistically significant differences were found in the content of citric acid in wheat and spelt bread formulations, being higher in WB10% (1957 ± 19 mg/100 g) and SB10% (1934 ± 22 mg/100 g) compared to the control samples, WB0% (1878 ± 7 mg/100 g) and SB0% (1600 ± 24 mg/100 g). Oxalic and fumaric acids were also found in lower amounts in the wheat and spelt bread samples, showing values around 0.24–1.98 mg/100 g for fumaric acid and 9.77–14.24 mg/100 g for oxalic acid. These results were in accordance with other studies that reported the presence of lactic, malic, and fumaric acids in wheat and spelt flours [21].

Furthermore, propionic acid was identified in wheat and spelt bread samples enriched with saffron flowers; this may have been generated as a result of the metabolic activity of microbiota that could metabolize molecules present in saffron floral by-products, producing different compounds. Moreover, the differences in the composition of organic acids between spelt and wheat breads could be related to the wide diversity in the presence of diverse microbial strains leading to the production of different enzymes and metabolites [22]. Therefore, the presence of organic acids in the different bread formulations could act as natural preservatives to improve their shelf life, avoiding microbiological spoilage.

The evaluation of sugar content in bakery products such as bread is essential due to its functional role. The natural composition of sugars in the initial ingredients is a major contributor to flavor in the final product by interacting with other ingredients due to two processes which are involved: fermentation and crust browning (caramelization and the Maillard reaction) during baking. Then, the sugars remaining after fermentation also contribute to the overall color and texture of the final product, since the affinity of sugars to bind to water will delay the development of gluten, which is essential for maintaining a tender bread texture [23]. The results of the analysis of the individual sugars are shown in Table 3.

Glucose was the major component in all bread formulations, followed by fructose and maltose. The disaccharide maltose is the first product released by the digestion of flour starch by the amylase enzyme that is further hydrolyzed to release glucose. Therefore, maltose is found in lower concentrations than glucose in the final product, ranging from 30 to 37 mg/100 g in all wheat and spelt breads. However, the glucose content is very high, showing statistically significant differences between each wheat and spelt formulation. The highest concentrations of this monosaccharide were found for WB10% (1917 ± 34 mg/100 g) and for SB10% (1829 ± 35 mg/100 g) and the lowest concentrations were found for the control samples, WB0% and SB0% (1303 ± 12 and 1210 ± 62 mg/100 g, respectively). These results were in accordance with other studies that reported the presence of maltose and glucose in wheat and spelt flours [21].

Table 3. Content of organic acids, soluble sugars, and inulin (mg/100 g dw) of the different wheat and spelt bread formulations enriched with saffron floral by-products [1].

		WB0%	WB2.5%	WB5%	WB10%	SB0%	SB2.5%	SB5%	SB10%
Organic acids	Phytic acid	n.d.	n.d.	n.d.	n.d.	178 ± 27	165 ± 5	168 ± 11	172 ± 3
	Lactic acid	241 ± 60 ^{d}	384 ± 22 ^{c}	503 ± 19 ^{b}	595 ± 20 ^{a}	301 ± 33 ^{d}	385 ± 6 ^{c}	448 ± 5 ^{b}	676 ± 11 ^{a}
	Citric acid	1878 ± 7 ^{c}	1895 ± 6 ^{bc}	1922 ± 14 ^{b}	1957 ± 19 ^{a}	1600 ± 24 ^{c}	1885 ± 5 ^{b}	1910 ± 9 ^{b}	1934 ± 22 ^{a}
	Malic acid	132 ± 5 ^{d}	214 ± 11 ^{c}	335 ± 30 ^{b}	516 ± 30 ^{a}	102 ± 10 ^{d}	189 ± 16 ^{c}	292 ± 26 ^{b}	521 ± 32 ^{a}
	Oxalic acid	14.24 ± 1.90	13.91 ± 0.62	13.84 ± 0.93	13.79 ± 0.93	11.33 ± 0.76	11.75 ± 0.22	9.77 ± 3.73	10.98 ± 0.26
	Fumaric acid	0.24 ± 0.07 ^{c}	0.04 ± 0.00 ^{c}	1.05 ± 0.08 ^{b}	1.98 ± 0.13 ^{a}	n.d.	n.d.	n.d.	1.71 ± 0.20
	Propionic acid	n.d.	835 ± 71 ^{b}	933 ± 33 ^{a}	982 ± 48 ^{a}	n.d.	845 ± 9 ^{c}	910 ± 16 ^{b}	1051 ± 48 ^{a}
Soluble sugars	Glucose	1303 ± 12 ^{d}	1411 ± 16 ^{c}	1617 ± 13 ^{b}	1917 ± 34 ^{a}	1210 ± 62 ^{d}	1317 ± 25 ^{c}	1512 ± 82 ^{b}	1829 ± 35 ^{a}
	Maltose	37.70 ± 0.14 ^{a}	33.32 ± 0.10 ^{c}	36.75 ± 0.16 ^{b}	33.11 ± 0.33 ^{c}	33.19 ± 0.24 ^{a}	33.03 ± 0.41 ^{a}	31.19 ± 0.32 ^{b}	30.49 ± 0.13 ^{b}
	Fructose	n.d.	n.d.	480 ± 40 ^{b}	913 ± 82 ^{a}	n.d.	n.d.	307 ± 48 ^{b}	813 ± 12 ^{a}
	Inulin	5011 ± 70 ^{c}	5377 ± 95 ^{b}	6504 ± 58 ^{a}	6511 ± 223 ^{a}	5464 ± 80 ^{c}	5682 ± 112 ^{c}	6231 ± 211 ^{b}	6846 ± 48 ^{a}

[1] Means ± standard deviation in the same row followed by different lowercase letters indicate statistically significant differences at ($p \leq 0.05$) for each wheat or spelt sample (n = 3); WB0%, WB2.5%, WB5%, and WB10%: wheat bread formulations with the addition of 0%, 2.5%, 5%, and 10% of dried saffron floral by-products, respectively; SB0%, SB2.5%, SB5%, and SB10%: spelt bread formulations with the addition of 0%, 2.5%, 5%, and 10% of dried saffron floral by-products, respectively; n.d.: not detected.

Fructose could be also released by hydrolysis of fructans, but this compound was only present in wheat and spelt bread formulations with 5 and 10% of saffron floral by-products. Thus, the content of this monosaccharide could be due to its natural occurrence in the saffron flowers ingredient [4]. Furthermore, the content of inulin as part of soluble dietary fiber was studied. The inulin concentration significantly increased with the content of saffron floral by-products in wheat and spelt breads. The highest amounts of inulin were present in WB10% and SB10%, with values around 6511 and 6846 mg/100 g, respectively. However, the lowest concentrations were found in samples without saffron floral by-products, being these amounts those present in wheat and spelt flours (around 5011 and 5464 mg/100 g for WB0% and SB0%, respectively). Therefore, this vegan ingredient is a natural source of dietary fiber to develop new bakery products with improved nutritional properties since inulin can be used as a prebiotic, stimulating the proliferation of the intestinal microbiota [24] At the same time, inulin contributed to an increase in TSS, since bread formulations with the highest inulin concentration (WB10% and SB10%) presented significantly higher TSS than the control samples, with values around 1.30 °Brix compared to WB0% and SB0% (0.77–0.93 °Brix) (Table 1). Moreover, the increase in TSS could be due to the presence of fructose in the composition of inulin in the WB10% and SB10% samples.

3.1.3. Texture

The textural properties are one of the quality parameters that have a crucial influence on the bread quality since they determine the shelf life and consumers' acceptability. Table 4 shows the textural parameters (hardness, cohesiveness, springiness, gumminess, chewiness) of the different wheat and spelt bread formulations. The results show that the control wheat and spelt breads, WB0% and SB0%, were characterized by the lowest hardness, whereas the bread enrichment with saffron floral by-products resulted in a statistically significant increase in this parameter, being around 28% in WB10% and around 50% in SB10% with respect to the control samples. All wheat- and spelt-enriched bread formulations presented a statistically significant decrease in cohesiveness when compared to control breads. Thus, these results revealed that the addition of saffron floral by-products significantly affected the textural properties of breads, which could be related to the interruption of the gluten network that leads to a low cohesiveness and the crumb disintegration [7]. These findings agree with those obtained in previous studies in which bread was enriched by adding chickpea and soy flour, also observing an increase in hardness [25].

Table 4. Parameters of the texture of the different wheat and spelt bread formulations enriched with saffron floral by-products [1].

	Hardness (N)	Cohesiveness	Springiness (mm)	Gumminess (N)	Chewiness (N)
WB0%	42.24 ± 2.45 ^{d}	0.93 ± 0.02 ^{a}	1.59 ± 0.01 ^{a}	39.47 ± 0.41 ^{d}	37.73 ± 0.74 ^{c}
WB2.5%	48.36 ± 3.32 ^{c}	0.88 ± 0.00 ^{b}	1.50 ± 0.08 ^{b}	42.58 ± 0.67 ^{c}	40.00 ± 0.49 ^{b}
WB5%	51.12 ± 1.12 ^{b}	0.87 ± 0.01 ^{b}	1.49 ± 0.03 ^{b}	44.68 ± 0.48 ^{b}	41.51 ± 0.23 ^{b}
WB10%	54.17 ± 0.53 ^{a}	0.86 ± 0.00 ^{b}	1.47 ± 0.02 ^{c}	46.91± 0.82 ^{a}	43.21 ± 0.15 ^{a}
SB0%	34.65 ± 1.66 ^{c}	0.89 ± 0.01 ^{a}	1.53 ± 0.04 ^{a}	31.14 ± 1.11 ^{d}	28.64 ± 0.16 ^{c}
SB2.5%	41.31 ± 2.24 ^{b}	0.84 ± 0.02 ^{b}	1.43 ± 0.00 ^{b}	34.76 ± 0.53 ^{c}	31.93 ± 0.27 ^{b}
SB5%	43.75 ± 2.35 ^{b}	0.82 ± 0.02 ^{bc}	1.40 ± 0.00 ^{c}	36.06 ± 0.14 ^{b}	32.94 ± 0.98 ^{b}
SB10%	52.12 ± 3.87 ^{a}	0.80 ± 0.00 ^{c}	1.37 ± 0.01 ^{d}	41.91 ± 1.14 ^{a}	38.22 ± 0.54 ^{a}

[1] Means ± standard deviation in the same column followed by different lowercase letters indicate statistically significant differences at ($p \leq 0.05$) for each wheat or spelt sample (n = 3); WB0%, WB2.5%, WB5%, and WB10%: wheat bread formulations with the addition of 0%, 2.5%, 5%, and 10% of dried saffron floral by-products, respectively; SB0%, SB2.5%, SB5%, and SB10%: spelt bread formulations with the addition of 0%, 2.5%, 5%, and 10% of dried saffron floral by-products, respectively.

The springiness parameter is associated with freshness, and high values are desired to extend the bread shelf life. However, the enriched bread samples had a statistically significant decrease in springiness values when compared to the controls, being linked to a high crumb brittleness [26]. Chewiness and gumminess increased significantly with the saffron floral by products' incorporation in wheat and spelt breads when compared to control samples. Similar results were found by García-Segovia, Igual and Martínez-Monzó [11] using pea protein or insect powder in the formulation of breads. These values might be related to the dietary fiber content, like inulin, which results in a strong chewiness and high water-swelling capacity [26].

3.1.4. Color

Bread crumb and crust color were also evaluated since it is an important factor for the consumers' choice of bread (Table 5). The L* values describing the brightness of breads decreased in bread crust and crumb prepared with saffron floral by-products.

Table 5. Color of the crust and crumb of the different wheat and spelt bread formulations enriched with saffron floral by-products [1].

	Crust			Crumb		
	L*	a*	b*	L*	a*	b*
WB0%	40.25 ± 3.15	12.49 ± 0.19 a	22.22 ± 2.19 b	61.86 ± 1.69 a	0.28 ± 0.10 b	18.45 ± 0.35 b
WB2.5%	37.21 ± 2.36	11.35 ± 0.63 ab	20.18 ± 2.51 b	51.72 ± 3.73 b	−0.38 ± 0.02 c	27.83 ± 0.60 a
WB5%	41.95 ± 2.37	10.72 ± 0.47 b	22.99 ± 1.56 a	47.72 ± 0.63 b	0.08 ± 0.04 b	27.44 ± 3.54 a
WB10%	34.78 ± 3.56	9.21 ± 0.53 c	17.05 ± 2.32 b	41.29 ± 1.47 c	1.52 ± 0.32 a	28.35 ± 1.89 a
SB0%	50.58 ± 1.71 a	11.05 ± 0.34 a	25.35 ± 1.16 a	63.15 ± 5.68 a	0.02 ± 0.00 d	19.08 ± 1.07 b
SB2.5%	43.54 ± 5.48 ab	10.60 ± 1.15 a	23.30 ± 2.87 a	58.45 ± 1.81 ab	0.30 ± 0.01 c	28.21 ± 1.40 a
SB5%	40.15 ± 0.42 bc	11.06 ± 0.56 a	21.42 ± 0.57 a	52.50 ± 2.54 c	0.59 ± 0.03 b	30.70 ± 0.81 a
SB10%	33.75 ± 1.42 c	8.73 ± 0.53 b	15.22 ± 1.14 b	47.01 ± 3.63 c	1.75 ± 0.09 a	30.57 ± 1.70 a

[1] Means ± standard deviation in the same column followed by different lowercase letters indicate statistically significant differences at ($p \leq 0.05$) for each wheat or spelt sample (n = 3); WB0%, WB2.5%, WB5%, and WB10%: wheat bread formulations with the addition of 0%, 2.5%, 5%, and 10% of dried saffron floral by-products, respectively; SB0%, SB2.5%, SB5%, and SB10%: spelt bread formulations with the addition of 0%, 2.5%, 5%, and 10% of dried saffron floral by-products, respectively.

Thus, the enrichment of the bread decreased crust and crumb L* values by about 13% and 33%, respectively, in WB10% compared to the control samples, WB0%. A similar tendency was followed by spelt breads, showing a darker color of the bread crumb and crust supplemented with saffron floral by-products, decreasing the crust and crumb L* values by about 33% and 25%, respectively, in SB10% when compared to SB0%.

These results suggested that the reduction of crust L* values could be related to an increase in Maillard browning reactions, and the decrease in L* crumb values may be explained by the intrinsic color of saffron floral by-products. Other authors reported similar observations in breads enriched with vegetal by-products [27]. Regarding the parameters, a* and b*, the crust bread showed similar values in enriched wheat and spelt bread formulations with respect to control samples. Nevertheless, the a* and b* values of the bread crumb increased in the wheat and spelt formulations with the addition of saffron flowers, resulting in redder-yellower loafs (Figure 2). Therefore, the color parameters of the crumb are related to those of the saffron flower by-products powder since the temperatures inside the loaves during baking did not exceed 100 °C [3,11].

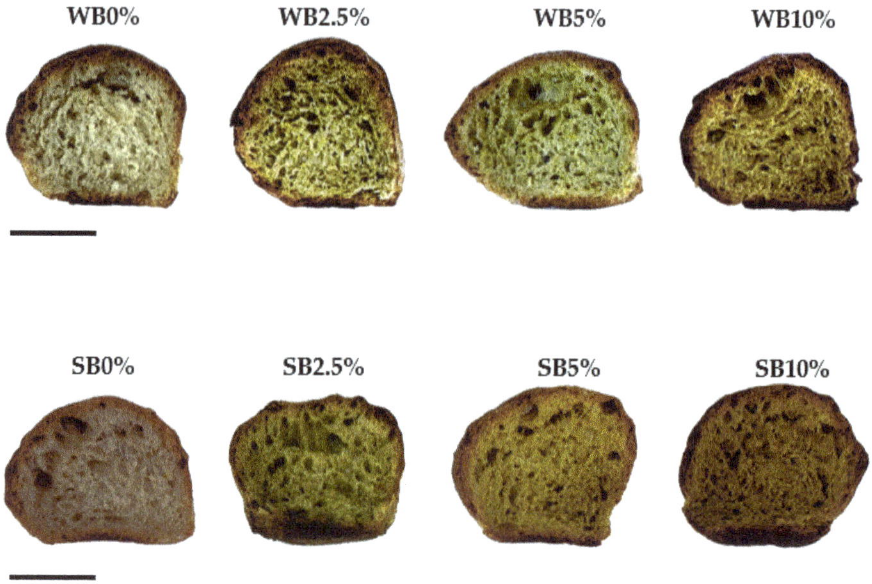

Figure 2. Wheat and spelt bread formulations obtained by adding different concentrations of dried saffron floral by-products. WB0%, WB2.5%, WB5%, and WB10%: wheat bread formulations with the addition of 0%, 2.5%, 5%, and 10% of dried saffron floral by-products, respectively; SB0%, SB2.5%, SB5%, and SB10%: spelt bread formulations with the addition of 0%, 2.5%, 5%, and 10% of dried saffron floral by-products, respectively. Scale bar: 2.5 cm.

3.2. Antioxidant Properties and Total Phenolic Content of Bread Formulations Enriched with Saffron Floral By-Products during Oral and Gastrointestinal In Vitro Digestion

The effect of the incorporation of saffron floral by-products into wheat and spelt bread matrices was also evaluated through the determination of the antioxidant activity and bioactive content at individual stages of simulated oral and gastrointestinal in vitro digestion. Plant bioactive compounds are very sensitive to several environmental factors, which means that their biological properties, such as the antioxidant power, could change significantly depending on the environmental conditions (pH, presence of digestive enzymes, temperature, etc.) [28]. Therefore, the changes in the antioxidant activity and in the total phenolic content at the different stages of in vitro digestion (oral, gastric, intestinal) of breads enriched with saffron floral by-products, as a rich source of phenolic compounds, were investigated.

The results of antioxidant activity ($ABTS^+$ and FRAP assays) and total phenolic compounds (TPCs) in the enriched wheat and spelt bread formulations at various stages of simulated in vitro digestion are presented in Table 6.

According to the values obtained for the ability to reduce free radicals by the $ABTS^+$ assay, Fe^{2+} ions by the FRAP assay, and TPC during the in vitro digestion, the addition of saffron floral by-products into wheat and spelt breads significantly improved the phenolic content and their antioxidant ability, especially in the formulations with 10% of the saffron floral ingredient. The enriched breads, WB10% and SB10%, exhibited a significantly higher antioxidant activity (ABTS and FRAP assays) compared to the control in all digestion phases. This could be due to a significant increase of phenolic compounds from saffron floral by-products with well-known antioxidant capacities. For the ABTS assay, the antioxidant capacity increased in all bread formulations in each digestion phase, reaching the highest values under intestinal digestion conditions, being the maximum values for WB10% and SB10% after 2 h of intestinal digestion (1576 ± 34 mmol Trolox/100 g and 1583 ± 83 mmol

Trolox/100 g, respectively). This increase could be due to intestinal conditions (neutral pH; presence of pancreatic enzymes and bile salts) which would facilitate an improvement in the release of bioactive compounds that may previously be bound or insolubilized due to their interaction with the bread matrix. These findings were in line with literature reports indicating that the free radical scavenging activity grew along with the increase of plant ingredients addition and with the progress of the in vitro digestion [12,28].

Table 6. Antioxidant properties (FRAP and ABTS assays) and Total Phenolic Content (TPC) of the different wheat and spelt bread formulations enriched with saffron floral by-products during the oral and gastrointestinal in vitro digestion process [1].

		In Vitro Digestion			
		Oral	Gastric	Intestinal (1 h)	Intestinal (2 h)
ABTS (mmol Trolox/100 g)	WB0%	3.47 ± 1.75	22.37 ± 1.99 c	421 ± 12 a	458 ± 6 b
	WB2.5%	3.94 ± 1.28	34.32 ± 8.47 b	437 ± 4 a	452 ± 10 b
	WB5%	3.61 ± 2.61	48.74 ± 2.14 a	435 ± 8 a	461 ± 5 b
	WB10%	4.94 ± 1.50	49.99 ± 0.98 a	399 ± 7 b	1576 ± 34 a
	SB0%	2.20 ± 0.16 b	30.74 ± 3.31 b	429 ± 11	437± 27 b
	SB2.5%	5.49 ± 0.67 a	43.34 ± 1.99 a	430 ± 12	1373 ± 59 a
	SB5%	4.91 ± 0.81 a	45.44 ± 6.40 a	426 ± 19	1458± 177 a
	SB10%	5.69 ± 0.20 a	50.54 ± 0.39 a	437 ± 27	1583 ± 83 a
FRAP (mmol Trolox/100 g)	WB0%	0.89 ± 0.50 c	8.07 ± 2.03 c	n.d.	n.d.
	WB2.5%	14.09 ± 4.08 b	17.12 ± 0.56 bc	1.40 ± 0.83 b	2.43 ± 0.73 b
	WB5%	28.05 ± 4.85 a	25.64 ± 6.04 b	9.83 ± 5.35 ab	15.00 ± 6.58 ab
	WB10%	38.77 ± 5.90 a	37.67 ± 4.67 a	20.71 ± 7.96 a	27.89 ± 6.99 a
	SB0%	6.99 ± 0.73 c	11.21 ± 1.01 b	n.d.	n.d.
	SB2.5%	18.08 ± 4.03 bc	21.56 ± 1.19 b	3.97 ± 2.81 b	7.60 ± 0.34 c
	SB5%	32.01 ± 9.90 b	26.69 ± 12.58 ab	14.26 ± 8.84 b	18.16 ± 4.89 b
	SB10%	61.86 ± 10.75 a	43.97 ± 6.96 a	31.79 ± 1.22 a	36.62 ± 3.130 a
TPC (mg GAE/100 g)	WB0%	76.99 ± 6.60 d	118 ± 8 d	109 ± 7 d	91.31 ± 5.25 c
	WB2.5%	111 ± 6 c	149 ± 16 c	140 ± 16 c	146 ± 13 b
	WB5%	198 ± 8 b	207 ± 1 b	222 ± 8 b	166 ± 14 b
	WB10%	221 ± 18 a	231 ± 11 a	272 ± 17 a	220 ± 17 a
	SB0%	102 ± 16 c	181 ± 92	137 ± 18 c	105 ± 15 c
	SB2.5%	124 ± 32 bc	236 ± 129	135 ± 10 c	135 ± 45 bc
	SB5%	175 ± 1 b	251 ± 71	187 ± 6 b	186 ± 5 b
	SB10%	261 ± 13 a	313 ± 63	258 ± 25 a	251 ± 17 a

[1] Means ± standard deviation in the same column followed by different lowercase letters indicate statistically significant differences at ($p \leq 0.05$) for each wheat or spelt sample ($n = 3$); WB0%, WB2.5%, WB5%, and WB10%: wheat bread formulations with the addition of 0%, 2.5%, 5%, and 10% of dried saffron floral by-products, respectively; SB0%, SB2.5%, SB5%, and SB10%: spelt bread formulations with the addition of 0%, 2.5%, 5%, and 10% of dried saffron floral by-products, respectively; n.d.: not detected.

For the FRAP assay, however, the maximum values were obtained after simulated oral digestion, which remained stable after simulated gastric digestion but decreased after the intestinal phase in all bread formulations. The lowest activity after the gastric digestion was detected in control samples and the highest in the samples obtained from wheat and spelt bread with 10% of saffron floral ingredients. Therefore, the influence of the bread food matrix and its interaction with bioactive compounds from saffron flower ingredients could have played an important role in the antioxidant activity of the bread along the oral and gastrointestinal in vitro digestion.

Regarding TPC values, in the wheat and spelt bread formulations, their concentrations remained stable in the oral, gastric, and intestinal digestion steps, reaching the statistically significant highest TPC content for the bread formulations, WB10% and SB10%. In spelt breads, a slight increase in TPC after gastric conditions was observed, which could be related to an increase in the solubility of certain phenolic compounds under acidic pH conditions, which may have been previously chelated, or in reduced form [29]. It should be noted that the TPC values were high after 2 h of the intestinal stage (220 ± 17 and 251 ± 17 mg GAE/100 g for WB10% and SB10%, respectively), so that the matrix enables these bioactive compounds to reach the colon after digestion and contribute to the maintenance of intestinal health [30]. However, further research is necessary to understand their mechanism of action at the physiological level in human health, since the bioavailability of phenolic compounds through the diet depends on many factors [28].

According to the results, saffron floral by-products have a high content of phenolic compounds that remained stable in the cereal food matrix during the in vitro digestion, so they could be absorbed and exert their beneficial physiological effects. In addition, these functional breads have improved antioxidant properties, especially in the formulations with a higher content of saffron floral by-products (5 and 10%).

The Principal Component Analysis (PCA) biplot graph (Figure 3) showed two different groups, one related to wheat and spelt bread formulations incorporating 10% of saffron floral by-products, WB10%, and SB10%, associated with a greater antioxidant power, revealed by FRAP and ABTS assays, after the in vitro digestion process, and with a higher total phenolic content. Therefore, the cereal matrix protected the bioactive content of saffron floral by-products, avoiding their oxidation and degradation and maintaining the antioxidant activity throughout the oral and gastrointestinal in vitro digestion. It is also observed in the figure that these samples were related to a high concentration of organic acids, inulin, and minerals improving the nutritional value when compared to SB0% and SB2.5%, which were not associated with any specific characteristic. The other group, WB0% and WB2.5%, was associated with pH, cohesiveness, and springiness.

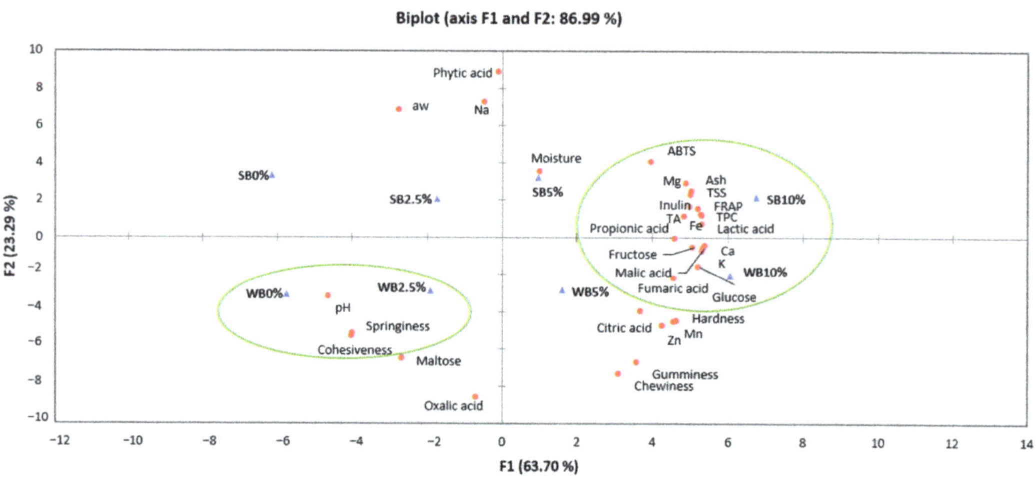

Figure 3. PCA biplot graph regarding physicochemical properties, minerals, organic acids, sugars content, texture, and antioxidant capacity (FRAP, ABTS) and Total Phenolic Content (TPC) after the in vitro digestion. TA: titratable acidity; TSS: total soluble sugars; WB0%, WB2.5%, WB5%, and WB10%: wheat bread formulations with the addition of 0%, 2.5%, 5%, and 10% of dried saffron floral by-products, respectively; SB0%, SB2.5%, SB5%, and SB10%: spelt bread formulations with the addition of 0%, 2.5%, 5%, and 10% of dried saffron floral by-products, respectively.

3.3. *Sensory Analysis of Bread Formulations Enriched with Saffron Floral By-Products*

Sensory evaluation plays an important role in the development of innovative and novel functional food products. Figure 4 summarizes the results of the sensory analysis of enriched bread formulations incorporating 5% of saffron floral by-products compared to the control samples. Appearance descriptors showed high scores regarding the uniformity of shape in all the formulations, but for the evenness of color of crust, as expected, enriched formulations such as WB5% and SB5% exhibited lower values when compared to the control samples WB0% and SB0%, respectively, due to the incorporation of saffron flowers which had an impact on the color of the crust.

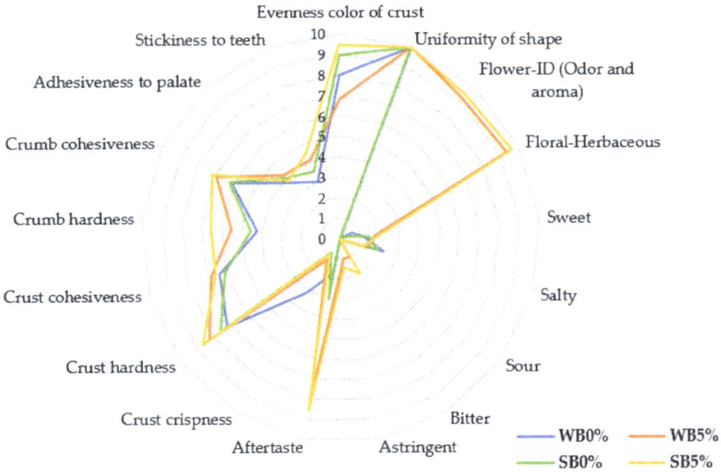

Figure 4. Results of the sensory analysis of selected bread formulations. Data are represented as means ± SD of three replicates. WB0% and WB5%: wheat bread formulations with the addition of 0% and 5% of dried saffron floral by-products, respectively; SB0% and SB5%: spelt bread formulations with the addition of 0% and 5% of dried saffron floral by-products, respectively.

As regards flavor evaluation, WB5% and SB5% had slightly higher values in terms of sweetness, probably due to the sugar concentration of saffron flowers. In the same way, flower-ID and floral–herbaceous attributes were related to the *Crocus sativus* L. flowers in enriched bread formulations, WB5% and SB5%. The intensity of astringency and bitterness, as expected, was higher in samples formulated with saffron floral by-products because of the content of polyphenolic compounds like anthocyanins. The astringency and bitterness could be derived from the precipitation of proline-rich salivary protein in the mouth caused by phenolic compounds [31,32]. Moreover, aftertaste may be linked to the high astringency of samples.

Moreover, the results of texture attributes were in line with the TPA analysis, according to which WB5% and SB5% had higher hardness than the control samples. The use of saffron floral by-products made breads softer than the control samples, reducing their crispness and affecting their cohesiveness, probably due to the addition of this vegetal ingredient that interrupts the gluten network, making breads more fragile [33]. However, with respect to adhesiveness to palate and stickiness to teeth, no important differences were observed between the studied bread formulations.

4. Conclusions

The present study aimed to evaluate the possibility of incorporating saffron floral by-products obtained from saffron spice production into wheat and spelt breads and to study their effect on physicochemical, functional, technological, and sensory properties. The enrichment of traditional breads with saffron flowers was an effective tool that allowed for obtaining functional food with significantly enhanced nutritional and functional potential, increasing the content of dietary fiber such as inulin, minerals, and organic acids, and showing a high antioxidant capacity and total phenolic content after the in vitro digestion process. Therefore, this study confirms that saffron floral by-products are a sustainable alternative for the development of novel bakery products suitable for vegans and vegetarians which could exert beneficial effects on human health after their intake. It also encompasses the valorization of a biomass that is currently unexploited, minimizing, at the same time, its environmental impact. Furthermore, this research provided new information for further studies that focus on the effects of the saffron floral ingredient at a physiological level after the intake of these functional breads.

Supplementary Materials: The following supporting information can be downloaded at: https://www.mdpi.com/article/10.3390/foods12122380/s1; Table S1: Lexicon for the sensory description of bread formulations.

Author Contributions: Conceptualization: D.C.-B. and M.J.F.; methodology: D.C.-B.; writing—original draft preparation: D.C.-B.; writing—review and editing: D.C.-B. and M.J.F.; supervision: M.J.F.; project administration: M.J.F. All authors have read and agreed to the published version of the manuscript.

Funding: This article is supported by the Partnership for Research and Innovation in the Mediterranean Area (PRIMA) program, project "SAFFROMFOOD". The PRIMA program is supported by the European Union.

Data Availability Statement: Data is contained within the article or supplementary material.

Acknowledgments: Thanks to the Spanish Ministry of Education, Culture, and Sport for the FPU-PhD fellowship (FPU18/02225) and thanks to the Spanish company Verdú Cantó Saffron, Spain, S.L for the supply of samples.

Conflicts of Interest: The authors declare no conflict of interest.

References

1. Tabanelli, G.; Pasini, F.; Riciputi, Y.; Vannini, L.; Gozzi, G.; Balestra, F.; Caboni, M.F.; Gardini, F.; Montanari, C. Fermented nut-based vegan food: Characterization of a home made product and scale-up to an industrial pilot-scale production. *J. Food Sci.* **2018**, *83*, 711–722. [CrossRef]
2. Sun, C.; Ge, J.; He, J.; Gan, R.; Fang, Y. Processing, quality, safety, and acceptance of meat analogue products. *Engineering* **2021**, *7*, 674–678. [CrossRef]
3. Cerdá-Bernad, D.; Clemente-Villalba, J.; Valero-Cases, E.; Pastor, J.-J.; Frutos, M.-J. Novel insight into the volatile profile and antioxidant properties of *Crocus sativus* L. flowers. *Antioxidants* **2022**, *11*, 1650. [CrossRef]
4. Serrano-Díaz, J.; Sánchez, A.M.; Martínez-Tomé, M.; Winterhalter, P.; Alonso, G.L. A contribution to nutritional studies on *Crocus sativus* flowers and their value as food. *J. Food Compos. Anal.* **2013**, *31*, 101–108. [CrossRef]
5. Cerdá-Bernad, D.; Costa, L.; Serra, A.T.; Bronze, M.R.; Valero-Cases, E.; Pérez-Llamas, F.; Candela, M.E.; Arnao, M.B.; Barberán, F.T.; Villalba, R.G.; et al. Saffron against neuro-cognitive disorders: An overview of its main bioactive compounds, their metabolic fate and potential mechanisms of neurological protection. *Nutrients* **2022**, *14*, 5368. [CrossRef]
6. Dewettinck, K.; Van Bockstaele, F.; Kühne, B.; Van de Walle, D.; Courtens, T.M.; Gellynck, X. Nutritional value of bread: Influence of processing, food interaction and consumer perception. *J. Cereal Sci.* **2008**, *48*, 243–257. [CrossRef]
7. Pycia, K.; Ivanišová, E. Physicochemical and antioxidant properties of wheat bread enriched with hazelnuts and walnuts. *Foods* **2020**, *9*, 1081. [CrossRef]
8. Świeca, M.; Gawlik-Dziki, U.; Dziki, D.; Baraniak, B. Wheat bread enriched with green coffee—In vitro bioaccessibility and bioavailability of phenolics and antioxidant activity. *Food Chem.* **2017**, *221*, 1451–1457. [CrossRef]
9. AOAC. *Official Methods of Analysis*, 16th ed.; Association of Official Analytical Chemists: Washington, DC, USA, 1995.
10. Cerdá-Bernad, D.; Valero-Cases, E.; Pastor, J.J.; Frutos, M.J.; Pérez-Llamas, F. Probiotic red quinoa drinks for celiacs and lactose intolerant people: Study of functional, physicochemical and probiotic properties during fermentation and gastrointestinal digestion. *Int. J. Food Sci. Nutr.* **2020**, *73*, 49–59. [CrossRef]
11. García-Segovia, P.; Igual, M.; Martínez-Monzó, J. Physicochemical properties and consumer acceptance of bread enriched with alternative proteins. *Foods* **2020**, *9*, 933. [CrossRef]
12. Gawlik-Dziki, U.; Dziki, D.; Baraniak, B.; Lin, R. The effect of simulated digestion in vitro on bioactivity of wheat bread with Tartary buckwheat flavones addition. *LWT—Food Sci. Technol.* **2009**, *42*, 137–143. [CrossRef]
13. Sicari, V.; Romeo, R.; Mincione, A.; Santacaterina, S.; Tundis, R.; Loizzo, M.R. Ciabatta bread incorporating goji (*Lycium barbarum* L.): A new potential functional product with impact on human health. *Foods* **2023**, *12*, 566. [CrossRef]
14. He, H.; Hoseney, R.C. Changes in bread firmness and moisture during long-term storage. *Cereal Chem.* **1990**, *67*, 603–605.
15. European Food Safety Authority. Dietary reference values for nutrients. *EFSA J.* **2017**, *2017*, e15121.
16. Fahim, N.; Sadat, S.; Janati, F.; Feizy, J. Chemical composition of agriproduct saffron (*Crocus sativus* L.) petals and its considerations as animal feed. *GIDA* **2012**, *37*, 197–201.

17. Fernández-Canto, M.N.; García-Gómez, M.B.; Boado-Crego, S.; Vázquez-Odériz, M.L.; Muñoz-Ferreiro, M.N.; Lombardero-Fernández, M.; Pereira-Lorenzo, S.; Romero-Rodríguez, M.Á. Element content in different wheat flours and bread varieties. *Foods* **2022**, *11*, 3176. [CrossRef]
18. Takeda, K.; Matsumura, Y.; Shimizu, M. Emulsifying and surface properties of wheat gluten under acidic conditions. *J. Food Sci.* **2001**, *66*, 393–399. [CrossRef]
19. Arendt, E.K.; Ryan, L.A.M.; Dal Bello, F. Impact of sourdough on the texture of bread. *Food Microbiol.* **2007**, *24*, 165–174. [CrossRef]
20. Munch-Petersen, A. Citric acid content in meal of certain cereals arid leguminous plants: A contribution to the question of exogenous citric acid. *Acta Physiol. Scand.* **2008**, *8*, 97–102. [CrossRef]
21. Shewry, P.R.; America AH, P.; Lovegrove, A.; Wood, A.J.; Plummer, A.; Evans, J.; van den Broeck, H.C.; Gilissen, L.; Mumm, R.; Ward, J.L.; et al. Comparative compositions of metabolites and dietary fibre components in doughs and breads produced from bread wheat, emmer and spelt and using yeast and sourdough processes. *Food Chem.* **2022**, *374*, 131710. [CrossRef]
22. De Luca, L.; Aiello, A.; Pizzolongo, F.; Blaiotta, G.; Aponte, M.; Romano, R. Volatile organic compounds in breads prepared with different sourdoughs. *Appl. Sci.* **2021**, *11*, 1330. [CrossRef]
23. Purlis, E. Browning development in bakery products—A review. *J. Food Eng.* **2010**, *99*, 239–249. [CrossRef]
24. Kolida, S.; Tuohy, K.; Gibson, G.R. Prebiotic effects of inulin and oligofructose. *Br. J. Nutr.* **2002**, *87* (Suppl. S2), S193–S1937. [CrossRef]
25. Lian, H.; Luo, K.; Gong, Y.; Zhang, S.; Serventi, L. Okara flours from chickpea and soy are thickeners: Increased dough viscosity and moisture content in gluten-free bread. *Int. J. Food Sci. Technol.* **2020**, *55*, 805–812. [CrossRef]
26. Tóth, M.; Kaszab, T.; Meretei, A. Texture profile analysis and sensory evaluation of commercially available gluten-free bread samples. *Eur. Food Res. Technol.* **2022**, *248*, 1447–1455. [CrossRef]
27. Frutos, M.J.; Guilabert-Antón, L.; Tomás-Bellido, A.; Hernández-Herrero, J.A. Effect of artichoke (*Cynara scolymus* L.) fiber on textural and sensory qualities of wheat bread. *Food Sci. Technol. Int.* **2008**, *14*, 49–55. [CrossRef]
28. Czubaszek, A.; Czaja, A.; Sokół-Łętowska, A.; Kolniak-Ostek, J.; Kucharska, A.Z. Changes in antioxidant properties and amounts of bioactive compounds during simulated in vitro digestion of wheat bread enriched with plant extracts. *Molecules* **2021**, *26*, 6292. [CrossRef]
29. Gião, M.S.; Gomes, S.; Madureira, A.R.; Faria, A.; Pestana, D.; Calhau, C.; Pintado, M.E.; Azevedo, I.; Malcata, F.X. Effect of *in vitro* digestion upon the antioxidant capacity of aqueous extracts of *Agrimonia eupatoria*, *Rubus idaeus*, *Salvia* sp. and *Satureja montana*. *Food Chem.* **2012**, *131*, 761–767. [CrossRef]
30. Laparra, J.M.; Sanz, Y. Interactions of gut microbiota with functional food components and nutraceuticals. *Pharmacol. Res.* **2010**, *61*, 219–225. [CrossRef]
31. Hoye, J.; Clifford Ross, C.F. Total phenolic content, consumer acceptance, and instrumental analysis of bread made with grape seed flour. *J. Food Sci.* **2011**, *76*, S428–S436. [CrossRef]
32. Soares, S.; Kohl, S.; Thalmann, S.; Mateus, N.; Meyerhof, W.; De Freitas, V. Different phenolic compounds activate distinct human bitter taste receptors. *J. Agric. Food Chem.* **2013**, *61*, 1525–1533. [CrossRef]
33. Altinok, E.; Kurultay, S.; Boluk, E.; Atik, D.S.; Kopuk, B.; Gunes, R.; Palabiyik, I.; Konar, N.; Toker, O.S. Investigation of using possibility of grape pomace in wafer sheet for wheat flour substitution. *Int. J. Food Sci. Technol.* **2022**, *57*, 3634–3642. [CrossRef]

foods

MDPI

Article

The Potential of Wine Lees as a Fat Substitute for Muffin Formulations

Federico Bianchi [1], Mariasole Cervini [2], Gianluca Giuberti [2] and Barbara Simonato [1,*]

[1] Department of Biotechnology, University of Verona, 37134 Verona, Italy; federico.bianchi_02@univr.it
[2] Department for Sustainable Food Process, Università Cattolica del Sacro Cuore, 29122 Piacenza, Italy; mariasole.cervini@unicatt.it (M.C.); gianluca.giuberti@unicatt.it (G.G.)
* Correspondence: barbara.simonato@univr.it

Abstract: The current study evaluates the prospect of wine lees (WL), a costless by-product from Amarone winemaking, as a fat replacer in muffin formulation. WL have elsewhere replaced sunflower oil, allowing the creation of 0, 25, 50, 75, and 100% fat-substituted muffins named ML0, ML25, ML50, ML75, and ML100, respectively. Batter rheology, in addition to the textural and colorimetric characteristics, the pore dimension, and the sensory aspect of the different formulations were evaluated. The batter consistency (K) of fat-replaced muffins was lower than that of the control, while the hardness and chewiness of the end products were higher. ML25 and ML50 samples reached the highest volume, while the baking loss decreased due to WL's fiber components. ML25, ML50, ML75, and ML100 accounted for caloric reductions of 9, 18, 22, and 26%, respectively, compared to full-fat muffins. Muffins with WL showed a darker crust and crumb as lightness (L^*) decreased. Moreover, a^* parameter increased with the increment of WL in the formulation, leading to a redder and less yellow-hued fat-replaced muffin. In conclusion, WL could effectively replace fat in the 25–50% range in muffins, achieving a final product with reduced calories, a higher dietary fiber content, higher volume, and promising sensory aspects.

Keywords: fat replacers; circular economy; wine lees; sustainable food products; rheology; sensory analyses

1. Introduction

Non-communicable diseases (NCDs), such as obesity and type 2 diabetes, are related to an unbalanced diet [1–3]. Therefore, consumers have become aware of the issue and changed their food habits by reducing sugar and fat consumption while increasing their dietary fiber intake. On the other hand, developing novel and healthier food products is a real challenge for bakers and food companies. In bakery goods, fats act as a texture improver, flavor retainer, and moistener, influencing the mouthfeel and delaying starch retrogradation [4]. Moreover, shortening/oils in cake or muffin formulations tend to plasticize, lubricate, and strengthen the dough, improve gas retention, and create higher volume and softness in final products. Even though fat is an essential ingredient in several bakery formulas, reducing fat intake helps to reduce body weight and the risk of NCDs [5].

Indeed, fat replacers have recently undergone significant development, driven by consumers' demand for low-fat foods [6]. Fat replacers can help to replace the whole or part of the fat within a food while maintaining the same functionality. The choice of the proper fat replacer is crucial and depends on the baked product's formulation. Fat replacers can be carbohydrate, protein, or lipid-based and impart fat-like properties such as texturization, stabilization, emulsification, gelling, and thickening. Many fat replacers have been proposed in the last decade, and several are available in the market [6].

Use of the discarded grape by-product, such as grape pomace and lees from the fermentation process, could promote a circular economy and responsible use of natural resources, and also prevent environmental degradation, following the principles of the ONU Agenda 2030 [7]. Moreover, the upcycling of agro-industrial by-products could represent an opportunity to exploit since they are inexpensive and rich in dietary fiber and bioactive compounds. The effect of wine by-product addition in various cereal-based foodstuffs has been widely reported [8–11]. Wine lees (WL) are one of the main by-products of vinification. According to Bonamente et al. (2015), grapes' by-products are comprised approximately 60% of pomace, 14% of stalks, and 25% of lees [12]. The WL are characterized mainly by mannoproteins from the yeast cell wall, mineral compounds, and dietary fibers, including pectin [13]. Pectin is a soluble fiber found in plants' cell walls that is commonly used in several food formulas to control their textures and rheologies. Lim et al. (2014) reported that making cake with 30% of its shortening replaced by pectin resulted in batter with higher viscosity and cakes with a slightly lower volume than the control [14]. In addition, mannoproteins have already been tested as effective emulsifiers in different food formulations [15,16]. Emulsifiers, mixes of emulsifiers, or emulsifiers blended with dietary fibers or other food components could represent an effective fat replacer [4,17,18]. For instance, emulsifiers can compensate for the volume losses incurred due to fat replacement in cakes [14,15]. Moreover, several authors have reported discrete success in replacing fat with vegetable puree in bakery products [19–22]. Therefore, WL could be a promising fat-mimetic ingredient used to produce low-fat muffins. To our knowledge, no studies on WL as a fat replacer in foods are present in the literature. This work aimed to evaluate the technological, sensory, and nutritional characteristics of muffins formulated with increasing levels of WL (25, 50, 75, and 100%) as a fat replacer.

2. Materials and Methods

2.1. Ingredient and Preparation

Wheat flour was supplied from Macinazione Lendinara (Arcole, Italy), and sunflower oil, salt, sugar, fresh eggs, and baking powder were commercially available. The wheat flour's composition was: total carbohydrates 71 g/100 g, protein 11 g/100 g, fat 1.2 g/100 g, and total dietary fiber 2.3 g/100 g (as reported on the label). The WL were obtained after the production of Amarone wine in Valpolicella and kindly provided by Cantina Sperimentale of Verona University (Verona, Italy). The WL were stored in plastic bottles in a −18 °C refrigerator and heated up to 25 °C upon use for WL characterization or muffin preparation. The muffins were prepared according to the formulation reported in Table 1. Briefly, wheat flour, salt, sugars, and baking powder were put in a planetary mixer set at medium speed for 5 min with tap water, sunflower oil, and whole eggs to obtain control sample ML0. Oil was replaced with WL in muffin preparation by 25, 50, 75, and 100% of its weight; samples were named ML25, ML50, ML75, and ML100, respectively. Fifty grams of muffin batter was poured into a pre-weighted silicone mold and then cooked at 180 °C for 20 min in a ventilated oven. Each formulation was prepared in triplicate.

Table 1. Experimental formulas of control and fat-replaced muffin samples with WL.

Sample	Sugar (g)	Sunflower Oil (g)	Wine Lees (g)	Salt (g)	Baking Powder (g)	Wheat Flour (g)	Water (g)	Fresh Eggs (g)
ML0	50	25	0	1	2	100	50	35
ML25	50	18.75	6.25	1	2	100	50	35
ML50	50	12.5	12.5	1	2	100	50	35
ML75	50	6.25	18.75	1	2	100	50	35
ML100	50	0	25	1	2	100	50	35

2.2. Proximate Composition of Muffins and WL

The proximate composition determination for the muffin and WL included DM (method 930.15), ash (method 942.05), total sugars (total of D-glucose, D-fructose, and sucrose; Megazyme cat. no. K-SUFRG), crude protein (method 976.05), crude lipid (method 954.02), and total starch (method 996.11), using thermostable amylase (Megazyme cat. no.

E-BSTAA), amyloglucosidase (Megazyme cat. no. E-AMGDF), and tartaric acids (Tartaric acid assay kit, Megazyme). The total, soluble and insoluble dietary fiber contents (TDF, SDF, and IDF, respectively) were determined according to method 991.43. [23]. Analyses were conducted in triplicate. In addition, WL was assessed for Brix, pH, and total phenol content (TPC). The TPC was determined according to Rainero et al. (2022) and expressed as gallic acid equivalent on DM, based on a 7-point standard curve [9]. The caloric contents of the muffin samples were calculated with the following equation:

$$\text{Calories}\left(\frac{\text{kcal}}{100\ \text{g}}\right) = 9 \times \text{F} + 4 \times \text{C} + 4 \times \text{P} \tag{1}$$

where F, C, and P represent the total amount of fat, carbohydrates, and protein in 100 g of muffin sample.

2.3. Rheology of Batters

A rheometer (DRS500 CP4000 PLUS, Lamy Rheology, France) equipped with an MSDIN-11 system was used for rheological analyses. Fifteen ml of batter were poured into the vessel and let rest for 1 min to reach a constant temperature of 25 °C. Shear stress was a function of shear rate over the 10–300 s^{-1} range. The results were fitted to the Power Law model Equation (2) by the software program Rheotex (Lamy Rheology, France).

$$\tau = \text{K}\cdot\gamma^{\text{n}} \tag{2}$$

where "τ" is shear stress (Pa), "K" is the consistency index (Pa), and "n" is the flow index. All measurements were in triplicate.

2.4. Technological Properties

The moisture content of the muffins was evaluated using AACC method 44-15A [24]. To determine the baking loss, the mass difference between the batter and the baked muffins was measured. The volume was determined using rapeseed displacement method AACC 10-05.01 (AACC, 2000), and the density was calculated by dividing the weight of the sample by its volume. The heights of the muffins were measured using a digital caliper [10].

2.5. Texture Profile Analyses

Once the muffins had cooled down, their texture profile analysis (TPA) was conducted using a texture analyzer TX700 (Lamy Rheology, Champagne-au-Mont-d'Or, France) with a flat probe measuring 50 mm in diameter. For the TPA analyses, two cm of each muffin was cut from the bottom and the upper part was discarded. Seven measurements were performed for each batch. Hardness, cohesiveness, chewiness, and resilience values were collected. Hardness was calculated as the maximum force of the 1st cycle of compression, cohesiveness was the ratio of the work of the first compression and the second, and chewiness was a product of hardness, cohesiveness, and springiness, while resilience was the ratio between the time to reach the maximum peak force of the second cycle and the time of the first cycle.

2.6. Physical Characterization of Muffins

Muffins were cut to obtain a slice of 1–2 cm in height and about 20 cm^2 in area. The scanner captured images of the muffin sections at a resolution of 600 DPI. Images were first converted to greyscale and binarized with ImageJ (version 1.8.0) to obtain the pore area fraction; the density of pores; the average pore diameter; D1,0 and D3,2; the pore perimeter; and the pore circularity [10]. Analyses were performed on five samples for each formulation.

2.7. Color Analysis

The color was measured by a reflectance colorimeter (Illuminant D65) (Minolta Chroma meter CR-300, Osaka, Japan) based on the color system CIE—L^* a^* b^*. Measurements were taken at five different points within the crumb and crust areas.

2.8. Sensory Analysis of Muffins

A quantitative descriptive sensory analysis (QDA) was conducted in the sensory test room of the University of Verona; the QDA was designed according to UNI ISO 8589 standards. The panelists were carefully selected based on their sensory sensitivity and ability to describe and communicate sensory perceptions. The panelists, consisting of 18 people (10 females and 8 males) between the ages of 22 and 28, recruited from the staff and students of the Department of Biotechnology at the University of Verona, were trained for 12 sessions of one hour each to evaluate the specific sensory attributes of muffins.

The panelists generated 10 sensory terms and were qualified to identify them. The following sensory attributes were evaluated: color intensity, egg taste, wine taste, wine odor, sweetness, acidity, chewiness, elasticity, hardness, and astringency. All sensory attribute definitions are reported in Table A1 (Appendix A). A nine-point scale, where 1 and 9 represented the lowest and highest intensities, respectively, was used [25]. The muffins were presented to panelists on covered plates, coded in a randomized and balanced manner, and then provided to the judges for the sensory analysis session. We asked the judges to rinse their mouths with water between each sample taste test. At the end of the session, the formats were taken, viewed, and statistically processed.

2.9. Statistical Analysis

All data are presented as mean values ± standard deviation, based on at least three measurements (n = 3). Analysis of variance (ANOVA) followed by a post-hoc Tukey test at a significance level of $p < 0.05$ was employed for comparing the means. Pearson's correlation tests and statistical analyses were performed using XLSTAT Premium Version (2021.1.1, Addinsoft SARL, Paris, France) software.

3. Results and Discussion

3.1. Proximate Composition of Wine Lees

The chemical composition of WL is reported in Table 2. The current composition of WL from Amarone production had content values (DM basis) of polyphenol, tartaric acid, ashes, pH, and dietary fiber content similar to the WL from Bustamante et al. (2008) and Gòmez et al. (2004) [26,27]. However, in Amarone WL, a higher content of proteins and a lower content of lipids were detected. The peculiar grapevine and unique vinification method of Amarone wine (such as the juice extraction from withered grapes) could explain these differences [28].

Table 2. Proximate composition, pH, Brix, and moisture content of Amarone wine lees. Protein, fat carbohydrates, ash, dietary fibers, phenol content and tartaric acid are reported on a DM basis.

Protein	Fat	Carbohydrates	Ash	Dietary Fibers	Phenols	Tartaric Acid	pH	Brix	Moisture
21.2 ± 3.4	0.07 ± 0	4.92 ± 0.11	10.27 ± 0.05	20.1 ± 3.0	1.53 ± 0.11	24.6 ± 1.2	3.41 ± 0.01	11.02 ± 0.21	87 ± 0.1

3.2. Rheological Properties of Muffins' Batter

Batter rheological data are reported in Table 3. All samples fitted the power law model equation, showing an r^2 higher than 0.997. All batter formulations showed a shear-thinning (pseudoplastic) behavior ($n < 1$) over the studied shear rate (10–300 s^{-1}), suggesting that the apparent viscosity decreased with the increase in shear rate. Bianchi et al. (2022) reported a similar performance in muffin batters made with grape pomace powder [10]. In the current study, the maximum flow index (n) value occurred in ML25 and ML50 samples, sequentially. In addition, ML75 and ML100 showed lower n than did the ML0, ML25, and ML50 samples, suggesting that their apparent viscosity increased less distinctly when the shear rate increased.

Table 3. Reported values for batter flow index n and batter consistency index k for each sample.

Sample	Flow Index n	Consistency Index K	r^2
ML0	0.558 ± 0.002 [c]	44.20 ± 0.44 [a]	0.997
ML25	0.593 ± 0.002 [a]	28.45 ± 0.11 [c]	0.992
ML50	0.577 ± 0.004 [b]	29.06 ± 0.30 [c]	0.989
ML75	0.529 ± 0.004 [d]	28.94 ± 0.63 [c]	0.991
ML100	0.407 ± 0.009 [e]	33.01 ± 1.14 [b]	0.974

Superscript numbers with different letters within the same column are statistically different ($p < 0.05$). ML0: control sample with no oil substitution; ML25: sample with 25% WL and 75% oil; ML50: sample with 50% WL and 50% oil; ML75: sample with 75% WL and 25% oil; ML100: sample with 100% WL.

ML25, ML50, and ML75 showed similar consistencies, but were all lower than the full-fat sample, while ML100 had a higher consistency, resembling the behavior of 100% pectin as a fat replacer in cakes, as reported by Psimouli & Oreopoulou et al. (2012) [29]. The consistency increase is possibly due to the high thickening capacity of pectins from WL, since WL contains about 20% DM of dietary fibers, mainly pectins [30]. However, ML25, ML50, and ML75 have almost 35% less consistency than does the full-fat sample, which could be attributed to mannoproteins contained in WL [13]. Indeed, emulsifiers could lower batter consistency and apparent viscosity [4,31].

The higher consistency indicated a higher ability to entrap air during the beating process, resulting in a lower density of batters and lower volume development. If the K value is too low, cake or muffin volumes shrink because batters lose effectiveness in retaining air bubbles [32,33]. This can explain the greater volume of the control muffins compared to ML75 and ML100. Indeed, ML0 had higher consistency and higher volume. However, ML25 and ML50 showed the opposite results. The flow index n was negatively correlated with fiber, hardness, and chewiness, whereas no relationships were found between consistency and textural attributes or fiber content (Table A2).

3.3. Proximate Composition of Muffins

The chemical composition of the samples is reported in Table 4. The protein and total carbohydrate content remained unchanged between the different formulations. However, the ash content increased with increasing levels of WL in the recipe. This increase is related to the presence of macro- and micro-elements such as P, K, Mn, Zn, Ca, and Mg in WL [28]. The TDF content significantly increased from 0.30 g/100 g in the control sample to 3.16 g/100 g in ML100. This increment is due to the dietary fiber fraction of Amarone WL. As a result, ML100 can benefit from the claim of "source of fiber" as it contains more than 3 g/100 g of TDF [34]. Other studies have reported similar findings for bread, breadsticks, cookies, and muffins fortified with grape by-products [9,35–38]. Muffin fat content was successfully reduced from 13.14% for sample ML0 to 1.82% in sample ML100. By replacing fat, the fat-reduced muffins achieved lower calories per gram, as fat is the most calorie-dense nutrient in all formulations. Accordingly, muffins ML25, ML50, ML75, and ML100 accounted for calorie reductions of 9, 18, 22, and 26%, respectively, as compared to ML0. The TDF and ash content were positively correlated with the density of muffins and negatively correlated with the fat content, due to the substitution of fat with the Amarone WL (Table A2).

3.4. Technological Properties of Muffins

The oil replacement in muffins influenced most of the technological parameters, as depicted in Table 5. Even though the heights of the muffins decreased with oil replacement in the formulations, the detected differences were low and perceptible mainly in ML0 and ML25 compared to ML100. Fat replacement slightly affected sample volume: ML25 and M50 showed the highest, and ML100 accounted for the lowest volume. Moreover, a small increase in muffin density was revealed in LM75 and LM100, showing a WL impact in the batter density, probably due to the water adsorption capacity and the high fiber content.

Table 4. The nutritional contents of muffins with varying levels of fat substitution. Data are expressed as grams per 100 g of dry matter (DM), and the calorie values are provided as kilocalories per 100 g of fresh product.

Sample	ML0	ML25	ML50	ML75	ML100
Protein	6.72 ± 0.23 [a]	6.34 ± 0.63 [a]	6.67 ± 0.46 [a]	6.67 ± 0.06 [a]	6.96 ± 0.52 [a]
Fat	13.14 ± 0.08 [a]	9.80 ± 0.07 [b]	6.59 ± 0.30 [c]	4.68 ± 0.13 [d]	1.82 ± 0.08 [e]
Carbohydrates	55.21 ± 0.62 [a]	54.64 ± 0.30 [a]	54.51 ± 1.11 [a]	54.12 ± 1.64 [a]	56.38 ± 0.66 [a]
Ash	1.36 ± 0.04 [b]	1.35 ± 0.03 [b]	1.44 ± 0.01 [ab]	1.49 ± 0.01 [ab]	1.54 ± 0.04 [a]
Dietary Fibers	0.30 ± 0.08 [d]	0.28 ± 0.00 [d]	1.14 ± 0.00 [c]	2.27 ± 0.32 [b]	3.16 ± 0.02 [a]
Calories	366 ± 24 [a]	332 ± 20 [b]	300 ± 17 [c]	285 ± 13 [d]	270 ± 12 [e]

Superscript numbers with different letters within the same line are statistically different ($p < 0.05$). ML0: control sample with no oil substitution; ML25: sample with 25% WL and 75% oil; ML50: sample with 50% WL and 50% oil; ML75: sample with 75% WL and 25% oil; ML100: sample with 100% WL.

Table 5. Technological properties of muffins with different levels of substitution of fat. Data are expressed on a DM basis.

Sample	Height (cm)	Baking Loss (%)	Volume (cm^3)	Density (g/cm^3)	Moisture (g/100 g)
ML0	4.53 ± 0.10 [a]	14.1 ± 1.0 [a]	87.8 ± 5.7 [b]	0.50 ± 0.04 [b]	22.20 ± 0.71 [d]
ML25	4.53 ± 0.08 [a]	9.15 ± 0.93 [bc]	95.7 ± 3.5 [a]	0.48 ± 0.00 [b]	28.25 ± 0.07 [c]
ML50	4.48 ± 0.12 [ab]	10.46 ± 0.78 [b]	92.2 ± 5.7 [ab]	0.50 ± 0.03 [b]	30.15 ± 0.49 [bc]
ML75	4.43 ± 0.19 [ab]	7.1 ± 1.1 [d]	73.9 ± 3.5 [c]	0.63 ± 0.02 [a]	31.90 ± 0.42 [ab]
ML100	4.37 ± 0.03 [b]	7.9 ± 1.2 [dc]	73.5 ± 2.3 [c]	0.63 ± 0.02 [a]	32.95 ± 0.07 [a]

Superscript numbers with different letters within the same column are statistically different ($p < 0.05$). ML0: control sample with no oil substitution; ML25: sample with 25% WL and 75% oil; ML50: sample with 50% WL and 50% oil; ML75: sample with 75% WL and 25% oil; ML100: sample with 100% WL.

In addition, the greatest volume was recorded for ML25 and ML50, increasing by 8.32 and 4.3% compared to the control, respectively, suggesting that the optimal substitution of fat with WL to achieve a better muffin volume should not exceed 50%. Higher fat replacement provoked a volume decrease of about 16% compared to the control. This could reflect the role of WL in capturing air during the beating step of batter preparation, stabilizing the air cell, and preventing their coalescence during the cooking, but only below the 50% threshold of replacement [22]. Even though parameters such K index should suggest a thicker batter and better gas retention ability, ML75 and ML 100 volumes were much lower than ML0, which was probably caused by an insufficient amount or the complete absence of oil in the formulations. Such conditions could promote bubble coalescence or a lack of gas retention and bubble expansion during cooking. For instance, Marchetti et al. (2018) reported that a low batter consistency led to cakes with lower volume [32]. The effect of dietary fibers on muffin volume has been widely reported. For instance, muffins have been enriched with brewer's spent grain, peach fibers, linseed, flaxseed, wheatgrass, and grape pomace [10,39–42].

ML0 accounted for the highest baking loss in the study, whereas ML100 and ML75 showed the lowest ones. Bender et al. (2017) argued that the hydration capacity of the fiber fractions in fortified muffins led to a decrease in the baking loss when Riesling skin or Tannat skin powder was added [38]. The presence of fiber components in the WL could affect the baking performances of the muffins even if no correlation was detectable. Indeed, the correlation analyses (Table A2) showed that fiber content was negatively correlated with the volume. The moisture content increased in the fat-replaced muffins due to water in the WL. High moisture was detected by other authors using vegetal-based fat replacers in baked goods [22,43]. However, the increase in the moisture content of a WL-containing muffin could presumably have a negative impact on its shelf-life. Consequently, further studies are warranted to explore the relationship between WL inclusion levels and the microbial stability of related food products during their shelf-life.

3.5. Image Analyses of Muffins

The image analyses showed that the muffin had no significative differences in pore area fraction and pore circularity (Table 6). Indeed, Abdul Manaf et al. (2017) obtained similar results replacing fat with *Persea* puree [20], while Rodríguez-García et al. (2012) showed how inulin increased pores' circularity and total cell area [44].

Table 6. Pore characterization of muffins' crumbs with different levels of substitution of fat.

Sample	Area Fraction (%)	Pore Density (Pores/cm^2)	Perimeter (mm)	Circularity (%)	D1,0	D3,2
ML0	10.76 ± 0.36 [a]	12.5 ± 2.0 [b]	2.84 ± 0.26 [a]	0.81 ± 0.01 [a]	1.41 ± 0.09 [a]	4.02 ± 0.33 [a]
ML25	10.91 ± 0.37 [a]	24.5 ± 5.1 [a]	2.13 ± 0.30 [b]	0.81 ± 0.01 [a]	1.13 ± 0.10 [ab]	3.39 ± 0.60 [b]
ML50	10.64 ± 0.30 [a]	23.7 ± 4.2 [a]	1.83 ± 0.26 [c]	0.81 ± 0.01 [a]	1.13 ± 0.10 [b]	3.07 ± 0.31 [b]
ML75	10.30 ± 0.20 [a]	25.1 ± 3.9 [a]	1.94 ± 0.23 [c]	0.82 ± 0.01 [a]	1.18 ± 0.09 [b]	2.81 ± 0.23 [b]
ML100	10.74 ± 0.29 [a]	14.1 ± 2.6 [b]	2.79 ± 0.27 [a]	0.81 ± 0.01 [a]	1.41 ± 0.07 [a]	3.94 ± 0.43 [a]

Superscript numbers with different letters within the same column are statistically different ($p < 0.05$). ML0: control sample with no oil substitution; ML25: sample with 25% WL and 75% oil; ML50: sample with 50% WL and 50% oil; ML75: sample with 75% WL and 25% oil; ML100: sample with 100% WL.

The ML25, ML50, and ML75 samples had higher pore densities than did the control or the ML100 sample. These data, for WL replacement < 100%, are in line with previous studies reporting that muffins with defatted sunflower seed flour, pecan nut expeller, and grape pomace powder had higher pore densities than did their controls [10,32,45]. Marchetti et al. (2018) stated that a higher pecan nut by-product amount promotes pore density in fortified muffins and a better gas retention capacity.

The pore perimeter values D1,0 and D3,2 showed a similar trend when replacing fat with Amarone WL. Indeed, ML100 led to a pore density similar to ML0, and thus to a more open crumb with a lower volume.

3.6. Color of Muffins

Consumers typically attribute darker crusts and crumbs to whole-grain baked goods. The addition of WL significantly changed ($p < 0.05$) the color of the muffins (Table 7). The color of the crust is formed during the baking process due to the caramelization of sugars and the Maillard reaction. Incorporating WL increases the Maillard reaction by increasing muffins' amino acid and carbohydrate contents. Additionally, the crumb color is affected by cake ingredients, and the pigments of WL increase the color parameters of muffins [15,36]. The L and b^* parameters decreased ($p < 0.05$), both in crust and crumb, following the substitution level of the fat with WL, while the a^* parameter significantly increased ($p < 0.05$). Bianchi et al. (2022), Rainero et al. (2022), and Acun et al. (2014) obtained similar results, respectively, in muffins with distilled grape pomace powder, breadsticks with grape pomace powder, and cookies fortified with different grape products [9,10,46]. Moreover, the color parameter of the bread and cake fortified with grape pomace from winemaking showed a similar trend [37,47,48].

3.7. Texture of Muffins

As depicted in Table 8, hardness increased as WL replaced the oil in the formulation. For instance, hardness in the ML100 was 2.6-fold higher than that in the M0. Arifin et al. (2019) explored the effect of pumpkin puree as a fat replacer in muffins [21]. The pumpkin-puree-added muffin had higher hardness than the full-fat sample because lipids interact with gluten fibrils, promoting a softer muffin. Moreover, our findings agree with Grigelmo-Miguel et al. (2001) and Bianchi et al. (2022), as they reported that fiber-increased muffins had higher hardness than their respective control samples [10,40]. ML25 was the only replaced muffin with no significant difference compared to control ML0. Similarly, Hussien et al. (2016) observed no differences in 25% fat-replaced cantaloupe and squash puree muffins [22].

Table 7. Cielab parameters of muffins with different levels of substitution of fat. Beneath, a picture of the cross-section of the experimental muffins.

Crumb/Crust	Cielab Parameters	ML0	ML25	ML50	ML75	ML100
Crumb	*L*	73.50 ± 0.95 [a]	61.1 ± 5.6 [b]	61.0 ± 2.4 [b]	55.5 ± 3.0 [c]	45.4 ± 3.4 [d]
	*a**	−3.63 ± 0.37 [e]	−1.03 ± 0.49 [d]	3.29 ± 0.39 [c]	5.44 ± 0.39 [b]	7.9 ± 0.51 [a]
	*b**	41.8 ± 3.1 [a]	21.3 ± 2.2 [b]	8.7 ± 1.39 [c]	7.9 ± 1.2 [c]	5.35 ± 0.23 [d]
Crust	*L*	70.0 ± 1.6 [a]	60.3 ± 6.0 [b]	59.9 ± 3.3 [b]	44.2 ± 2.4 [c]	36.8 ± 3.6 [d]
	*a**	−2.17 ± 0.50 [e]	−1.33 ± 0.48 [d]	2.39 ± 0.34 [c]	5.80 ± 0.91 [b]	7.46 ± 0.83 [a]
	*b**	45.6 ± 3.4 [a]	24.0 ± 1.3 [b]	8.7 ± 1.2 [c]	9.0 ± 1.4 [cd]	7.0 ± 1.3 [d]

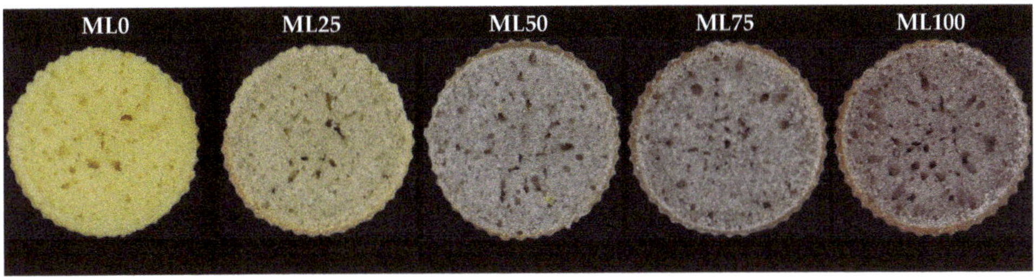

Superscript numbers with different letters within the same line are statistically different ($p < 0.05$). ML0: control sample with no oil substitution; ML25: sample with 25% WL and 75% oil; ML50: sample with 50% WL and 50% oil; ML75: sample with 75% WL and 25% oil; ML100: sample with 100% WL.

Table 8. Nutritional content of muffins with or without different levels of substitution of fat. Data are expressed on a DM basis.

Sample	Hardness (N)	Cohesiveness	Chewiness (N)	Resilience
ML0	38.0 ± 4.6 [c]	0.52 ± 0.01 [c]	19.8 ± 2.0 [e]	0.32 ± 0.05 [d]
ML25	35.3 ± 4.80 [c]	0.64 ± 0.01 [b]	29.0 ± 3.0 [d]	0.40 ± 0.05 [c]
ML50	46.4 ± 6.1 [c]	0.63 ± 0.01 [b]	32.43 ± 0.51 [c]	0.42 ± 0.06 [c]
ML75	73 ± 7.5 [b]	0.71 ± 0.05 [a]	49.7 ± 8.7 [b]	0.52 ± 0.05 [b]
ML100	99.1 ± 9.3 [a]	0.72 ± 0.02 [a]	67.8 ± 5.0 [a]	0.59 ± 0.05 [a]

Superscript numbers with different letters within the same column are statistically different ($p < 0.05$). ML0: control sample with no oil substitution; ML25: sample with 25% WL and 75% oil; ML50: sample with 50% WL and 50% oil; ML75: sample with 75% WL and 25% oil; ML100: sample with 100% WL.

Cohesiveness increased with the rate of fat replacement level. These results agree with Rodriguez-Garcia et al. (2012), who stated that this increase might be due to a denser crumb cell structure in fat-replaced cakes [44]. Indeed, below the 75% level of fat replacement with WL there are positive correlations with pore density, perimeter of cells, D3,2, and specific volume. Grigelmo-Miguel et al. (2001) reported that peach fibers, when used as fat replacers, had little effect on either cohesiveness or springiness in muffins. However, the fat replacement was lower than 10%, and this reported effect could depend on the fat replacer concentration [40].

Chewiness values for fat-replaced samples were statistically higher than that of the control sample. Martínez-Cervera et al. (2011) obtained the opposite results in a muffin with cocoa fibers used as a fat substitute [49]. However, cocoa fiber as a fat replacer still contained fat and had a higher fraction of soluble dietary fibers, which could promote softness in bakery products [4]. The chewiness and hardness were positively correlated with the density of muffins, the ash, and the fiber content, while the flow index "n" and the fat content were negatively correlated (Table A2).

In the present study, all the samples had resilience values lower than 0.6, reflecting the presence of softening ingredients, such as sugars and fat, in the formulations. Indeed, resilience increased in muffins with lower fat content, underlying their role as tenderizers in bakery products.

Overall, hardness, cohesiveness, chewiness, and resilience increased with the incremental increase of WL used as a fat replacer in muffins.

3.8. Sensory Analysis of Selected Muffins

The replacement of oil with WL significantly influenced most of the selected sensory attributes, as depicted in Figure 1 and Table A2.

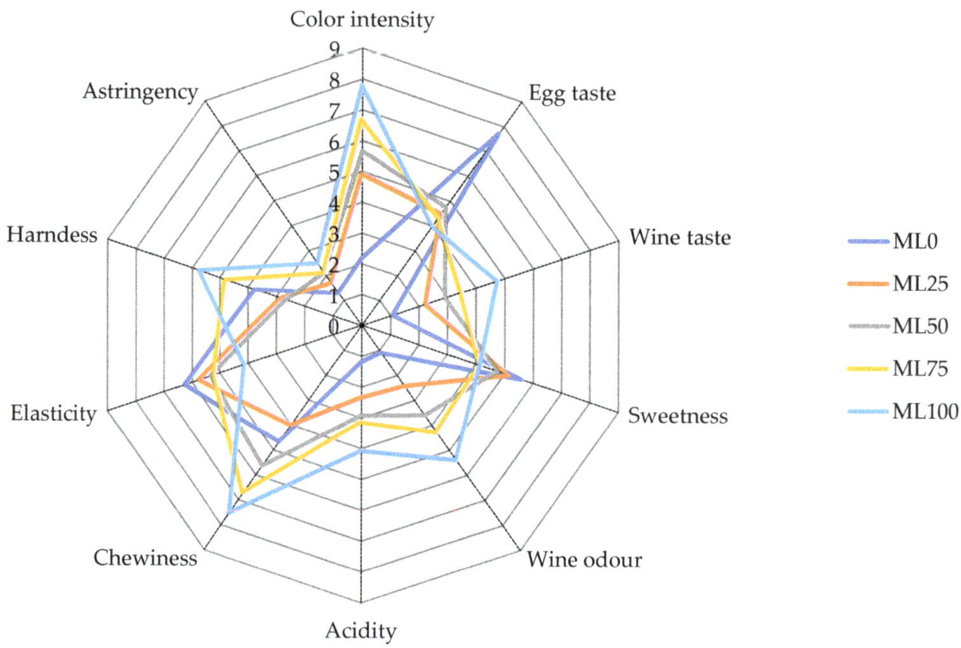

Figure 1. Spider plot of the sensory attributes of the muffin samples. ML0: control sample with no oil substitution; ML25: sample with 25% WL and 75% oil; ML50: sample with 50% WL and 50% oil; ML75: sample with 75% WL and 25% oil; ML100: sample with 100% WL.

Color intensity changed with the rate of fat replacement in muffins and resulted in higher values for the ML100. Similarly, Bianchi et al. (2022) reported an increase in the perceived color of muffins fortified with distilled grape pomace powder, due to grape pigments. The WL's inclusion affected almost all aroma- and taste-based sensory attributes ($p < 0.05$). The WL substitution reduced the egg taste perception and raised the wine taste. Noticeably, the egg perception was strongly affected by the WL addition, decreasing from a score of 7.9 for ML0 to less than 4.8 for the fat-reduced muffin. Wine odor and acidity were influenced by the oil replacement with WL as well. Bianchi et al. (2022) reported a similar effect in muffins fortified with grape pomace powder. The chewiness and hardness increased in muffins with the increased use of fat replacer. Indeed, the chewiness and hardness attributes of ML100 and ML75 were higher and statistically significant (Table A3) compared to ML0. For all samples, stringency and sweetness were not statistically different.

4. Conclusions

Reducing dietary fat in cereal-based food products represents a key focus of food industries. This study showed the potential of Amarone wine lees (WL) as a sustainable and effective ingredient used to partially replace fat ingredients in muffin formulations.

Overall, nutritional, technological, and sensorial analyses indicated that the suggested amount of Amarone WL as a fat replacer in wheat-based muffins was between 25% and 50%, causing a calorie reduction and a higher dietary fiber content of the products without drastically compromising technological and sensorial aspects.

However, further optimization of the muffin formulas with 75% and 100% of WL by the addition of ingredients or alternative mixing and whipping strategies may improve their texture and sensory properties. Overall, this study highlights the potential of using WL as a sustainable fat-substitute in baked goods. Further research is also needed to explore the potential of WL as fat-replacer in different cereal-based food formulations.

Author Contributions: Conceptualization: B.S. and F.B.; formal analysis: F.B. and M.C.; methodology: B.S. and G.G.; data curation: F.B., B.S., G.G. and M.C.; original draft preparation: F.B., B.S. and M.C.; review and editing: B.S., F.B. and G.G. All authors have read and agreed to the published version of the manuscript.

Funding: This research received no external funding.

Data Availability Statement: Data is contained within the article.

Conflicts of Interest: The authors declare no conflict of interest.

Appendix A

Table A1. Definitions of Attributes used in the sensory analysis.

Attributes	Definition
Color intensity	Intensity of perceived color
Egg taste	Intensity of egg taste perceived during the mastication of muffins
Wine taste	Intensity of wine taste perceived during the mastication of muffins
Sweetness	Intensity of sweet taste perceived during the mastication of muffins
Wine odor	Intensity of wine odor perceived by smelling the muffins
Acidity	Intensity of acid taste perceived during the mastication of muffins
Chewiness	Number of chews (at 1 chew/s) needed to masticate the sample to a consistency suitable for swallowing
Elasticity	Energy required to disintegrate a semi-solid food to a state ready for swallowing
Hardness	Force required to deform the product to a given distance, i.e., force to compress between molars, bite through with incisors, or compress between tongue and palate
Astringency	The intensity of drying-out/abrading sensation felt on the tongue or the palate during mastication.

Table A2. Correlation analyses table. The reported values represent the Pearson's coefficients, and the bold numbers represent significative differences for p-value < 0.05. On the right is depicted a scale used as a reference for the color in the main table.

	BL	Density	n	Fat	Ash	TDF	Crumb L	Crust L	Hardness	Chewiness
Volume	0.539	**−0.981**	0.818	0.712	−0.853	**−0.894**	0.657	0.824	**−0.906**	−0.832
BL		−0.688	0.437	0.835	−0.677	−0.709	0.867	0.865	−0.658	−0.781
Density			−0.808	−0.814	**0.901**	**0.939**	−0.769	**−0.908**	**0.934**	**0.898**
n				0.737	−0.823	**−0.881**	0.776	0.821	**−0.941**	**−0.894**
TDF					**−0.949**	**−0.941**	**0.963**	**0.949**	**−0.895**	**−0.946**
Ash						**0.987**	−0.869	**−0.915**	**0.953**	**0.938**
Fibers							**−0.898**	**−0.955**	**0.987**	**0.975**
Crumb L								**0.964**	**−0.884**	**−0.960**
Crust L									**−0.947**	**−0.987**
Hardness										**0.978**
Chewiness										

Scale
1.000
0.750
0.500
0.250
0.000
-0.250
-0.500
-0.750
-1.000

BL: Baking loss; TDF: total dietary fiber content; Crumb L: crumb lightness; Crust L: crust lightness. The bold numbers are statistically significative for $p < 0.05$.

Table A3. Values of the sensory attributes for all experimental muffin samples.

Attribute	ML0	ML25	ML50	ML75	ML100
Color Intensity	2.2 ± 1.9 [c]	4.9 ± 1.4 [b]	5.7 ± 1.1 [b]	6.7 ± 1.2 [a]	7.8 ± 1.0 [a]
Egg taste	7.7 ± 1.9 [a]	4.5 ± 2.1 [b]	4.8 ± 2.4 [b]	4.4 ± 1.8 [b]	4.0 ± 2.0 [b]
Wine taste	1.1 ± 0.3 [c]	2.3 ± 1.4 [bc]	2.9 ± 1.3 [b]	3.5 ± 1.0 [b]	4.8 ± 1.2 [a]
Sweetness	5.6 ± 1.6 [a]	5.2 ± 1.6 [a]	4.8 ± 1.8 [a]	4.2 ± 2.1 [a]	4.2 ± 1.8 [a]
Wine odor	1.1 ± 0.3 [c]	2.4 ± 1.7 [bc]	3.6 ± 1.8 [b]	4.3 ± 1.7 [b]	5.4 ± 1.7 [a]
Acidity	1.2 ± 1.4 [c]	2.3 ± 1.3 [bc]	2.9 ± 2.0 [b]	3.1 ± 2.1 [b]	4.1 ± 1.8 [a]
Chewiness	4.7 ± 1.6 [bc]	4.0 ± 1.3 [c]	5.6 ± 1.4 [b]	6.7 ± 1.1 [a]	7.5 ± 1.4 [a]
Elasticity	6.3 ± 1.8 [b]	5.8 ± 1.8 [b]	5.3 ± 1.1 [b]	5.2 ± 1.4 [ab]	4.1 ± 1.0 [a]
Hardness	3.8 ± 2.0 [b]	2.8 ± 1.0 [b]	2.7 ± 1.1 [b]	4.8 ± 1.5 [a]	5.8 ± 1.9 [a]
Astringency	1.3 ± 1.6 [a]	1.7 ± 0.8 [a]	2.1 ± 1.3 [a]	2.1 ± 1.4 [a]	2.5 ± 1.8 [a]

Superscript numbers with different letters within the same line are statistically different ($p < 0.05$). ML0: control sample with no oil substitution; ML25: sample with 25% WL and 75% oil; ML50: sample with 50% WL and 50% oil; ML75: sample with 75% WL and 25% oil; ML100: sample with 100% WL.

References

1. Lustig, R.H.; Schmidt, L.A.; Brindis, C.D. The Toxic Truth about Sugar. *Nature* **2012**, *482*, 27–29. [CrossRef]
2. Palmer, J.R.; Boggs, D.A.; Krishnan, S.; Hu, F.B.; Singer, M.; Rosenberg, L. Sugar-Sweetened Beverages and Incidence of Type 2 Diabetes Mellitus in African American Women. *Arch. Intern. Med.* **2008**, *168*, 1487–1492. [CrossRef]
3. Malik, V.S.; Schulze, M.B.; Hu, F.B. Intake of Sugar-Sweetened Beverages and Weight Gain: A Systematic Review. *Am. J. Clin. Nutr.* **2006**, *84*, 274–288. [CrossRef]
4. Yazar, G.; Rosell, C.M. Fat Replacers in Baked Products: Their Impact on Rheological Properties and Final Product Quality. *Crit. Rev. Food Sci. Nutr.* **2022**, *14*, 1–24. [CrossRef]
5. Mozaffarian, D. Dietary and Policy Priorities for Cardiovascular Disease, Diabetes, and Obesity. *Circulation* **2016**, *133*, 187–225. [CrossRef] [PubMed]
6. Syan, V.; Kaur, J.; Sharma, K.; Patni, M.; Rasane, P.; Singh, J.; Bhadariya, V. An Overview on the Types, Applications and Health Implications of Fat Replacers. *J. Food Sci. Technol.* **2022**, *59*, 1–12. [CrossRef]
7. ONU *70/1*. Transforming Our World: The 2030 Agenda for Sustainable Development. 2015. Available online: https://www.un.org/en/development/desa/population/migration/generalassembly/docs/globalcompact/A_RES_70_1_E.pdf (accessed on 20 October 2015).
8. Tolve, R.; Pasini, G.; Vignale, F.; Favati, F.; Simonato, B. Effect of Grape Pomace Addition on the Technological, Sensory, and Nutritional Properties of Durum Wheat Pasta. *Foods* **2020**, *9*, 354. [CrossRef] [PubMed]
9. Rainero, G.; Bianchi, F.; Rizzi, C.; Cervini, M.; Giuberti, G.; Simonato, B. Breadsticks Fortification with Red Grape Pomace: Effect on Nutritional, Technological, and Sensory Properties. *J. Sci. Food Agric.* **2021**, *102*, 2545–2552. [CrossRef]
10. Bianchi, F.; Cervini, M.; Giuberti, G.; Rocchetti, G.; Lucini, L.; Simonato, B. Distilled Grape Pomace as a Functional Ingredient in Vegan Muffins: Effect on Physicochemical, Nutritional, Rheological and Sensory Aspects. *Int. J. Food Sci. Technol.* **2022**, *57*, 4847–4858. [CrossRef]
11. Bianchi, F.; Lomuscio, E.; Rizzi, C.; Simonato, B. Predicted Shelf-Life, Thermodynamic Study and Antioxidant Capacity of Breadsticks Fortified with Grape Pomace Powders. *Foods* **2021**, *10*, 2815. [CrossRef]
12. Bonamente, E.; Scrucca, F.; Asdrubali, F.; Cotana, F.; Presciutti, A. The Water Footprint of the Wine Industry: Implementation of an Assessment Methodology and Application to a Case Study. *Sustainability* **2015**, *7*, 12190–12208. [CrossRef]
13. De Iseppi, A.; Marangon, M.; Lomolino, G.; Crapisi, A.; Curioni, A. Red and White Wine Lees as a Novel Source of Emulsifiers and Foaming Agents. *LWT* **2021**, *152*, 112273. [CrossRef]
14. Lim, J.; Ko, S.; Lee, S. Use of Yuja (*Citrus junos*) Pectin as a Fat Replacer in Baked Foods. *Food Sci. Biotechnol.* **2014**, *23*, 1837–1841. [CrossRef]
15. Hedayati, S.; Niakousari, M.; Seidi Damyeh, M.; Mazloomi, S.M.; Babajafari, S.; Ansarifar, E. Selection of Appropriate Hydrocolloid for Eggless Cakes Containing Chubak Root Extract Using Multiple Criteria Decision-Making Approach. *LWT* **2021**, *141*, 110914. [CrossRef]
16. Felix, M.; Martínez, I.; Sayago, A.; Recamales, M.Á.F. Wine Lees: From Waste to O/W Emulsion Stabilizer. *Innov. Food Sci. Emerg. Technol.* **2021**, *74*, 102810. [CrossRef]
17. Sanchez, C.; Klopfenstein, C.F.; Walker, C.E. Use of Carbohydrate-Based Fat Substitutes and Emulsifying Agents in Reduced-Fat Shortbread Cookies. *Cereal. Chem.* **1995**, *72*, 25–29.
18. Dogan, I.S.; Yıldız, O. Multiple Response Optimisations for the Development of Reduced-Fat Cake. *Food Manuf. Effic.* **2010**, *3*, 35–40. [CrossRef]
19. de Souza, N.C.O.; de Lacerda de Oliveira, L.; Rodrigues de Alencar, E.; Moreira, G.P.; dos Santos Leandro, E.; Ginani, V.C.; Zandonadi, R.P. Textural, Physical and Sensory Impacts of the Use of Green Banana Puree to Replace Fat in Reduced Sugar Pound Cakes. *LWT* **2018**, *89*, 617–623. [CrossRef]

20. Abdul Manaf, M.; Othman, N.A.; Harith, S.; Wan Ishak, W.R. Thermal Properties of Batter and Crumb Structure of Muffin Incorporated with Persea Americana Puree. *J. Culin. Sci. Technol.* **2017**, *15*, 259–271. [CrossRef]
21. Arifin, N.; Siti Nur Izyan, M.A.; Huda-Faujan, N. Physical Properties and Consumer Acceptability of Basic Muffin Made from Pumpkin Puree as Butter Replacer. *Food Res.* **2019**, *3*, 840–845. [CrossRef]
22. Hussien, H.A. Using Vegetable Puree as a Fat Substitute in Cakes. *Int. J. Food Sci. Nutr.* **2016**, *5*, 284. [CrossRef]
23. AOAC. *Official Methods of Analysis*, 18th ed.; The Association of Official Analytical Chemist: Gaithersburg, MD, USA, 2007.
24. AACC. *Approved Method of the AACC*, 10th ed.; American Association of Cereal Chemists: St. Paul, MN, USA, 2000.
25. Bianchi, F.; Giuberti, G.; Cervini, M.; Simonato, B. Fortification of Durum Wheat Fresh Pasta with Maqui (*Aristotelia Chilensis*) and Its Effects on Technological, Nutritional, Sensory Properties, and Predicted Glycemic Index. *Food Bioprocess Technol* **2022**, 1563–1572. [CrossRef]
26. Bustamante, M.A.; Moral, R.; Paredes, C.; Pérez-Espinosa, A.; Moreno-Caselles, J.; Pérez-Murcia, M.D. Agrochemical Characterisation of the Solid By-Products and Residues from the Winery and Distillery Industry. *Waste Manag.* **2008**, *28*, 372–380. [CrossRef]
27. Gómez, M.E.; Igartuburu, J.M.; Pando, E.; Rodríguez Luis, F.; Mourente, G. Lipid Composition of Lees from Sherry Wine. *J. Agric. Food Chem.* **2004**, *52*, 4791–4794. [CrossRef] [PubMed]
28. de Iseppi, A.; Lomolino, G.; Marangon, M.; Curioni, A. Current and Future Strategies for Wine Yeast Lees Valorization. *Food Res. Int.* **2020**, *137*, 109352. [CrossRef]
29. Psimouli, V.; Oreopoulou, V. The Effect of Alternative Sweeteners on Batter Rheology and Cake Properties. *J. Sci. Food Agric.* **2012**, *92*, 99–105. [CrossRef]
30. Spinei, M.; Oroian, M. The Potential of Grape Pomace Varieties as a Dietary Source of Pectic Substances. *Foods* **2021**, *10*, 867. [CrossRef]
31. Sakiyan, O.; Sumnu, G.; Sahin, S.; Bayram, G. Influence of Fat Content and Emulsifier Type on the Rheological Properties of Cake Batter. *Eur. Food Res. Technol.* **2004**, *219*, 635–638. [CrossRef]
32. Marchetti, L.; Califano, A.N.; Andrés, S.C. Partial Replacement of Wheat Flour by Pecan Nut Expeller Meal on Bakery Products. Effect on Muffins Quality. *LWT* **2018**, *95*, 85–91. [CrossRef]
33. Aydogdu, A.; Sumnu, G.; Sahin, S. Effects of Addition of Different Fibers on Rheological Characteristics of Cake Batter and Quality of Cakes. *J. Food Sci. Technol.* **2018**, *55*, 667–677. [CrossRef]
34. European Parliament. REGULATION (EU) No 1169/2011 of the EUROPEAN PARLIAMENT and of the 454 COUNCIL of 25 October 2011 on the provision of food information to consumers. *Off. J. Eur. Union* **2011**. Available online: https://eur-lex.europa.eu/legal-content/EN/ALL/?uri=celex%3A32011R1169 (accessed on 20 January 2023).
35. Smith, I.N.; Yu, J. Nutritional and Sensory Quality of Bread Containing Different Quantities of Grape Pomace from Different Grape Cultivars. *EC Nutr.* **2015**, *2*, 291–301.
36. Walker, R.; Tseng, A.; Cavender, G.; Ross, A.; Zhao, Y. Physicochemical, Nutritional, and Sensory Qualities of Wine Grape Pomace Fortified Baked Goods. *J. Food Sci.* **2014**, *79*, S1811–S1822. [CrossRef] [PubMed]
37. Tolve, R.; Simonato, B.; Rainero, G.; Bianchi, F.; Rizzi, C.; Cervini, M.; Giuberti, G. Wheat Bread Fortification by Grape Pomace Powder: Nutritional, Technological, Antioxidant, and Sensory Properties. *Foods* **2021**, *10*, 75. [CrossRef] [PubMed]
38. Bender, A.B.B.; Speroni, C.S.; Salvador, P.R.; Loureiro, B.B.; Lovatto, N.M.; Goulart, F.R.; Lovatto, M.T.; Miranda, M.Z.; Silva, L.P.; Penna, N.G. Grape Pomace Skins and the Effects of Its Inclusion in the Technological Properties of Muffins. *J. Culin. Sci. Technol.* **2017**, *15*, 143–157. [CrossRef]
39. Shih, Y.T.; Wang, W.; Hasenbeck, A.; Stone, D.; Zhao, Y. Investigation of Physicochemical, Nutritional, and Sensory Qualities of Muffins Incorporated with Dried Brewer's Spent Grain Flours as a Source of Dietary Fiber and Protein. *J. Food Sci.* **2020**, *85*, 3943–3953. [CrossRef]
40. Grigelmo-Miguel, N.; Carreras-Boladeras, E.; Martín-Belloso, O. Influence of the Addition of Peach Dietary Fiber in Composition, Physical Properties and Acceptability of Reduced-Fat Muffins. *Food Sci. Technol. Int.* **2001**, *7*, 425–431. [CrossRef]
41. Sudha, M.L.; Begum, K.; Ramasarma, P.R. Nutritional Characteristics of Linseed/Flaxseed (*Linum usitatissimum*) and Its Application in Muffin Making. *J. Texture Stud.* **2010**, *41*, 563–578. [CrossRef]
42. Rahman, R.; Hiregoudar, S.; Veeranagouda, M.; Ramachandra, C.; Kammar, M.; Nidoni, U.; Roopa, R. Physico-Chemical, Textural and Sensory Properties of Muffins Fortified with Wheat Grass Powder. *Karnataka J. Agric. Sci.* **2015**, *28*, 79–82.
43. Caggia, C.; Palmeri, R.; Russo, N.; Timpone, R.; Randazzo, C.L.; Todaro, A.; Barbagallo, S. Employ of Citrus By-Product as Fat Replacer Ingredient for Bakery Confectionery Products. *Front. Nutr.* **2020**, *7*, 46. [CrossRef]
44. Rodríguez-García, J.; Puig, A.; Salvador, A.; Hernando, I. Optimization of a Sponge Cake Formulation with Inulin as Fat Replacer: Structure, Physicochemical, and Sensory Properties. *J. Food Sci.* **2012**, *77*, C189–C197. [CrossRef]
45. Grasso, S.; Liu, S.; Methven, L. Quality of Muffins Enriched with Upcycled Defatted Sunflower Seed Flour. *LWT* **2020**, *119*, 108893. [CrossRef]
46. Acun, S.; Gül, H. Effects of Grape Pomace and Grape Seed Flours on Cookie Quality. *Qual. Assur. Saf.* **2014**, *6*, 81–88. [CrossRef]
47. Šporin, M.; Avbelj, M.; Kovač, B.; Možina, S.S. Quality Characteristics of Wheat Flour Dough and Bread Containing Grape Pomace Flour. *Food Sci. Technol. Int.* **2018**, *24*, 251–263. [CrossRef] [PubMed]

48. Nakov, G.; Brandolini, A.; Hidalgo, A.; Ivanova, N.; Stamatovska, V.; Dimov, I. Effect of Grape Pomace Powder Addition on Chemical, Nutritional and Technological Properties of Cakes. *LWT* **2020**, *134*, 109950. [CrossRef]
49. Martínez-Cervera, S.; Salvador, A.; Mugerza, B.; Moulay, L.; Fiszman, S.M. Cocoa Fibre and Its Application as a Fat Replacer in Chocolate Muffins. *LWT* **2011**, *44*, 729–736. [CrossRef]

foods

MDPI

Article

Development of Fortified Breads Enriched with Plant-Based Bioactive Peptides Derived from the Chia (*Salvia hispanica* L.) Expeller

Brenda Ozón [1], Juliana Cotabarren [1], Florencia R. Geier [1], M. Paula Kise [2], Javier García-Pardo [3,*], Mónica G. Parisi [2,*] and W. David Obregón [1]

1 Departamento de Ciencias Biológicas, Centro de Investigación de Proteínas Vegetales (CIProVe), Facultad de Ciencias Exactas, Universidad Nacional de La Plata, 47 y 115 s/N, La Plata B1900, Buenos Aires, Argentina; brendaozon@gmail.com (B.O.); cotabarren.juliana@biol.unlp.edu.ar (J.C.); florgei.94.fg@gmail.com (F.R.G.); davidobregon@biol.unlp.edu.ar (W.D.O.)

2 Departamento de Ciencias Básicas, Instituto de Ecología y Desarrollo Sustentable (INEDES, CONICET-UNLu), Universidad Nacional de Luján, Ruta 5 y Avenida Constitución, Luján B6700, Buenos Aires, Argentina; paulakise@yahoo.com.ar

3 Institut de Biotecnologia i de Biomedicina (IBB) and Departament de Bioquímica i Biologia Molecular, Universitat Autònoma de Barcelona, Bellaterra, 08193 Barcelona, Spain

* Correspondence: javiergarciapardo@msn.com (J.G.-P.); mparisi@unlu.edu.ar (M.G.P.); Tel.: +34-93-5868936 (J.G.-P.); +54-2323-423171 (M.G.P.)

Abstract: By-products from the industrialization of oilseeds, particularly chia, can be sustainably used for the development of new functional products. In this work, wheat breads supplemented with up to 10 mg of chia expeller hydrolysate/g of flour were prepared, obtaining fortified breads with acceptability for consumption, according to a preliminary consumer research study based on an affective test employing a five-point hedonic scale of global acceptance. In this context, protein hydrolysates of the chia expeller were produced using Alcalase, reaching a degree of hydrolysis of 54.3 ± 1.6% with an antioxidant activity of 55.8 ± 0.4% after 6 h incubation at 25 °C in the presence of the enzyme. These peptides showed appropriate techno-functional properties and chemical compositions suitable for the further development of bakery products. Taken together, our approach and the development of a fortified bread with plant-based bioactive peptides provide a novel and eco-friendly alternative for the recovery of nutrients from agro-industrial waste. More importantly, these enriched breads could exert beneficial effects on human health by exploiting the antioxidant properties of functional peptides derived from the chia expeller.

Keywords: antioxidant; bioactive peptides; bread; chia expeller; chia by-product; enzymatic hydrolysates; functional ingredient; plant-derived peptides; *Salvia hispanica* L.

1. Introduction

Wheat bread stands as a global staple food and is the main source of food and energy in most developing countries [1]. However, wheat-based diets are typically poor in essential amino acid content and in micro- and macronutrients; therefore, these essential amino acids and micro- and macronutrients must be incorporated through other dietary sources. Moreover, with the constant increase in the human population, food production must increase substantially to meet future demands. In this context, the incorporation of plant-based functional ingredients in wheat bread formulations has been shown to be an ideal approach to improve its organoleptic and nutritional properties [2–5].

Chia (*Salvia hispanica* L.) is a traditional crop native to Central and Southern Mexico and Guatemala, and nowadays, it is a cultivated plant across the landscapes of South and Central America [6,7]. Notably, the oil extracted from chia seeds has emerged as an outstanding reservoir of α-linolenic acid and phytosterols, surpassing many other vegetable oils in their concentrations. Of remarkable significance is the abundant presence of α-linolenic acid, constituting the predominant component of the fatty acid composition within chia seeds (approximately 60% of the total) compared with the lower proportion of linoleic acid (only about 10%) [6]. Moreover, chia seeds can be considered as a "functional superfood" because they contain a high concentration of protein (15–26%) of a good nutritional quality and digestibility and a good balance of essential amino acids that are higher than most grains, especially due to the high content of cysteine and methionine amino acids [8–10]. Another important characteristic of chia seeds is the absence of gluten, which makes it more desirable to coeliac patients. It also helps to increase the satiety index and prevent cardiovascular diseases, inflammatory and nervous system disorders, and diabetes, along with contributing to human nutrition [11].

A possible approach to increase the amino acid content and the nutritional quality of wheat bread is to incorporate food processing by-products [12,13]. It is worth mentioning that between a third and a quarter of food is lost or wasted annually along the food chain. This is also the case in the industrialization of chia seeds, in which the production of cold-pressed chia oil results in large amounts of protein-rich by-products. Among them, the chia expeller can be processed to yield functionally derived peptides that can be effectively used for the development of high value-added food products [13,14]. Typically, these bioactive peptides are inactive when bound inside the parent protein, and they should be enzymatically activated to fully display its beneficial biological properties [14].

In this study, we attempted to develop a novel functional product from bread by incorporating bioactive peptides produced via the enzymatic hydrolysis of the chia expeller by-product. These peptides display antioxidant properties and are a high-nutritional-value component, improving the amino acid content in the final product. In this context, we also investigated whether the incorporation of chia expeller hydrolysates into the bread formulation may alter its functional and sensory properties. Taken together, our results demonstrate that the enrichment of bread fortified with these chia-derived peptides may be a useful strategy to improve the protein content of the bread product and simultaneously generate more economic profitability for the chia industry.

2. Materials and Methods

2.1. Materials and Reagents

Protease from *Bacillus licheniformis* (Alcalase), sodium chloride, tris (hydroxymethyl) aminomethane, sodium dodecyl sulphate (SDS), β-mercaptoethanol (βME), Coomassie Blue G-250, N,N,N′,N′-tetramethyl ethylene diamine (TEMED), 1,1-diphenyl-2-picrylhydrazyl (DPPH), 2,2′-Azino-bis(3-ethylbenzothiazoline-6-sulf,onic acid) diammonium salt (ABTS), o-phthaldialdehyde (OPA), and bovine serum albumin were purchased from Sigma-Aldrich (St Louis, MO, USA). Ingredients such as wheat flour, yeast, sugar, salt, milk, butter, and oil were purchased locally.

2.2. Obtention of the Chia Expeller Protein Hydrolysate

Chia expeller, provided by Greenborg S.R.L. (Lincoln, Buenos Aires, Argentina), was defatted and then grinded to produce some flour with a high protein content [4]. Chia expeller protein concentrate (PC) was obtained by mixing 40 g of defatted flour with 1000 mL of 0.01 M phosphate buffer, 0.1 M NaCl, pH 7.4, before the mixture was subjected to subsequent homogenization in a cool blender and, after 1 h at room temperature, stored at −20 °C until analysis. A fraction of the PC was subjected to heat treatment at 50 °C for 1 h (called "HT") to evaluate the degree of hydrolysis (DH%) after Alcalase treatment.

The enzymatic hydrolysis of PC and HT (PCH and HTH) was performed with the food-grade microbial protease Alcalase (enzyme/substrate ratio of 0.3 U g^{-1}). The reaction mixture was adjusted to pH 7.0 to achieve the optimal enzymatic activity and incubated at 25 °C for 2, 4, 6, 8, 10, and 24 h at 80 rpm in a shaker (PCH25 and HTH25 samples) or 50 °C for 1, 2, 3, 4, 5, 6, 7, and 8 h at 80 rpm in a shaker (PCH50 and HTH50 samples) (Figure S1). All samples were withdrawn in triplicate, and the hydrolysis was stopped at the selected intervals via heating at 85 °C for 15 min. The hydrolysates were then lyophilised and kept at 4 °C until use.

2.3. Characterization of Hydrolysates

2.3.1. Determination of the Degree of Hydrolysis

The degree of hydrolysis (DH%) was estimated by using the o-phthalaldehyde (OPA) method [15]. The OPA reagent was prepared as described by Ozón et al. [16]. A total of 10 μL of the sample and 200 μL of the OPA reagent were mixed in a 96-well flat bottom plate. The assay was performed by triplicate, including a negative control using water instead of the sample, and the absorbance at 340 nm was recorded after 2 min of incubation. A Tecan Infinite M200 PRO spectrophotometer (Männedorf, Switzerland) was used for the determinations. The DH was calculated as follows:

$$DH\% = h/htot \times 100$$

where h is the number of hydrolysed bonds, and htot is the total number of peptide bonds per protein equivalent (which depends on the amino acid composition of the raw material). Complete hydrolysis of the sample (100%) was achieved by incubating the sample with 6 N HCl at 100 °C for 24 h.

2.3.2. Evaluation of Antioxidant Activity

The antioxidant activity of the chia expeller hydrolysates was evaluated by using the ABTS+ method [17] with slight modifications. The ABTS•+ solution was prepared through reacting an aqueous ABTS solution (5 mL) and potassium persulfate ($K_2S_2O_8$) solution in a 2:1 ratio. After storage in the dark for 16–17 h, the blue–green radical cation solution was further diluted in phosphate buffer (pH 7.5) until its absorbance reached 1 ± 0.01 AU at 734 nm. The antioxidant compound content in PC, HT, and the hydrolysates were analysed by adapting the method to a 96-well flat bottom plate in a Tecan Infinite M200 PRO spectrophotometer (Tecan Group Ltd., Männedorf, Switzerland). A total of 190 μL of diluted ABTS•+ solution was added to a 10 μL sample or Trolox standard according to Ozón et al. [6]. The mixture was incubated for 10 min at room temperature in dark conditions, and after that, the absorbance was measured at 734 nm. ABTS+ radical scavenging activity was estimated as follows:

$$\text{ABTS radical scavenging activity } (\%) = 100 - 100 \times \frac{A1 - A2}{A0}$$

The absorbance of the control without sample was expressed as A0; the absorbance in the presence of the sample and ABTS•+ and the absorbance of the sample blank without ABTS•+ was expressed as A1 and A2, respectively.

To evaluate the antioxidant capacity of the breads, 1 g of ground bread crumbs were homogenised in 5 mL of a phosphate buffer (0.1 M, pH 7.0). The suspensions were homogenised in a vortex mixer for 2 min and centrifuged at 7000× *g* for 10 min, and the supernatant from each sample was evaluated via the ABTS method, as previously described [17].

2.3.3. Chemical and Techno-Functional Properties of the Chia Expeller Hydrolysates

PC hydrolysates produced by Alcalase hydrolysis at 25 °C for 6 h (PCH25,6) were selected for subsequent analysis. The moisture (AOAC 934.01), ashes (AOAC 942.05), lipids (AOAC 920.39), dietary fibre (AOAC 985.29), and protein (micro-Kjeldahl method, AOAC 984.13) contents of the raw materials were determined according to Association of Official Analytical Chemists (2000) protocols [18], and the carbohydrate content was obtained by subtracting 100 from the sum of other components.

Water retention capacity (WRC) and oil retention capacity (ORC) were determined according to Chau et al. [19] with modifications. Thus, 0.5 g of PCH25,6 were homogenised with 5 mL of water/oil via vortex for 2 min and centrifuged at 2200× *g* for 30 min. Water/oil on the supernatant was removed, and the remaining fibrous suspension was weighed to determine the water/oil weight gain. The WRC/ORC was expressed as grams of water/oil retained per gram of sample, considering oil density as 0.92 g/mL.

$$\text{WRC or ORC} = \frac{\text{final weight} - \text{initial weight}}{\text{initial weight}}$$

Water adsorption capacity (WAC) was determined according to Chen et al. [20]. A total of 1 g of PCH25,6 was placed in a microenvironment of 98% relative humidity in equilibrium, generated by placing 20 mL of saturated potassium sulphate saline solution in a hermetically sealed glass container at 25 °C. The sample was incubated until it achieved a constant weight, reporting this capacity as the weight gain expressed in grams of water per gram of dry sample.

$$\text{WAC} = \frac{\text{g of adsorbed water}}{\text{g of CEH}}$$

Emulsifying activity (EA) and emulsion stability (ES) were determined according to Chau et al. [19]. To evaluate EA, 100 mL of distilled water were added to 2 g of hydrolysates, and the mixture was homogenised at 11,000 rpm. Then, 100 mL of refined corn oil (density = 0.92 g/mL) were added and homogenised again at 11,000 rpm for 3 min in a disperser (Ultra-Turrax®, T25 Basic, IKA®, Werke, The Netherlands). The samples were placed in graduated centrifuge tubes (50 mL) and centrifuged at 170× *g* for 10 min, and the volume of the remaining emulsion was measured to calculate EA as follows:

$$\text{EA} = \frac{\text{volume of remanent emulsion (mL)} \times 100}{\text{initial volume of emulsion (mL)}}$$

ES was determined by heating the obtained emulsions at 80 °C for 30 min. The samples were cooled to room temperature and centrifuged at 170× *g* for 10 min. ES was expressed as the ratio of the remaining emulsion volume (mL) with respect to the original volume of the emulsion.

$$\text{ES} = \frac{\text{volume of remanent emulsion (mL)} \times 100}{\text{initial volume of emulsion (mL)}}$$

2.4. *Ingredients and Breadmaking Procedure*

Wheat breads were prepared with the addition of different PCH:flour ratios (i.e., 1, 3, 5, and 10 mg of PCH25,6/g flour (Table S1) using the N° 11 program of a commercial bread oven (ATMA Easy Cook, HP4031E). Wet ingredients were first added to the recipient, and then dry ingredients were placed on the top. The mixture was homogenised for 15 min at medium speed, fermented at 30 °C for 110 min with 80% relative humidity, and cooked for 50 min at 220 °C. After 12 h, the breads were sliced for further analysis.

2.5. *Technological and Physicochemical Parameters*

2.5.1. Specific Volume Determination

Specific volume (SV) was determined according to the millet seed displacement method [21]. Millet seeds were poured into a container of known volume until the bottom was covered. The test bread was then placed inside the container, followed by more millet seeds, which were levelled across the top with a spatula. The displacement of the millet seeds that were not required to fill the container were measured in a graduated cylinder and used to express the specific volume.

2.5.2. Colour Analysis

The breads were cut into 2 cm wide slices, and bread colour analysis was performed using a colour spectrophotometer (HunterLab MiniScan EZ, Reston, VA, USA) according to the colour system in the L*, a*, and b* space defined by CIE-L*a*b* [22]. Calibration with black and white tiles was performed before colour measurement.

The colour appearance of bread slices was quantified using three-dimensional colour spaces: lightness, colour, and hue. Lightness is described as the brightness of a surface, and colour indicates the intensity of the colour content (light green, dark green, etc.). Hue is the visual perception according to which an area appears to be similar to one of the colours: red, yellow, green, or blue. To calculate the hue, the CIE L*a*b* (CIELAB) colour space is taken into account, where the hue angle is calculated as h = arctan (b/a), between 0 and 360 degrees, corresponding to the hue circle, where (a) is the red–green component and (b) is the yellow-bluish component. In this way, h = 0 approximates the appearance of red object surfaces; h = 90 to the yellow ones, h = 180 to the green ones, and h = 270 to the blue ones. Colour difference values (ΔE), which measure the changes in colour of the total colour differences, were calculated as previously described [23] by using the following equation:

$$\sqrt{\Delta L2 + \Delta a^2 + \Delta b^2}$$

2.5.3. Texture Profile Analysis

The textural properties of the fortified breads were determined on three central slices (2 cm thickness) using a TA-XTplus Texture Analyser (Stable Micro Systems, Surrey, UK) with a 36 mm cylindrical probe (p/36 R) according to the 74-09 methodology proposed by the Approved Methods of the American Association of Cereal Chemists (AACC) [21]. The pre-test speed, test speed, and post-test speed values were 1, 5, and 5 mm/s, respectively, with a trigger force of 5 g. The data were measured as force (N) versus time (s) curves. The hardness (g), adhesiveness (g.s), resilience (%), cohesiveness, and springiness (%) were recorded as well. Gumminess (G) was calculated as G = hardness × cohesiveness, and chewiness (Ch) was calculated as Ch = gumminess × springiness.

2.5.4. Sensory Evaluation

The sensory analysis of freshly baked bread samples (CB and B10) involved 37 untrained panellists aged between 20 and 60 years who were regular consumers of wheat bread. This panel consisted of 18 women and 17 men, including undergraduate and graduate students, as well as faculty members from the Faculty of Exact Sciences at the National University of La Plata in La Plata, Argentina.

To assess the organoleptic characteristics of the breads, the panellists participated in a descriptive sensory analysis panel. The breads were baked the day before the evaluation. On the day of the evaluation, the untrained panellists were randomly divided into groups of nine people. These groups were accommodated in a room that was equipped with individual cabinets and specifically arranged for the sensory test. The panellists received basic instructions regarding the evaluation procedure, which was conducted in a single day. During the sensory evaluation, the bread samples (CB and B10) were presented as 2 cm cubes that were placed on different dishes, each labelled with random three-digit numbers. Water was provided for rinsing between tasting each sample.

Following the instructions, training, and discussions related to product-oriented affective testing, each panellist received a rating form score sheet. This score sheet was based on a five-point hedonic scale for assessing the overall acceptance of the bread. The scale ranged from 5, indicating "I liked it very much", to 1, indicating "I disliked it very much". Additionally, panellists performed a preference test where they selected their preferred bread out of all of the options.

The untrained panellists also performed a triangular test. This test is useful for evaluating overall bread differences between control samples and upon supplementation with the chia expeller-derived peptides. To perform the evaluation, three cubes were presented in random order: (i) CB-CB-B10, (ii) CB-B10-CB, (iii) B10-CB-CB. After sensory evaluation, the panellists opined on the different tastes of the bread cubes, and their answers were interpreted as "right", "wrong", or "do not know".

The results were interpreted according to the relationship between the number of positive judgments and the total number of judgments. A significant difference between the samples at the corresponding level of probability was considered when the number of positive judgments was greater than or equal to the value of the two-sided table [24].

2.6. Statistical Analysis

Statistical analyses (ANOVA) were performed using GraphPad Prism (v9.4.1, GraphPad Software Inc., San Diego, CA, USA, 2012). All experiments were conducted in triplicate of two independent experiments. Data are expressed as the mean ± standard deviation, and significant differences were determined by using Tukey's post hoc test ($p < 0.05$).

3. Results and Discussion

3.1. Enzymatic Hydrolysis of Chia Expeller Using Alcalase

Defatted chia expeller is a by-product with a high protein content of about 37% [14]. The optimization of the enzymatic hydrolysis of the chia expeller protein concentrate (PC) and the heat-treated sample (HT) was carried out to find the optimal digestion conditions. As observed in Figure 1, when the enzymatic hydrolysis was carried out at 50 °C, a gradual increase in the degree of hydrolysis (DH) was observed as the incubation time increased in both samples (PCH and HTH), reaching a DH of 70% after 3 h of incubation (Figure 1A). No significant increases in DH were observed at longer times, reaching 90% after 8 h, indicating a certain degree of enzyme inactivation when the samples were incubated at this temperature. However, when the enzymatic hydrolysis was carried out at 25 °C, the incubation time required to reach the same DH as in the test at 50 °C was longer, achieving 70% DH after 10 h of incubation for both samples (Figure 1B). Furthermore, by maintaining the process for 24 h, DH only increased up to 80%. Under our experimental conditions, the faster Alcalase hydrolysis of chia proteins was achieved at 50 °C incubation [16,25]. This observation aligns with the optimal temperature range for Alcalase activity, which typically falls between 50 and 65 °C [26].

Additionally, we also evaluated the effect that thermal pre-treatment had on the samples. Interestingly, thermally treating the samples at 50 °C for 1 h before the hydrolysis process (HTH) increased the DH by 10% in the first 4 h when the hydrolysis was carried out at 50 °C, suggesting that thermal treatment can also change substrate accessibility for hydrolysis due to substrate unfolding. No significant differences were observed between the DH obtained at higher hydrolysis times. On the other hand, when the hydrolysis was carried out at 25 °C, no important differences were observed in the resulting DH for both the thermally treated samples and the PC.

Next, we evaluated the antioxidant activity of the chia expeller protein concentrate hydrolysates (PCHs) and the corresponding heat-treated Alcalase hydrolysates (HTHs) produced at 50 °C, which was found to be the faster digestion condition. As shown in Figure 1C, the antioxidant capacity was similar for both sample groups (PCHs and HTHs) and was obtained after 2 to 8 h of hydrolysis, with values in the range 65–80%. Furthermore, the antioxidant activities produced by the chia expeller samples digested at 25 °C were also comparable (Figure 1D). When we compared the antioxidant activity of the hydrolysates obtained at 25 °C to those obtained at 50 °C, the values obtained after 2 h of hydrolysis at 50 °C were equivalent, which is in agreement with the DH values obtained. These results demonstrated the possibility of producing hydrolysates with Alcalase at room temperature, since they only require a longer hydrolysis time to obtain the same antioxidant effect as at 50 °C. Furthermore, no significant differences were observed in DH and the antioxidant activity when using PC and HT as substrates. Thus, we selected the PCH sample produced after 6 h at 25 °C (PCH25,6) for further characterization since this treatment showed a DH of 54.3 ± 1.6% and a high antioxidant activity (55.8 ± 0.4%).

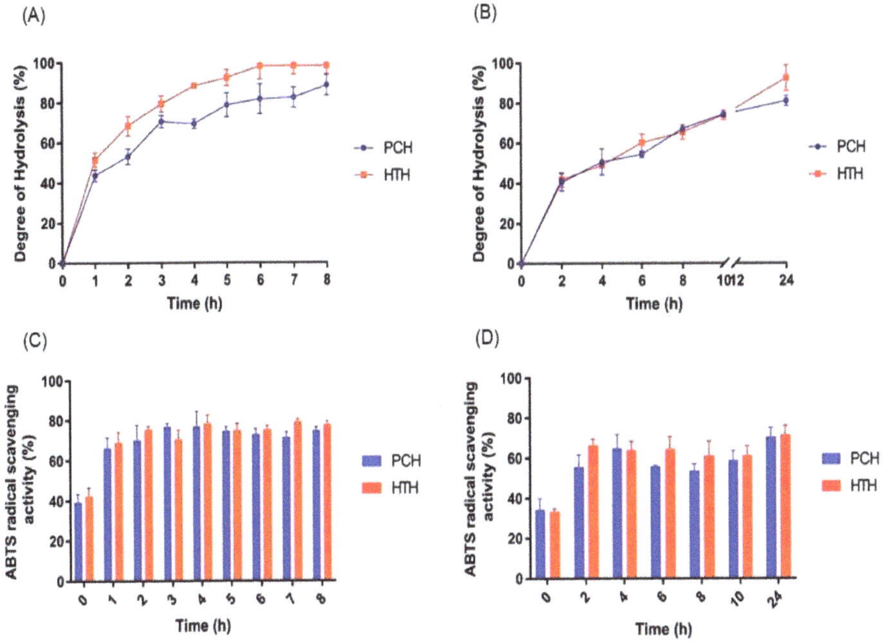

Figure 1. Characterization of chia expeller hydrolysates. Degree of hydrolysis (DH%) for enzymatic hydrolysis with Alcalase at 50 °C (**A**) and 25 °C (**B**) carried out using the OPA method. (**C**,**D**) ABTS radical cation decolorization assay to improve antioxidant activity for Alcalase hydrolysis at 50 °C (**C**) and 25 °C (**D**). In (**A**–**D**), the abbreviations are as follows: PCH, chia expeller protein concentrate hydrolysate; HTH, Heat-treated chia expeller protein concentrate hydrolysate.

3.2. Physicochemical and Techno-Functional Properties of the Chia Expeller Hydrolysates

The techno-functional properties of hydrolysates are important factors that need to be considered when developing novel ingredients as food additives [27]. For this reason, the proximal composition and the techno-functional properties of the hydrolysates (PCH25,6) produced in this study were characterised (Tables S2 and S3). As shown in Table S2, a protein content of 19.5% was determined for the PCH25,6 sample. This is slightly higher than the protein content reported in the literature for the chia seeds (16.5%, according to the USDA National Nutrient Database for Standard Reference) [28].

As expected, the total lipid content found in the PCH25,6 sample derived from the defatted expeller was dramatically reduced to 1.6%, a value that is substantially lower compared to the reference values for the chia seeds (30.7% of lipid content). Moreover, the ash content was higher compared to the values found in chia seeds (4.8% compared to 33%, respectively). The amount of assimilable carbohydrates were found to be 5.9%, and the soluble fibre content stayed in the range of 30–35%, with both values being comparable to those commonly found in the chia seeds (7.7% and 34.4%, respectively). Importantly, the protein content observed for PCH25,6 suggests that the chia expeller hydrolysate maintains protein and soluble fibre contents comparable to those found in the chia seeds, which suggest that chia expeller hydrolysates are excellent candidates for use as dietary supplements.

Extensive hydrolysates with a DH higher than 10% are commonly used for the generation of functional foods, including hypoallergenic and bioactive foods. In this study, we obtained PCH25,6 derived from the chia expeller, which has the particularity of containing a high protein and fibre content. In agreement with this, the Water Retention Capacity (WRC) of this was found to be 1.92 g H_2O/g sample (Table S3). This value is in the same range to those observed in previous studies for chia seed protein isolates obtained via three distinct drying methods [29] and from flour obtained via solvent extraction or by pressing seeds [30]. In contrast, the oil retention capacity (ORC) of PCH25,6 was determined to be 2.36 g oil/g hydrolysate. This value is slightly increased compared to that reported by Capitani et al. [30], presumably due to the balance between the release of hydrophobic groups and the decrease in the molecular mass and release of ionizable groups, which probably have a negative effect on the ORC. Finally, the emulsifying activity of PCH25,6 was also determined. As shown in Table S3, we found values reported by Villanueva-Lazo et al. [31] for PCH25,6 that were slightly lower for the Alcalase hydrolysates derived from chia seeds (DH: 36.2%) than that reported for a defatted meal [30]. It is widely accepted that protein hydrolysis increases the emulsifying activity of food samples, as this has been previously reported for samples of corn, lupine, soybean, rice bran, pumpkin, oats, sesame, chickpea, peanut, rapeseed, and potato, among others [32].

3.3. Technological Parameters of the Fortified Breads Enriched with Functional Peptides Derived from the Chia Expeller

Plant-based proteins represent a good alternative to animal proteins in the production of functional peptides due to their low cost and ease of large-scale production. As described above, we have obtained chia hydrolysates produced via the enzymatic digestion of the chia expeller with Alcalase (i.e., PCH25,6). The resultant chia expeller hydrolysates displayed excellent values of emulsifying activity and water and oil retention capacities and an adequate proximal composition. Based on these results, breads enriched with different amounts of PCH25,6 (i.e., 1, 3, 5, and 10 mg of chia expeller hydrolysate/g of wheat flour) were formulated and analysed.

3.3.1. External Analysis: Crust Colour

The surface browning of bread is an indicator of bread quality. As shown in Figure 2A, the crust colour of breads containing increased concentrations of the chia expeller hydrolysates did not show significant differences compared to the bread made with wheat flour (control). In a previous study, Xing et al. [33] attributed a change in crust colour to a higher content of amino acids (specially lysine), which react with carbonyl compounds (reducing sugar in the Maillard reaction) during baking. We believe that the lack of darkening of the crust colour in our study is in agreement with the overall low lysine content (0.97 g/100 g) found in chia proteins [28].

3.3.2. Alveolate and Specific Volume Analysis

The specific volume and the alveolate of bread crumbs are common indices of quality as they reflect the ability of the dough to occlude air. Regarding the alveolate, bread slices were scanned, and a qualitative analysis of images was performed (see Figure 2B). The addition of increasing amounts of hydrolysates revealed a positive effect on the formation of alveoli, with larger alveoli being observed in the breads supplemented with 3, 5, and 10 mg of hydrolysate per gram of flour than those observed for the control bread (CB). In addition, a considerable increase in height was also observed in the supplemented samples, showing an increase of 1 cm compared to the CB.

Interestingly, the specific volume was not significantly affected by the level of wheat flour substitution (Table 1). Only slightly higher values were recorded for the breads fortified with the chia hydrolysates (B10, B5, and B3, respectively), with mean values of 2.27 ± 0.05, 2.29 ± 0.08 and 2.25 ± 0.06 mL/g. The lowest value was observed for the control sample (CB with 2.07 ± 0.04). These results showed a minor loss of dough upon baking in the enriched formulations compared to the control bread, probably due to the high water retention capacity of the remaining mucilage in the expeller [34]. Previously, Silveira Coelho and de las Mercedes Salas-Mellado and coworkers [35] produced breads enriched with 11% of hydrated chia seeds, and they found that the enriched bread showed a significantly smaller volume than the control bread. This discrepancy was attributed to a decrease in the share of gluten in the bread's dough due to a decreased ability to hold CO_2 during fermentation and baking.

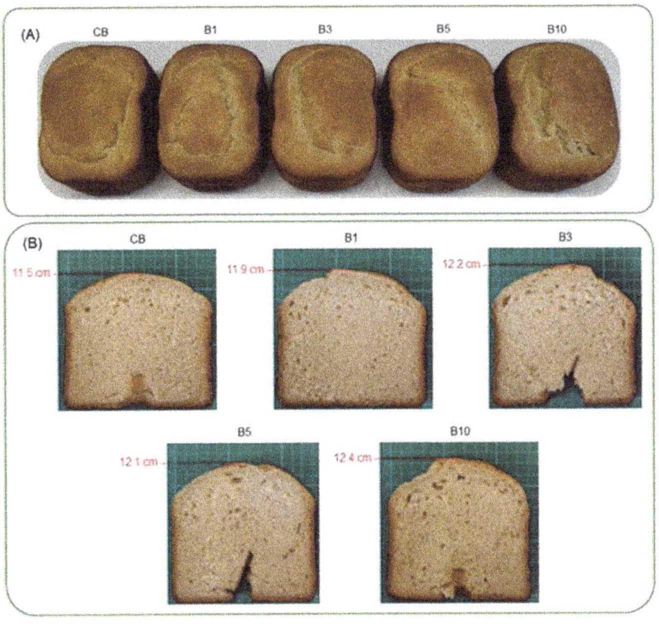

Figure 2. Photographs of wheat bread supplemented with chia expeller functional peptides. (**A**) Photographs of whole bread loaves. (**B**) Photographs of central transversal slices of each loaf positioned on a graduated cutting board. Bread slices produced with wheat flour (CB) or different blends of wheat flour via chia hydrolysates (B1, B3, B5 and B10).

Table 1. Technological characteristics of breads supplemented with different concentrations of chia expeller hydrolysates compared to wheat flour.

Bread	SV (mL/g)	H (g)	A (g. sec)	R (%)	Co	S (%)	G	Ch
CB	2.07 ± 0.04	2986.4 ± 936	−0.90 ± 0.60	15.31 ± 2.34	0.44 ± 0.04	88.03 ± 2.48	1304.2 ± 357	1145.1 ± 306
B1	2.11 ± 0.03	2696.2 ± 546	−0.83 ± 0.61	15.11 ± 1.24	0.44 ± 0.03	86.76 ± 2.95	1170.3 ± 194	1016.2 ± 175
B3	2.25 ± 0.06 ^{b}	2478.5 ± 687	−0.71 ± 0.61	15.71 ± 1.67	0.45 ± 0.03	87.87 ± 2.09	1116.6 ± 297	981.2 ± 261
B5	2.29 ± 0.08 ^{b}	2197.4 ± 449	−0.32 ± 0.57	18.89 ± 2.45	0.50 ± 0.04	88.51 ± 1.62	1086.2 ± 180	960.4 ± 154
B10	2.27 ± 0.05 ^{b}	2160.9 ± 342	−0.59 ± 0.62	19.16 ± 1.87	0.51 ± 0.02	91.05 ± 2.63	1102.1 ± 153	1003.0 ± 140

Abbreviations are as follows: CB—Control bread; B1, B3, B5, and B10—Breads supplemented with 1, 3, 5, and 10 mg of chia expeller hydrolysate/g of flour; SV—Specific volume; H—Hardness; A— Adhesiveness; R—Resilience; Co—Cohesion; S—Springiness; G—Gumminess (G = H × Co); Ch— Chewiness (Ch = G × S = H × Co × S). Values show the average of three values ± the standard deviation (SD). ^{b} indicates a significant difference between the means ($p < 0.05$; Tukey test).

3.3.3. Texture Profile Analysis

A detailed analysis of the textural properties (Table 1) showed no significant differences between the hardness parameters (H) of the CB, B1, and B3 bread formulations, while the values for B5 and B10 were considerably lower. This trend suggests that the addition of 5 and 10 mg of hydrolysates (with DH = 54.3 ± 1.6%) of chia expeller/g of flour may have improved the textural properties of the enriched bread compared to the control bread, resulting in a higher specific volume, lower firmness, and increase in the alveolate. In addition to changes in the Specific Volume (SV) and the hardness (H), bread fortification with 5 and 10 mg of hydrolysates also induced slight changes in the other textural properties of the bread, such as the resilience (R) and cohesiveness (Co) of the final baked product. In a previous report, Madruga et al. [5] reported a considerable reduction in the specific volume of the bread when 3 mg of chia seed hydrolysate (with DH = 30%/g of flour) was added to the bread formulation. The authors attributed these changes to a different formation of the "gluten network" due to the increase in the concentration of hydrolysate used in the formulation of the breads. In our study, the improvement in the textural parameters may be caused by the use of a chia expeller derivative with a superior degree of hydrolysis (DH = 54.3 ± 1.6%).

3.3.4. Crumb Colour Analysis

Crumb colour is a key property of baked products since, together with volume and texture, it influences consumer acceptance. In our analysis, the crumb colour of the breads supplemented with hydrolysates from the chia expeller were expressed in CIE L*, a*, and b* values—corresponding to brightness, green (−)/red (+), and blue (−)/yellow (+), respectively—and compared to the standard bread (control). Table 2 summarises the colorimetric parameters obtained for the crumb of breads formulated with the chia expeller hydrolysates. In essence, no significant differences were observed in the brightness (L*) of breads B1, B3, and B5 compared to the control bread. Only breads with 10 mg of hydrolysate/g of flour showed a value of 69.39 ± 1.26, which is slightly lower compared to the control bread, which showed a value of 73.26 ± 0.74. This decrease in brightness can be attributed to the darker colour of the chia expeller hydrolysates compared to the wheat flour. However, the effect produced by baking represents only a 5% reduction compared to the control bread. A similar darkening effect was previously observed in another study after the addition of chia flour to the bread formulation [36–38].

Table 2. Colour parameters of the crumbs of breads supplemented with chia expeller hydrolysates.

Bread Formulation	L*[2]	a*[2]	b*[2]	h (°)	ΔE
CB	73.26 ± 0.74	4.88 ± 0.39	27.33 ± 0.39	79.88 ± 0.65	0.92
B1	73.07 ± 0.50	4.13 ± 0.13	26.40 ± 0.59	81.11 ± 0.08	0.78
B3	73.54 ± 1.05	4.27 ± 0.10	26.41 ± 0.18	80.82 ± 0.15	1.07
B5	71.12 ± 1.19	4.68 ± 0.28	26.94 ± 0.33	80.15 ± 0.46	1.27
B10	69.39 ± 1.26 ^{b}	4.58 ± 0.13	26.15 ± 0.29	80.07 ± 0.17	1.30

Abbreviations are as follows: CB—Control bread; B1, B3, B5, and B10—Breads supplemented with 1, 3, 5, and 10 mg of chia expeller hydrolysate/g of flour. L*—Lightness; a*—Tendency for green–red coloration; b*—Tendency for blue–yellow coloration; h—Hue angle; ΔE—colour difference values. Values show the average of three values ± the standard deviation (SD). ^{b} indicates a significant difference between the means ($p < 0.05$; Tukey test).

The values of the a* and b* parameters showed a slight tendency towards reddish and yellow coloration in all the formulations, including the one with the highest concentration of chia expeller hydrolysates. Despite such small changes in coloration upon baking, no substantial differences were observed between the different bread formulations. Typically, the crumb colour of the bread crumbs tended to be yellow (h close to 90°). In this case, the value of the h parameter did not show significant differences between the CB (79.88 ± 0.65) and the bread formulation with the highest level of chia expeller hydrolysate supplementation, B10 (80.07 ± 0.17). Regarding the colour difference values (ΔE), the addition of different hydrolysate amounts slightly increased this parameter compared to the control bread. In summary, the results obtained for the chromatic parameters showed no considerable differences between the breads containing chia expeller hydrolysates and the control bread.

3.4. Antioxidant Activity of Baked Breads Fortified with Chia Expeller Hydrolysates

The antioxidant activity of the different bread formulations was evaluated using the ABTS method after the breads were subjected to long-term storage (up to 75 days) at −20 °C. As shown in Figure 3 all fortified breads tested maintained or increased their antioxidant activity over time. Interestingly enough, the B10 bread formulation, which had 10 mg of chia expeller hydrolysate/g of flour, showed the highest antioxidant activity scores and a marked time-dependent increase in antioxidant capacity up to a value of 56 ± 3.9% after 75 days of storage at −20 °C. Under the same conditions, the control bread showed a much lower antioxidant activity (44 ± 2.5%). These results are consistent with those reported by Sayed-Ahmad et al. [37] and Romankiewicz et al. [39], who demonstrated that the addition of chia seed improved the antioxidant activity of wheat bread. Similar results were reported by Zdybel et al. [38], who evaluated the antioxidant potential against DPPH free radical scavenging of breads added with a waste product from the pressing of chia seed oil. In contrast, Segura-Campos et al. [40] demonstrated that the partial replacement of wheat flour by chia protein hydrolysates had no effect on the antioxidant activity of white bread, probably due to the effect of high temperatures during baking, causing the oxidation of tryptophan and histidine or the methionine desulphuration. However, in this study, the formulation contained only between 1 and 3 mg hydrolysates/g of flour.

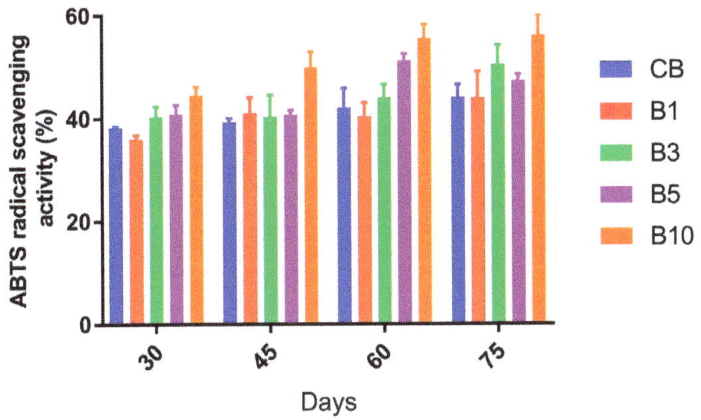

Figure 3. Antioxidant activity of breads supplemented with chia expeller hydrolysates. CB—Control bread; B1, B3, B5, and B10—Breads with 1, 3, 5, and 10 mg of chia expeller hydrolysate/g of flour.

In our study, the bread formulation supplemented with 10 mg of chia expeller hydrolysate/g of flour achieved increased antioxidant activity. Therefore, it is tempting to speculate that chia expeller hydrolysates could be a beneficial additive for improving the antioxidant properties of a novel generation of functional breads.

3.5. Sensory Test: Consumer Acceptability of Selected Bread Fortified with Chia Expeller Hydrolysate (B10)

To gain insights into consumer preferences and acceptability criteria, consumer-oriented testing methods and untrained sensory panels have proven to be a valuable approach [41,42]. In order to assess the consumer acceptance and preference information for the formulations treated with chia protein hydrolysates in a wheat bread, the B10 formulation, which had highest amount of hydrolysate per gram of flour, was selected for a further preliminary sensory analysis [43]. This bread formulation showed a higher protein content and the best textural properties compared to the control bread. Moreover, this bread formulation exhibited a superior antioxidant capacity upon long-term storage. As summarised in Figure 4, the sensory analysis of the breads supplemented with 10 mg of hydrolysate/g of flour, 17 untrained panellists awarded the B10 formulation with the best possible rating ("I liked it very much"), while 19 panellists gave it the second-best possible rating, stating "I liked it". Compared to the control wheat bread, where the rating of "I liked it" predominated, a positive trend was found regarding the bread fortified with 10 mg of hydrolysate, which received the best score (Figure 4A). In a similar manner, when carrying out the preference test (Figure 4B), 21 panellists chose bread B10 over the control bread (which was preferred by 9 panellists), while 7 panellists did not present a preference for any particular bread. Out of the 37 panellists, 24 positive responses had to have been obtained for one of the samples to ensure that there was a significant difference ($p < 0.05$). Therefore, from the 21 positive responses received for the bread fortified with chia hydrolysates, it can be inferred that the addition of 10 mg of hydrolysates/g of flour did not show significant differences in the preferences of the panellists with respect to the control bread.

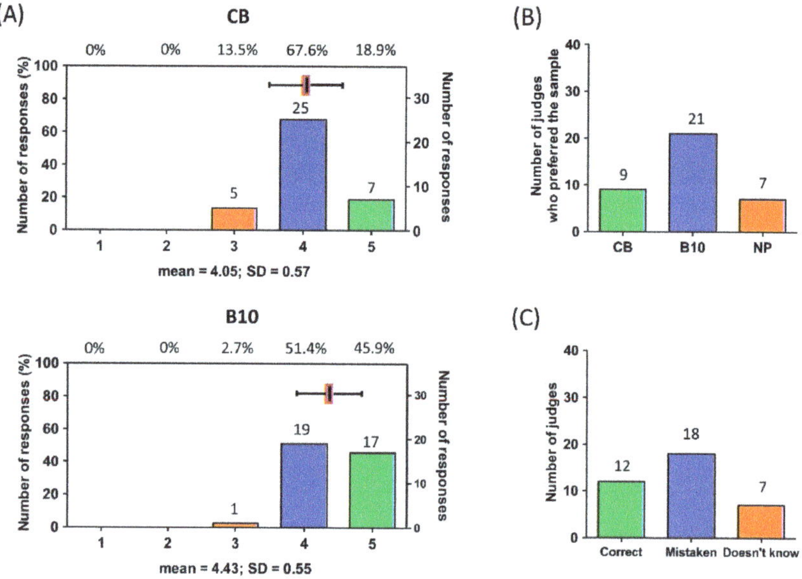

Figure 4. Sensory analysis of control breads (CB) and breads supplemented with 10 mg of chia expeller hydrolysate/g of flour (B10). (**A**) Five-point hedonic sensory evaluation. (**B**) Preference test. NP—No preference. (**C**) Triangular test. In (**A**), the mean (red square) ± Standard Deviation (SD) is displayed over the bars.

In a recent study, Madruga and coworkers [5] prepared wheat and rice breads supplemented with chia hydrolysates. In this study, the authors concluded that the addition of chia hydrolysate flour did not affect the consumer preference. The similarities in the observations between our work and previously reported work could be due to the reduced impact of sensory attributes introduced by the addition of hydrolysates at the tested concentration. On the contrary, a previous study by da Silva et al. [44] found that the addition of *ora-pro-nobis* (*Pereskia aculeata*) to salt bread had a substantial impact on consumer preference. In a similar manner, Robles-Ramírez et al. [45] recently reported that the substitution of wheat flour with 60% barley flour caused a reduction in consumer preference. Interestingly, the sensory qualities of this bread were improved by adding sesame seeds to the bread formulation.

Finally, a triangular test was also carried out to investigate whether the panellists were able to identify fine differences between the three bread samples. As shown in Figure 4C, only 12 panellists were able to identify the different bread samples, while 18 of them pointed out an erroneous sample as different, and 7 of them found no differences. These results are in agreement with the preference test since there were no significant differences between the preference of the panellists towards a particular sample.

Taken together, the results from the sensory evaluation suggest that both the control and fortified breads present similar consumer preferences and overall sensory characteristics since consumers could not identify which one corresponded to the formulation fortified with the chia expeller hydrolysates.

4. Conclusions

This study represents the first attempt to develop a wheat bread product enriched with functional peptides derived from the chia expeller. For this purpose, we first aimed to prepare a protein hydrolysate from defatted chia expeller via enzymatic hydrolysis with Alcalase, reaching a degree of hydrolysis of 54.3 ± 1.6% after 6 h of hydrolysis at 25 °C. This chia expeller hydrolysate (PCH25,6) showed a notable antioxidant activity of 55.8 ± 0.4%, as determined via the ABTS method. Moreover, the proximate composition values of the PCH25,6 hydrolysate suggested that this additive contains a notable amount of protein and fibre, which is comparable to the values found in raw chia seeds. In this context, the techno-functional characterization of the hydrolysate showed adequate values of emulsifying activity, as well as water and oil retention capacities, rendering it compatible as a supplemental component in bread formulations.

Notably, the fortification of wheat breads with different amounts of chia expeller hydrolysates (i.e., 1, 3, 5, and 10 mg of chia expeller hydrolysate/g of flour) did not affect the bread's textural properties and provided a higher antioxidant activity even after 75 days of storage at −20 °C. The most substantial enhancement in antioxidant activity was observed when 10 mg of chia expeller hydrolysate was added to the bread formulation.

More importantly, the bread prepared with the highest content of hydrolysates showed acceptable sensorial attributes that were comparable to the control bread. In a preliminary consumer study, the participants enjoyed trying the bread product supplemented with the chia expeller hydrolysates, but no significant differences in consumer preference were identified between the control and fortified bread samples. This finding is in agreement with the results from a previous study by Madruga and coworkers [5], in which the authors demonstrated that the sensory characteristics of wheat and rice bread supplemented with 3 mg of chia hydrolysate/g flour were comparable to those of the control bread.

Taken together, the results presented herein demonstrate that the incorporation of chia expeller hydrolysates into a novel wheat bread formulation could be a valuable way of increasing the protein content of bread. Furthermore, it positions the fortified bread as a functional wheat bread product that offers increased protein content and enhanced antioxidant activity. Ultimately, this approach facilitates the utilisation of a by-product derived from the chia oil industry, offering both significant economic and environmental advantages.

Supplementary Materials: The following supporting information can be downloaded at: https://www.mdpi.com/article/10.3390/foods12183382/s1, Figure S1: Process flow diagram for obtaining and characterizing chia expeller protein hydrolysates; Table S1: Summary of the different bread formulations prepared in this study; Table S2: Proximate composition of chia expeller hydrolysates produced with Alcalase at 25 °C for 6 h (PCH25,6) compared to chia seeds; Table S3: Techno-functional characterization of chia expeller hydrolysates produced with Alcalase at 25 °C for 6 h (PCH25,6).

Author Contributions: Conceptualization, J.C., J.G.-P., M.G.P. and W.D.O.; methodology, B.O., J.C., M.G.P. and W.D.O.; validation, B.O., F.R.G., M.P.K., J.C., M.G.P. and W.D.O.; formal analysis, B.O., J.C., M.G.P. and W.D.O.; investigation, B.O., F.R.G., M.P.K. and J.C.; resources, J.C., J.G.-P., M.G.P. and W.D.O.; writing—original draft preparation, B.O., J.C., J.G.-P., M.G.P. and W.D.O.; writing—review and editing, J.C., J.G.-P., M.G.P. and W.D.O.; visualization, B.O., J.C., F.R.G., M.P.K., J.G.-P., M.G.P. and W.D.O.; supervision, J.C., J.G.-P., M.G.P. and W.D.O.; project administration, J.C., M.G.P. and W.D.O.; funding acquisition, M.G.P. and W.D.O. All authors have read and agreed to the published version of the manuscript.

Funding: This research study was funded via Departamento de Ciencias Básicas of the Universidad Nacional de Luján (UNLu) and C82 Project, Ciencia y Tecnología contra el Hambre Program—MINCYT (RESOL-2021-289-APN-MINCYT) grants to MGP and PICT-2018-03271 (2019-2021) and UNLP I+D 2020 project Code X-920, grants to WDO.

Institutional Review Board Statement: Not applicable.

Informed Consent Statement: All sensory panellists were informed of the procedure, and their consent for participation was obtained before the start of the sensory test.

Data Availability Statement: The data used to support the findings of this study can be made available by the corresponding author upon request.

Acknowledgments: This study is part of the DSc degree thesis of Brenda Ozón carried out at the UNLP. B. Ozón is a doctoral fellow from the National Agency of Scientific and Technological Promotion (ANPCyT—Argentina); J. Cotabarren and W.D. Obregón are members of the Researcher Career Program of the CONICET, and M.G. Parisi is an established researcher of UNLu. The authors acknowledge Eng. Gustavo Kreymborg from Greenborg Industries S.R.L. for his substantial contribution and valuable support (providing us the starting material—the chia expeller—required for the development of this work).

Conflicts of Interest: The authors declare no conflict of interest.

References

1. Kiran, A.; Wakeel, A.; Mahmood, K.; Mubaraka, R.; Hafsa; Haefele, S.M. Biofortification of Staple Crops to Alleviate Human Malnutrition: Contributions and Potential in Developing Countries. *Agronomy* **2022**, *12*, 452. [CrossRef]
2. Tolve, R.; Simonato, B.; Rainero, G.; Bianchi, F.; Rizzi, C.; Cervini, M.; Giuberti, G. Wheat Bread Fortification by Grape Pomace Powder: Nutritional, Technological, Antioxidant, and Sensory Properties. *Foods* **2021**, *10*, 75. [CrossRef] [PubMed]
3. Czarnowska-Kujawska, M.; Starowicz, M.; Barišić, V.; Kujawski, W. Health-Promoting Nutrients and Potential Bioaccessibility of Breads Enriched with Fresh Kale and Spinach. *Foods* **2022**, *11*, 3414. [CrossRef] [PubMed]
4. Cabello-Olmo, M.; Krishnan, P.G.; Araña, M.; Oneca, M.; Díaz, J.V.; Barajas, M.; Rovai, M. Development, Analysis, and Sensory Evaluation of Improved Bread Fortified with a Plant-Based Fermented Food Product. *Foods* **2023**, *12*, 2817. [CrossRef]
5. Madruga, K.; Rocha, M.; Fernandes, S.S.; Salas-Mellado, M.d.l.M. Properties of wheat and rice breads added with chia (*Salvia hispanica* L.) protein hydrolysate. *Food Sci. Technol.* **2020**, *40*, 596–603. [CrossRef]
6. Rabail, R.; Khan, M.R.; Mehwish, H.M.; Rajoka, M.S.R.; Lorenzo, J.M.; Kieliszek, M.; Khalid, A.R.; Shabbir, M.A.; Aadil, R.M. An overview of chia seed (*Salvia hispanica* L.) bioactive peptides' derivation and utilization as an emerging nutraceutical food. *Front. Biosci.* **2021**, *26*, 643–654. [CrossRef]
7. Adamczyk, G.; Ivanišová, E.; Kaszuba, J.; Bobel, I.; Khvostenko, K.; Chmiel, M.; Falendysh, N. Quality Assessment of Wheat Bread Incorporating Chia Seeds. *Foods* **2021**, *10*, 2376. [CrossRef]
8. Grancieri, M.; Stampini Duarte Martino, H.; Gonzalez de Mejia, E. Chia Seed (*Salvia hispanica* L.) as a Source of Proteins and Bioactive Peptides with Health Benefits: A Review. *Compr. Rev. Food Sci. Food Saf.* **2019**, *18*, 480–499. [CrossRef] [PubMed]
9. Muñoz, L.A.; Cobos, A.; Diaz, O.; Aguilera, J.M. Chia seed (*Salvia hispanica*): An ancient grain and a new functional food. *Food Rev. Int.* **2013**, *29*, 37–41. [CrossRef]
10. Sandoval-Oliveros, M.R.; Paredes-Lopez, O. Isolation and characterization of proteins from chia seeds (*Salvia hispanica* L.). *J. Agric. Food Chem.* **2013**, *61*, 193–201. [CrossRef]
11. Gupta, E.; Mishra, P. Functional food with some health benefits, so called superfood: A review. *Curr. Nutr. Food Sci.* **2021**, *17*, 144–166. [CrossRef]
12. Martins, Z.E.; Pinho, O.; Ferreira, I.M.P.L.V.O. Food industry by-products used as functional ingredients of bakery products. *Trends Food Sci. Technol.* **2017**, *67*, 106–128. [CrossRef]
13. Amoah, I.; Taarji, N.; Johnson, P.-N.T.; Barrett, J.; Cairncross, C.; Rush, E. Plant-Based Food By-Products: Prospects for Valorisation in Functional Bread Development. *Sustainability* **2020**, *12*, 7785. [CrossRef]
14. Cotabarren, J.; Rosso, A.M.; Tellechea, M.; García-Pardo, J.; Rivera, J.L.; Obregón, W.D.; Parisi, M.G. Adding value to the chia (*Salvia hispanica* L.) expeller: Production of bioactive peptides with antioxidant properties by enzymatic hydrolysis with Papain. *Food Chem.* **2019**, *274*, 848–856. [CrossRef] [PubMed]
15. Nielsen, P.M.; Petersen, D.; Dambmann, C. Improved Method for Determining Food Protein Degree of Hydrolysis. *Food Chem. Toxicol.* **2001**, *66*, 642–646. [CrossRef]
16. Ozón, B.; Cotabarren, J.; Valicenti, T.; Parisi, M.G.; Obregón, W.D. Chia expeller: A promising source of antioxidant, antihypertensive and antithrombotic peptides produced by enzymatic hydrolysis with Alcalase and Flavourzyme. *Food Chem.* **2022**, *380*, 132185. [CrossRef]

17. Pukalskas, A.; van Beek, T.A.; Venskutonis, R.P.; Linssen, J.P.H.; van Veldhuizen, A.; de Groot, A. Identification of Radical Scavengers in Sweet Grass (*Hierochloe odorata*). *J. Agric. Food Chem.* **2002**, *50*, 2914–2919. [CrossRef] [PubMed]
18. AOAC International. *Official Methods of Analysis*, 19th ed.; Horwitz, W., Latimer, G.W., Eds.; Association of Official Analytical Chemists Internationa: Washington, DC, USA, 2012.
19. Chau, C.F.; Cheung, P.C.K.; Wong, Y.S. Functional properties of protein concentrates from three Chinese indigenous legume seeds. *J. Agric. Food Chem.* **1997**, *45*, 2500–2503. [CrossRef]
20. Chen, J.; Piva, M.; Labuza, T.P. Evaluation of water binding capacity (WBC) of food fiber sources. *J. Food Sci.* **1984**, *49*, 59–63. [CrossRef]
21. AACC. *Approved Methods of the American Association of Cereal Chemists*, 11th ed.; Cereals & Grains Association: St. Paul, MI, USA, 2012.
22. Carter, E.C.; Schanda, J.; Hirschler, R.; Jost, S.; Luo, M.R.; Melgosa, M.; Ohno, Y.; Pointer, M.R.; Rich, D.C.; Viénot, F.; et al. *CIE 015:2018 Colorimetry*, 4th ed.; The International Commission on Illumination: Vienna, Austria, 2018.
23. Zain, N.M.; Ghani, M.A.; Kasim, Z.M. Effects of Water Addition Level on Physical Properties, Rheological Profile and Sensory Evaluation of Gluten-Free Bread: A Preliminary Approach. *Sains Malays.* **2023**, *52*, 487–499. [CrossRef]
24. Utrilla-Coello, R.G.; Osorio-Díaz, P.; Bello-Pérez, L.A. Alternative Use of Chickpea Flour in Breadmaking: Chemical Composition and Starch Digestibility of Bread. *Food Sci. Technol. Int.* **2007**, *13*, 323–327. [CrossRef]
25. Tacias-Pascacio, V.G.; Morellon-Sterling, R.; Siar, E.H.; Tavano, O.; Berenguer-Murcia, Á.; Fernandez-Lafuente, R. Use of Alcalase in the production of bioactive peptides: A review. *Int. J. Biol. Macromol.* **2020**, *165*, 2143–2196. [CrossRef] [PubMed]
26. Hunsakul, K.; Laokuldilok, T.; Sakdatorn, V.; Klangpetch, W.; Brennan, C.S.; Utama-Ang, N. Optimization of enzymatic hydrolysis by alcalase and flavourzyme to enhance the antioxidant properties of jasmine rice bran protein hydrolysate. *Sci. Rep.* **2022**, *12*, 12582. [CrossRef] [PubMed]
27. Halim, N.R.A.; Yusof, H.M.; Sarbon, N.M. Functional and bioactive properties of fish protein hydolysates and peptides: A comprehensive review. *Trends Food Sci. Technol.* **2016**, *51*, 24–33. [CrossRef]
28. Kulczyński, B.; Kobus-Cisowska, J.; Taczanowski, M.; Kmiecik, D.; Gramza-Michałowska, A. The Chemical Composition and Nutritional Value of Chia Seeds-Current State of Knowledge. *Nutrients.* **2019**, *11*, 1242. [CrossRef]
29. Timilsena, Y.P.; Adhikari, R.; Barrow, C.J.; Adhikari, B. Physicochemical and functional properties of protein isolate produced from Australian chia seeds. *Food Chem.* **2016**, *212*, 648–656. [CrossRef]
30. Capitani, M.I.; Spotorno, V.; Nolasco, S.M.; Tomás, M.C. Physicochemical and functional characterization of by-products from chia (*Salvia hispanica* L.) seeds of Argentina. *LWT-Food Sci. Technol.* **2012**, *45*, 94–102. [CrossRef]
31. Villanueva-Lazo, A.; Montserrat-De la Paz, S.; Rodríguez-Martin, N.M.; Millán, F.; Carrera, C.; Pedroche, J.J.; Millán-Linares, M.D.C. Antihypertensive and antioxidant activity of chia protein techno-functional extensive hydrolysates. *Foods* **2021**, *10*, 2297. [CrossRef]
32. Wouters, A.G.B.; Rombouts, I.; Fierens, E.; Brijs, K.; Delcour, J.A. Relevance of the Functional Properties of Enzymatic Plant Protein Hydrolysates in Food Systems. *Compr. Rev. Food Sci. Food Saf.* **2016**, *15*, 786–800. [CrossRef]
33. Xing, Q.; Kyriakopoulou, K.; Zhang, L.; Boom, R.M.; Schutyser, M.A.I. Protein fortification of wheat bread using dry fractionated chickpea protein-enriched fraction or its sourdough. *LWT* **2021**, *142*, 110931. [CrossRef]
34. Roncolini, A.; Milanović, V.; Cardinali, F.; Osimani, A.; Garofalo, C.; Sabbatini, R.; Clementi, F.; Pasquini, M.; Mozzon, M.; Foligni, R.; et al. Protein fortification with mealworm (*Tenebrio molitor* L.) powder: Effect on textural, microbiological, nutritional and sensory features of bread. *PLoS ONE* **2019**, *14*, e0211747. [CrossRef]
35. Silveira Coelho, M.; de las Mercedes Salas-Mellado, M. Chemical Characterization of CHIA (*Salvia hispanica* L.) for Use in Food Products. *J. Food Nutr. Res.* **2014**, *2*, 263–269. [CrossRef]
36. Steffolani, E.; de la Hera, E.; Pérez, G.; Gómez, M. Effect of Chia (*Salvia hispanica* L) Addition on the Quality of Gluten-Free Bread. *J. Food Qual.* **2014**, *37*, 309–317. [CrossRef]
37. Sayed-Ahmad, B.; Talou, T.; Straumite, E.; Sabovics, M.; Kruma, Z.; Saad, Z.; Hijazi, A.; Merah, O. Evaluation of nutritional and technological attributes of whole wheat based bread fortified with Chia flour. *Foods* **2018**, *7*, 135. [CrossRef]
38. Zdybel, B.; Różyło, R.; Sagan, A. Use of a waste product from the pressing of chia seed oil in wheat and gluten-free bread processing. *J. Food Process Preserv.* **2019**, *43*, e14002. [CrossRef]
39. Romankiewicz, D.; Hassoon, W.H.; Cacak-Pietrzak, G.; Sobczyk, M.; Wirkowska-Wojdyła, M.; Ceglińska, A.; Dziki, D. The Effect of Chia Seeds (*Salvia hispanica* L.) Addition on Quality and Nutritional Value of Wheat Bread. *J. Food Qual.* **2017**, *2017*, 7352631. [CrossRef]
40. Segura-Campos, M.R.; Salazar-Vega, I.M.; Chel-Guerrero, L.A.; Betancur-Ancona, D.A. Biological potential of chia (*Salvia hispanica* L.) protein hydrolysates and their incorporation into functional foods. *LWT-Food Sci. Technol.* **2013**, *50*, 723–731. [CrossRef]
41. Watts, B.M.; Ylimaki, G.L.; Jeffery, L.E.; Elias, L.G. *Basic Sensory Methods for Food Evaluation*; IDRC: Ottawa, ON, Canada, 1989.
42. Olusanya, R.N.; Kolanisi, U.; Ngobese, N.Z. Mineral Composition and Consumer Acceptability of Amaranthus Leaf Powder Supplemented Ujeqe for Improved Nutrition Security. *Foods* **2023**, *12*, 2182. [CrossRef]
43. Mamat, H.B.; Wan Chen, Y.; Abdul Hamid, M.; Md Haque Akanda, J.; Pusiran, A.K.; Zainol, M.K. Assessment of dough rheological characteristics and soft bread roll quality of wheat flour incorporated with seaweed powder. *Br. Food J.* **2021**, *123*, 3888–3901. [CrossRef]

44. da Silva, D.O.; Di Primio, E.M.; Botelho, F.T.; Gularte, M.A. Valor nutritivo e análise sensorial de pão de sal adicionado de Pereskia aculeata. *DEMETRA Aliment. Nutr. Saúde* **2014**, *9*, 1027–1040. [CrossRef]
45. del Carmen Robles-Ramírez, M.; Ortega-Robles, E.; Monterrubio-López, R.; Mora-Escobedo, R.; del Carmen Beltrán-Orozco, M. Barley bread with improved sensory and antioxidant properties. *Int. J. Gastron. Food Sci.* **2020**, *22*, 100279. [CrossRef]

foods

MDPI

Article

Comprehensive Analyses of Breads Supplemented with Tannic Acids

Yanbin Guan [1,†], Xun Yang [1,†], Chuang Pan [1], Jie Kong [1], Ruizhe Wu [1], Xueli Liu [1], Yuesheng Wang [1], Mingjie Chen [1], Miao Li [2], Qiong Wang [3], Guangyuan He [1], Guangxiao Yang [1], Junli Chang [1], Yin Li [1] and Yaqiong Wang [1,*]

1 The Genetic Engineering International Cooperation Base of Chinese Ministry of Science and Technology, The Key Laboratory of Molecular Biophysics of Chinese Ministry of Education, College of Life Science and Technology, Huazhong University of Science & Technology, Wuhan 430074, China; gyb@hust.edu.cn (Y.G.); u202112936@hust.edu.cn (X.Y.); d202280833@hust.edu.cn (C.P.); u202112929@hust.edu.cn (J.K.); u202112943@hust.edu.cn (R.W.); wysh@hust.edu.cn (Y.W.); cmj@hust.edu.cn (M.C.); hegy@hust.edu.cn (G.H.); ygx@hust.edu.cn (G.Y.)

2 Grain Storage and Security Engineering Research Center of Education Ministry, School of Food and Strategic Reserves, Henan University of Technology, Zhengzhou 450052, China; limiao@haut.edu.cn

3 College of Life Sciences and Health, Wuhan University of Science and Technology, Wuhan 430065, China

* Correspondence: wangyq@hust.edu.cn; Tel.: +86-27-87792271

† These authors contributed equally to this work.

Abstract: Tannic acid (TA) has been recently considered as a new dough additive for improving the bread-making quality of wheat. However, the effects of TA supplementation on the sensory quality parameters (color, crumb grain structure, and sensory properties) of bread have not been studied. Further, the potential of TA supplementation in bread-making quality improvement has not been evaluated by using commercial flour. In the present study, three commercial wheat flours (namely, XL, QZG, and QZZ) with different gluten qualities were used to evaluate the effects of TA supplementation (in concentrations of 0.1% and 0.3%, respectively). TA supplementation did not change the proximate composition of the breads but increased the volumes and specific volumes of XL and QZG breads. TA supplementation enhanced antioxidant activities, with 0.3% TA significantly increasing the antioxidant capacities of bread made from all three flour samples by approximately four-fold (FRAP method)/three-fold (ABTS method). Positive effects of TA on the reduction in crumb hardness, gumminess, and chewiness were observed in the XL bread, as determined by the texture profile analysis. For the analyses on visual and sensory attributes, our results suggest that TA did not affect the crust color, but only slightly reduced the L* (lightness) and b* (yellowness) values of the crumb and increased the a* (redness) value. TA supplementation also increased the porosity, total cell area, and mean cell area. Satisfactorily, the sensory evaluation results demonstrate that TA-supplemented breads did not exhibit negative sensory attributes when compared to the non-TA-added breads; rather, the attributes were even increased. In summary, TA-supplemented breads generally had not only better baking quality attributes and enhanced antioxidant activities, but, more importantly, presented high consumer acceptance in multiple commercial flour samples. Our results support the commercial potential of TA to be used as a dough improver.

Keywords: tannic acid; commercial wheat flour; antioxidant activity; bread volume; crust and crumb color; sensory evaluation

1. Introduction

White wheat bread is a major staple food in many countries and is also popular and consumed to varying degrees in other countries and regions of the world. Bread and other wheat flour products provide about 20% of the energy (calories) and 21% of plant-sourced proteins for human beings [1]. Moreover, wheat dough has unique viscoelasticity not possessed in other cereal flours, mainly due to gluten proteins [2]. Consequently, breads and other wheat-flour-derived food products hold both nutritional values and diverse culinary properties.

White flour represents a refined product of wheat grain. During the process of grain milling, the endosperm is separated from the germ and bran and refined to produce white flour [3]. Wheat brans and germs are removed during grain milling, which are rich in fiber (such as beta-glucan and arabinoxylan) and phytochemicals (such as sterols, phenols, vitamins and tocols) [4]. These compounds are thought to have positive effects on many health-related metabolic processes [5]. Consequently, the overconsumption of white flour and its food products may contribute to the increasing prevalence of chronic diseases (e.g., chronic kidney disease, cardiovascular diseases, neurodegenerative diseases, chronic obstructive pulmonary disease, and cancer) that are known to be associated with oxidative stress [6–9]. Accordingly, sensory quality and health benefits are increasingly important considerations for consumers when choosing breads and other bakery products [10]. Therefore, the development of innovative breads containing functional ingredients is a new important research area and echoes the consumers' demands for healthy bakery products [11].

In addition, consumers expect bakery products to have a high sensory quality [12]. For bread, the volume and crumb grain structure are important quality parameters. Nevertheless, some studies have demonstrated that the addition of nutritious ingredients to bread, particularly those from coarse cereals enriched with nutrients, while improving the antioxidant capacity and/or the content of dietary fibers, significantly decreases the bread-making quality, therefore, impacting the appearance, texture or taste of the bread and consequently its acceptability [13–16]. For example, some polyphenols have also been reported to improve antioxidant capacity but reduce bread quality [17,18]. Therefore, it is essential to explore additives that enhance the antioxidant capacity without negatively affecting the sensory attributes of bread.

Tannins have been studied as potential fortifying ingredients for bread-making [19–22], as tannin compounds possess excellent antioxidant properties and are widely presented in many kinds of plants. Tannins can be separated into two groups following their distinct structural features: condensed and hydrolyzable tannins. Tannic acid (TA), a pale solid powder with a slightly astringent taste, is one typical compound of hydrolyzable tannins. TA consists of a central glucose unit and ten attached gallic acid molecules. TA is a safe food additive approved by the U.S. Food and Drug Administration (FDA) and the Joint FAO/WHO Expert Committee on Food Additives (JECFA). In the U.S.A., TA is a common additive widely used in beverages (wine, beer, and coffee) and pharmaceuticals, with an estimated daily use about 1 g of TA per person [23]. Previously, our group discovered that TA significantly improves the farinograph-based dough parameters [20]. Later, we confirmed that TA improves the dough quality by using a mixograph and identified that TA promotes glutenin aggregation and preferentially interacts with high-molecular-weight glutenin subunits (HMW-GS) through hydrogen bonds [21].

While the above-mentioned studies reveal the potential of TA as a new dough improver, key issues regarding the commercial application of TA in the baking industry remain to be addressed, and these include the following: (1) whether TA supplementation could affect sensory attributes and consumer acceptability of bread; (2) whether TA remains effective in bread quality improvement when supplemented into commercially used bread flour, since the previous studies examined TA's positive effects in certain wheat cultivars that are grown locally (e.g., Zhengmai 9023) and not particularly bred for bread-making [20,21]. To address these issues, the present work aims to provide comprehensive analyses on several chemical and physical characteristics of the TA-supplemented bread made from commercial flour samples (including proximate compositions, baking parameters, antioxidant activity and bread texture), visual attributes (color analysis and crumb image analysis) and the sensory evaluation results.

2. Materials and Methods

2.1. Materials

Food-grade TA (Catalog No. 58454) was purchased from Henan Yiduoyuan Biotechnology Co., Ltd. (Zhengzhou, China). Three commercial wheat flours of different gluten strengths were purchased from the local market (Wuhan, China), including a high-gluten white bread flour Xinliang (XL), a bread flour Qizi (QZG) and a plain flour Qizi (QZZ).

2.2. Preparation of Bread

Pan bread samples were made according to the AACC-approved method (AACC 10-10B) with slight modifications. TA was added at 0.1% and 0.3% of the dry weight of the wheat flour, respectively. Doughs were prepared at room temperature by mixing 40 g of wheat flour with 23.2 g of tap water, 0.4 g of dry yeast, 2.4 g of caster sugar, 0.6 g of table salt and 1.2 g of margarine. The baking temperatures and times were optimized for the above-mentioned scale. The dough was first fermented for 90 min in a controlled fermentation cabinet at 28 °C and 85% relative humidity. Then, the dough was formed into a cylindrical shape, placed in a baking mold (10 cm × 5.6 cm × 3.4 cm) and again had a second fermentation in the controlled fermentation cabinet for 60 min. Finally, the bread was baked for 20 min at 180 °C in an electric top-and-bottom heated oven. After cooling to room temperature, the bread was analyzed for loaf weight, loaf volume and specific volume. The baking experiment was performed in three batches.

2.3. Proximate Analysis

The fresh bread was used as a sample for moisture. The freeze-dried bread samples were analyzed to determine the ash, protein, fat and fiber. The moisture content was determined by the oven drying method according to 925.10 AOAC [24]. The ash content was performed by conductivity following ICC Standard No. 157 [25]. The protein content was determined by the Kjeldahl method using a KjeltecTM 8400 analyzer (Foss, Hilleroed, Denmark) following GB5009.5—2016. The nitrogen content was multiplied by 6.25 to obtain the protein content. The fat content was analyzed by hydrolysis in hydrochloric acid at 75 °C for 45 min followed by extraction in petroleum ether and evaporation according to GB5009.6—2016. The fiber content was evaluated using FIWE6 Fiber Analyzer (VELP, Usmate, Italy) according to GB5009.88—2014.

2.4. Volume and Specific Volume Measurement and Texture Profile

The loaf volume was determined according to the AACC-approved method 10-05 (rapeseed displacement test). The bread sample was placed in a measuring cylinder. The measuring cylinder was filled with rapeseed to the maximum line. The volume of the rapeseed after removing the bread sample was measured. The volume of the bread was calculated by subtracting the volume of rapeseed from the volume of the measuring cylinder. The specific volume was calculated by dividing the loaf volume by the loaf weight.

Three slices of bread including the middle slice and one on either side were used for texture profile analysis (TPA) at room temperature using a Texture Analyzer (TA-XT plus, Stable Micro Systems, Godalming, UK) according to a previous study [26]. The probe used was the P/36 aluminum cylinder (36.0 mm in diameter). The rates for the pre-test, test, and post-test were 1.0, 0.5, and 10.0 mm/s, respectively, the trigger force was 5.0 g, and the distance was 5.0 mm. The parameters analyzed were investigated in terms of their hardness, springiness, cohesiveness, gumminess, chewiness, and resilience.

2.5. Antioxidant Capacity Assays

A ferric-reducing antioxidant power assay (FRAP) kit (Catalog #G0115F) and a total antioxidant capacity assay kit with ABTS method (Catalog #G0142W) were purchased from Suzhou Grace Biotechnology Co., Ltd. (Suzhou, China). Antioxidant capacity was determined by the FRAP kit and ABTS kit. An amount of 10 mg of each of the freeze-dried bread samples were used for the FRAP assay and ABTS assay with three biological replicates. The results were expressed in μmol Trolox/g dry bread.

2.6. Determination of Crust and Crumb Color

The color of bread crust and crumb was measured using a Minolta colorimeter (Minolta CR- 400, Konica Minolta Sensing, Inc., Osaka, Japan). Color values were recorded in ten areas of each loaf. The color analysis was expressed in L*, a*, and b* values. L* value is a measure of lightness (100: perfect white, zero: black), whereas a* (+red/−green) and b* (+yellow/−blue) values are the chromaticity values [27].

2.7. Image Analysis

Digital image analysis was used to evaluate crumb structure according to a previous study [28], as the image-analysis-based fineness score correlated well with the visual evaluation of fineness. The cross-sections of bread slices were scanned at 300 dpi on a grey scale with a scanner (CanoScan LiDE110, Tokyo, Japan). The images obtained were analyzed using ImageJ2 v1.0 software (National Institutes of Health, Bethesda, MD, USA). Sub-image areas of 30 mm × 30 mm were obtained at the same position on each slice (center of the bread) and then binarized for analysis. Five crumb features were determined, including the total number of cells, total area of cells, average area of cells, porosity, and percentage of the number of cells < 4 mm^2.

2.8. Sensory Evaluation of Bread

A specially designed, odor-free meeting room with separate tables was kept at 22 ± 2 °C. The sensory evaluation was performed by a group of 20 evaluators (consumers) consisting of 11 males and 9 females, aged 18−45, all of whom are non-smokers. The experimental design of sensory evaluation was based on the sensory analyses of bread samples previously published in the Foods journal [29–33]. They were recruited from students and teachers working on the improvement of wheat processing quality at the Genetic Engineering International Cooperation Base of the Chinese Ministry of Science and Technology. The panelists are consumers of bread. To facilitate the description of the sensory characteristics of the bread, appearance, crumb characteristics, crust and crumb color, aroma, and taste were chosen based on previous report [14]. Appearance, crumb characteristics, crust and crumb color, aroma, and taste were evaluated using a 9-point Hedonic scale (1: dislike extremely, 5: neither like nor dislike, and 9: like extremely).

2.9. Statistical Analysis

All of the statistical analyses were conducted using the SPSS 26.0 software. All experiments were performed in triplicate. In order to determine the significance of differences between the flours, a two-way ANOVA was performed at a significance level of $p < 0.05$. The sensory data were checked for the homogeneity of variance and normality before performing ANOVA analysis with SPSS.

3. Results

3.1. Proximate Analysis

Three commercial wheat flours (namely XL, QZG and QZZ) were used in the present study. The proximate analysis of bread, measuring ash, protein, fat, and fiber, is shown in Table 1. Of the three flours, XL has the highest content of nutrients. QZZ has lower ash and protein content than QZG. The two-way statistical analysis showed a significant statistical effect ($p < 0.05$) of the type of flour on the composition of breads. As TA is a polyphenol, predictably, the addition of TA had no significant ($p < 0.05$) effect on the nutrient content of bread.

Table 1. Proximate value and antioxidant activity of breads supplemented with or without TA.

Breads	Samples	Ash Content (g/100 g DM)	Protein Content (g/100 g DM)	Fat Content (g/100 g DM)	Fiber Content (g/100 g DM)	ABTS (μmol TE/g DM)	FRAP (μmol TE/g DM)
XL	0	6.17 ± 0.14 [a]	13.74 ± 0.09 [a]	3.23 ± 0.10 [a]	3.23 ± 0.10 [a]	4.65 ± 0.75 [a]	4.27 ± 0.34 [a]
	0.10%	6.02 ± 0.05 [a]	13.76 ± 0.07 [a]	3.46 ± 0.09 [a]	3.46 ± 0.09 [a]	10.11 ± 0.88 [b]	9.58 ± 0.05 [b]
	0.30%	6.10 ± 0.08 [a]	13.44 ± 0.18 [a]	3.52 ± 0.03 [a]	3.59 ± 0.04 [a]	15.03 ± 1.51 [c]	21.97 ± 0.08 [c]
QZG	0	5.85 ± 0.06 [b]	11.05 ± 0.06 [a]	2.94 ± 0.04 [a]	2.94 ± 0.04 [a]	4.14 ± 0.07 [a]	4.24 ± 0.08 [a]
	0.10%	5.97 ± 0.09 [b]	11.07 ± 0.02 [a]	2.97 ± 0.08 [a]	2.97 ± 0.08 [a]	7.34 ± 0.21 [b]	10.09 ± 0.16 [b]
	0.30%	5.37 ± 0.16 [a]	11.21 ± 0.08 [a]	2.91 ± 0.10 [a]	2.91 ± 0.10 [a]	16.71 ± 0.54 [c]	22.30 ± 0.11 [c]
QZZ	0	5.36 ± 0.08 [a]	10.28 ± 0.10 [a]	3.02 ± 0.14 [a]	3.02 ± 0.14 [a]	4.43 ± 0.74 [a]	3.97 ± 0.13 [a]
	0.10%	5.19 ± 0.08 [a]	10.41 ± 0.09 [a]	2.88 ± 0.03 [a]	2.88 ± 0.03 [a]	6.68 ± 0.22 [a]	10.15 ± 0.20 [b]
	0.30%	5.52 ± 0.09 [a]	10.46 ± 0.07 [a]	2.68 ± 0.08 [a]	2.68 ± 0.08 [a]	15.52 ± 1.49 [b]	22.57 ± 0.17 [c]
Two-Factor ANOVA—p							
Factor 1 (flour)		<0.05	<0.05	<0.05	>0.05	>0.05	>0.05
Factor 2 (TA concentration)		>0.05	>0.05	>0.05	<0.05	<0.05	<0.05
Factor 1 × factor 2		<0.05	>0.05	<0.05	>0.05	>0.05	>0.05

Lower case letters indicate the statistical results of a bread proximate component or antioxidant activity (corresponding to each of the columns) between control and TA treatments. Different letters indicate values that are significantly different ($p < 0.05$, Tukey's HSD test).

3.2. Baking Parameters

Next, we investigated whether TA addition to commercial flour could lead to improved dough quality. Without TA supplementation, the bread made from XL had the highest bread volume and specific volume, followed by QZG and QZZ (Figure 1A,B). TA supplementation showed different effects in improving bread volume. TA addition had no effect on QZZ breads, whereas TA supplementation significantly increased the bread volume and specific volume of XL and QZG breads in a dosage-dependent manner (from 0.1% to 0.3%). Indeed, TA improved the bread-making properties in commercial flour-made breads, and such improving effects became quite evident in high-gluten flour samples. The two-way statistical analysis showed a significant statistical effect ($p < 0.05$) of the type of flour on the baking parameters.

As a natural food additive, the effects of TA supplementation on bread-making properties are distinct from many other nutritional components. The addition of nut seeds, dietary fiber, polyphenols, plant extracts, or polyphenol-rich ingredients and vegetables all significantly reduced the bread volume [14,15,34–36]. The baking characteristics of the bread are correlated to the degree of crosslinking of gluten proteins and the addition of the above-mentioned ingredients likely led to weakened gluten networks, explaining why the bread-making quality was reduced. Unlike the above-mentioned supplements, TA has phenolic hydroxyl and gallic acyl groups that interact with gluten proteins via hydrogen bonding and hydrophobic interactions [20,21], leading to the dough containing well-distributed gluten polymers with a higher degree of cross-linking that could help improve the air retention ability of the dough during baking, thus improving the bread-making quality.

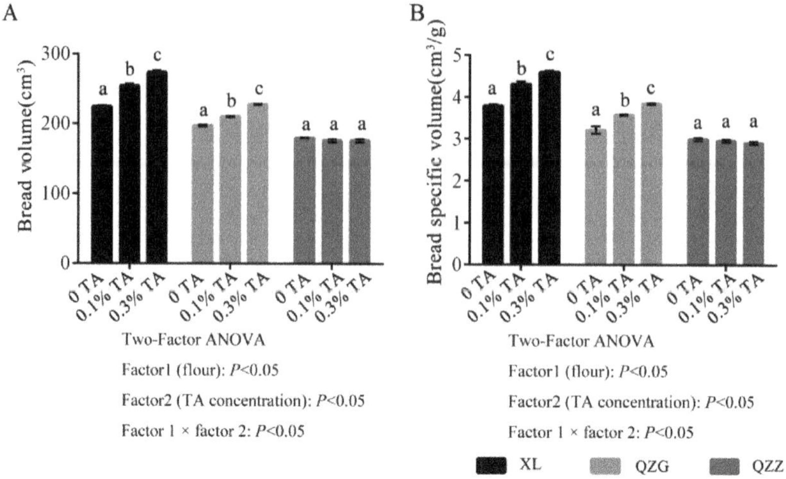

Figure 1. Effect of TA on baking characteristics of breads: bread volume (**A**) and bread specific volume (**B**). (**A**) The bread volume was determined according to the AACC-approved method (AACC 10-05); (**B**) The specific volume was calculated by dividing the bread volume by the bread weight. Different letters indicate values that are significantly different ($p < 0.05$, Tukey's HSD test).

3.3. Antioxidant Activity of the TA-Supplemented Breads

The FRAP assay and ABTS assay were performed to test whether the TA-supplemented bread samples had enhanced antioxidant activity (Table 1). In the bread without TA, both FRAP and ABTS were about 4.5 µmol Trolox g^{-1} bread. TA supplementation dramatically increased the antioxidant capacity of the breads. The antioxidant capacity was increased by one-fold (approximately 10 µmol Trolox g^{-1} bread) in the FRAP assay and 0.5-fold in the ABTS assay (approximately 7.5 µmol Trolox g^{-1} bread) when supplemented with 0.1% TA and further increased by four-fold (~22 µmol Trolox g^{-1} bread) in the FRAP assay and three-fold in the ABTS assay (approximately 15 µmol Trolox g^{-1} bread) when supplemented with 0.3% TA. There was no statistical difference ($p < 0.05$) between the flours on the antioxidant activity according to the two-way ANOVA analysis.

Our results showed that the supplementation of low-level TA (i.e., 0.3%) can significantly increase the antioxidant activity of the bread samples. Previous studies have also reported improved antioxidant activity with the addition of nutritional ingredients. For instance, the addition of 10% sea buckthorn pulp, 8% flaxseed, and even 70% orange-fleshed sweet potato could increase the antioxidant activity of breads by two-fold, one-fold, and three-fold, respectively, as determined by the FRAP assay [34,37]. Although 6% apple polyphenols and 10% grape pomace powder (rich in polyphenols) improved the antioxidant capacity by about nine-fold and eight-fold, respectively [15,35], their amount of addition far exceeded the amount of TA added in our study. Previously, our work has shown that TA affects the dough's macroscopic properties and microstructure through non-covalent interactions with gluten proteins, thus improving the dough processing parameters [21]. In the present study, TA supplementation improved not only the baking performance of breads (volume and specific volume of bread) (Figure 1A,B) but also the antioxidant capacity by around four-fold (FRAP assay)/three-fold (ABTS assay) (Table 1).

3.4. Texture Evaluation

The bread texture was evaluated after baking using TPA (Table 2). Without the addition of TA, XL bread had the least hardness, gumminess and chewiness among the three breads, followed by QZG and QZZ. After the addition of TA, TA significantly affected ($p < 0.05$) the hardness, gumminess and chewiness of XL bread, while there was no significant effect

($p < 0.05$) on the other parameters. TA did not produce a significant effect ($p < 0.05$) on the TPA parameters of either QZG or QZZ. The two-way statistical analysis showed a significant statistical effect ($p < 0.05$) by the type of flour on the hardness, cohesiveness, gumminess and chewiness.

Table 2. Textural parameters of breads supplemented with or without TA.

Breads	Samples	Hardness (N)	Springiness (−)	Cohesiveness (−)	Gumminess (−)	Chewiness (−)	Resilience (−)
XL	0 TA	950.75 ± 9.11 [c]	0.95 ± 0.01 [a]	0.74 ± 0.01 [a]	705.87 ± 9.32 [b]	672.74 ± 9.79 [b]	0.43 ± 0.01 [a]
	0.1% TA	858.99 ± 14.28 [b]	0.97 ± 0.02 [a]	0.78 ± 0.03 [a]	667.46 ± 16.42 [b]	645.86 ± 25.17 [b]	0.47 ± 0.03 [a]
	0.3% TA	595.73 ± 7.14 [a]	0.96 ± 0.01 [a]	0.78 ± 0.02 [a]	464.50 ± 8.30 [a]	447.60 ± 11.50 [a]	0.47 ± 0.01 [a]
QZG	0 TA	1243.19 ± 32.00 [ab]	0.97 ± 0.01 [a]	0.68 ± 0.01 [a]	840.05 ± 15.14 [a]	814.07 ± 16.15 [a]	0.40 ± 0.01 [a]
	0.1% TA	1272.64 ± 9.04 [b]	0.96 ± 0.01 [a]	0.74 ± 0.03 [a]	934.70 ± 31.12 [a]	896.29 ± 22.03 [b]	0.43 ± 0.02 [a]
	0.3% TA	1143.08 ± 38.36 [a]	0.97 ± 0.01 [a]	0.74 ± 0.01 [a]	840.94 ± 16.66 [a]	812.32 ± 15.91 [a]	0.45 ± 0.01 [a]
QZZ	0 TA	2359.12 ± 88.59 [a]	0.97 ± 0.01 [a]	0.71 ± 0.06 [a]	1653.40 ± 77.11 [a]	1610.66 ± 75.72 [a]	0.45 ± 0.06 [a]
	0.1% TA	2221.69 ± 95.21 [a]	0.98 ± 0.01 [a]	0.67 ± 0.01 [a]	1495.30 ± 60.60 [a]	1459.18 ± 61.83 [a]	0.42 ± 0.01 [a]
	0.3% TA	2167.52 ± 107.32 [a]	0.97 ± 0.01 [a]	0.66 ± 0.03 [a]	1414.10 ± 21.97 [a]	1371.37 ± 10.89 [a]	0.40 ± 0.03 [a]
Two-Factor ANOVA—p							
Factor 1 (flour)		<0.05	>0.05	<0.05	<0.05	<0.05	>0.05
Factor 2 (TA concentration)		<0.05	>0.05	>0.05	<0.05	<0.05	>0.05
Factor 1 × factor 2		>0.05	>0.05	>0.05	<0.05	<0.05	>0.05

Lower case letters indicate the statistical results of a textural parameter (corresponding to each of the columns) between control and TA treatments. Different letters indicate values that are significantly different ($p < 0.05$, Tukey's HSD test).

Texture properties determine product durability and consumer acceptance. A denser crumb structure leads to greater hardness [34]. To maximize consumer acceptance of the bread, minimizing the hardness value is the highest priority [38]. The effect of TA on the hardness of bread meets this requirement, which is reflected in the fluffier crumb structure of TA-added bread. The dextran produced by the sourdough fermentation of whole and sprouted lentil flour affects the crumb structure of white bread, resulting in less hardness [39]. Additionally, chewiness is a dimensionless size and characterizes the energy needed to chew the food [34]. A positive effect of TA on the reduction in adhesion and chewiness of bread crumb was observed, which may be attributed to the effect of TA on the gluten protein network.

3.5. *Visual Attributes of the TA-Supplemented Breads*

The visual attributes analyzed in the present study include crumb color, crust color and crumb structure parameters [40].

3.5.1. Color Analysis

Since TA is a pale-yellow powder, we also investigated the crumb color of TA-supplemented bread samples. Visual comparison did not reveal obvious changes in the crumb color between the bread slices supplemented with or without TA (Figure 2). The crust colors of the bread samples were then analyzed by using a colorimeter (Table 3). In the non-TA-added crust samples, the L* and b* values of the crust color were different between the flour samples, ranking from high to low as QZZ, QZG, and XL, while no difference was detected for the a* value. After TA supplementation, a slight increase in L* values for QZZ and XL bread crusts and a slight decrease for QZG was detected. TA supplementation led to a slight decrease in b* values of the QZZ and QZG bread crusts but a slight increase in the XL bread crust. Generally, TA addition did not affect a* values.

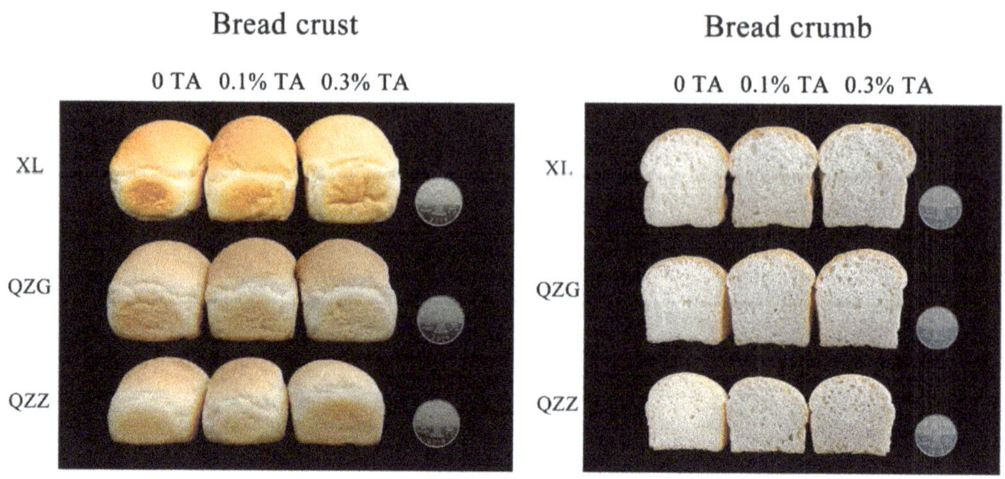

Figure 2. Effect of TA on crust and crumb color.

Table 3. The crust and crumb color parameters of the bread samples supplemented with or without TA.

Breads	Samples	Crust Color			Crumb Color		
		L*	a*	b*	L*	a*	b*
XL	0 TA	68.31 ± 0.75 [a]	16.98 ± 0.53 [ab]	29.91 ± 0.98 [a]	90.64 ± 0.24 [a]	0.71 ± 0.03 [a]	10.58 ± 0.18 [a]
	0.1% TA	70.77 ± 0.54 [b]	17.85 ± 0.35 [b]	34.21 ± 0.49 [b]	83.55 ± 0.27 [b]	2.05 ± 0.01 [b]	4.98 ± 0.15 [b]
	0.3% TA	72.99 ± 0.22 [c]	15.71 ± 0.24 [a]	34.40 ± 0.42 [b]	82.17 ± 0.19 [c]	3.03 ± 0.01 [c]	5.32 ± 0.09 [b]
QZG	0 TA	79.91 ± 0.80 [a]	18.68 ± 0.27 [a]	39.87 ± 0.50 [a]	90.90 ± 0.21 [a]	0.98 ± 0.02 [a]	10.09 ± 0.24 [a]
	0.1% TA	78.79 ± 0.18 [a]	17.37 ± 0.87 [ab]	37.77 ± 0.61 [a]	85.41 ± 0.15 [b]	1.95 ± 0.02 [b]	7.13 ± 0.11 [b]
	0.3% TA	74.57 ± 0.89 [b]	16.40 ± 0.38 [b]	34.54 ± 0.78 [b]	84.01 ± 0.25 [c]	2.79 ± 0.01 [c]	6.84 ± 0.21 [b]
QZZ	0 TA	83.75 ± 0.89 [ab]	16.78 ± 0.41 [ab]	42.10 ± 1.02 [a]	85.94 ± 0.17 [a]	0.97 ± 0.02 [a]	11.00 ± 0.24 [a]
	0.1% TA	82.36 ± 1.50 [b]	19.01 ± 1.30 [a]	42.11 ± 0.34 [a]	81.71 ± 0.26 [b]	1.75 ± 0.03 [b]	7.68 ± 0.24 [b]
	0.3% TA	86.36 ± 0.77 [a]	13.99 ± 0.66 [b]	39.47 ± 0.37 [b]	78.75 ± 0.19 [c]	2.74 ± 0.02 [c]	8.63 ± 0.17 [c]
Two-Factor ANOVA—*p*							
Factor1 (flour)		<0.05	>0.05	<0.05	<0.05	<0.05	<0.05
Factor 2 (TA concentration)		<0.05	<0.05	<0.05	<0.05	<0.05	<0.05
Factor 1 × factor 2		>0.05	<0.05	<0.05	<0.05	<0.05	<0.05

Lower case letters indicate the statistical results of a crust or crumb color parameter (corresponding to each of the columns) between control and TA treatments. Different letters indicate values that are significantly different ($p < 0.05$, Tukey's HSD test).

Although a two-way ANOVA analysis showed significant differences ($p < 0.05$) in color between flours, the differences in crumb color values were not large between the three bread samples (i.e., XL, QZG and QZZ). After TA addition, as the concentration of TA increased, the crumb L* and b* values gradually decreased in all three breads, while the crumb a* values increased gradually. Overall, the crust colors were not changed significantly by TA supplementation, and the crumb colors became slightly deepened based on colorimeter analysis.

Previous studies have shown that bread color can be influenced by additives in a dosage-dependent manner [13,35,41]. When millet flour was added to the bread at 10%, 30% and 50% percent, respectively, the crumb color darkened as the amount of millet flour increased [13]. Another example is that gluten-free bread made with pigmented rice flour (containing 19% anthocyanins) shows a purple color [41]. Here, color analysis results showed that the effect of TA on crumb color values was dosage-dependent (Table 3). Because TA is only pale yellow in color and its effective amount of addition is pretty low (0.3%), TA supplementation into breads only leads to a subtle (albeit detectable for some parameters by the colorimeter) change in crust color values and may not affect consumer acceptability.

3.5.2. Crumb Image Analysis

Flour additives can influence the crumb characteristics of bread, which are related to the end-use quality of the bread [42]. To obtain a detailed and quantitative evaluation of the crumb structure, image analysis was performed on slices of breads supplemented with/without TA (Figure 3). The total area of cells, average area of cells, and porosity were correlated with the bread volume and specific volume, and the percentage of total cells and those of the cells < 4 mm^2 were correlated with the gas–cell regularity (Figure 3) [43].

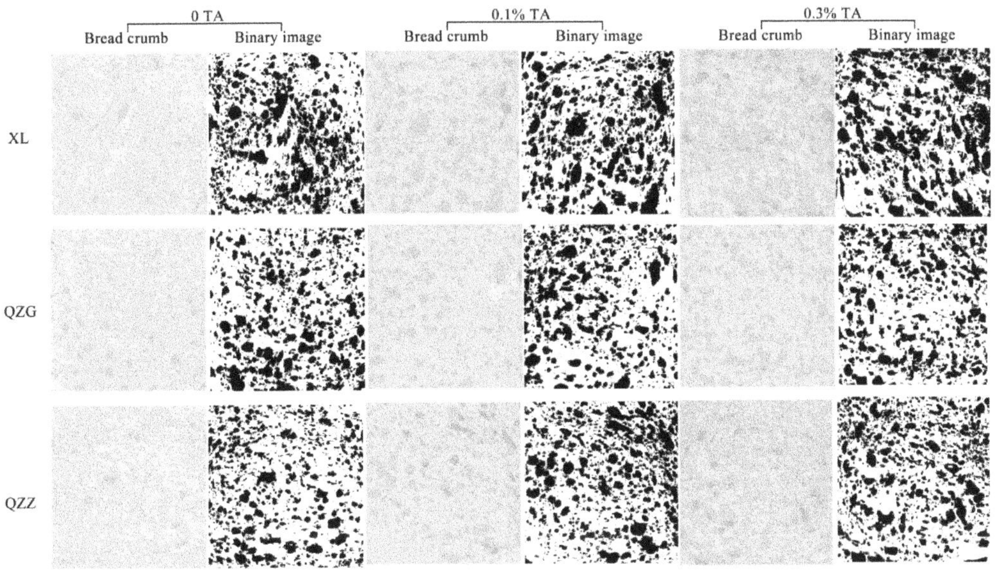

Figure 3. Determination of the visual attributes of bread samples with or without TA supplementation. Cross-section of bread crumb and its binary image with different levels of TA. The binary image was obtained by analyzing a 30 mm × 30 mm bread crumb area with ImageJ2 software (National Institutes of Health, USA).

The two-way statistical analysis showed a significant statistical effect ($p < 0.05$) of the type of flour on the gas cell characteristic. Without TA addition, the porosity, total area of cells, and average area of cells were highest in the XL bread, followed by the QZG and QZZ bread (Table 4). Although the total number of cells of the QZG bread was lower than that of the QZZ bread, the total area of cells, average area of cells, and porosity of the QZG bread was higher than that of the QZZ bread. Theoretically, large values of the average area of cells, porosity, and total area of cells imply high bread volumes [44]. Our results of crumb structure parameters were consistent with the bread volume results (Figure 1). Overall, TA supplementation significantly altered crumb structures, and those changes were reflected in the image analysis results. TA supplementation significantly increased ($p < 0.05$) the total area of cells, average area of cells, and porosity of the XL breads, while TA addition to QZG breads led to a significantly increased ($p < 0.05$) total area of cells and porosity. Further, TA supplementation into the QZZ breads resulted in significantly increased ($p < 0.05$) total area of cells, average area of cells, and porosity compared to those of non-TA breads. However, the percentage of the number of cells < 4 mm^2 of the XL or QZG breads decreased significantly along with the increased TA dosage. These results indicate that TA addition increased the number of large cells, while we did not detect statistical differences in the total number of cells with or without TA supplementation. In summary, TA supplementation improved the crumb structure parameters. Moreover, TA addition had more pronounced effects on the XL bread than those made from QZG and QZZ flour samples.

Table 4. The gas cell characteristics of the bread samples supplemented with or without TA.

Breads	Samples	Total Number of Cells	Total Area of Cells (mm^2)	Average Area of Cells (mm^2)	Porosity (%)	Percentage of Number of Cells < 4 mm^2 (%)
XL	0 TA	1419.33 ± 43.43 [a]	323.07 ± 16.04 [a]	0.23 ± 0.02 [a]	35.23 ± 1.57 [a]	99.00 ± 0.15 [a]
	0.1% TA	1242.33 ± 65.58 [a]	353.30 ± 10.23 [a]	0.29 ± 0.02 [ab]	37.77 ± 1.14 [a]	98.43 ± 0.16 [ab]
	0.3% TA	1322.33 ± 18.17 [a]	411.43 ± 2.91 [b]	0.31 ± 0.00 [b]	43.38 ± 0.70 [b]	98.33 ± 0.12 [b]
QZG	0 TA	1223.00 ± 39.51 [a]	262.60 ± 14.88 [a]	0.22 ± 0.02 [a]	28.54 ± 1.52 [a]	99.10 ±0.13 [a]
	0.1% TA	1290.33 ± 52.60 [a]	278.55 ± 2.01 [ab]	0.22 ± 0.01 [a]	30.32 ± 0.49 [ab]	99.10 ± 0.01 [a]
	0.3% TA	1184.33 ± 20.76 [a]	313.25 ± 4.44 [b]	0.26 ± 0.01 [a]	33.87 ± 0.59 [b]	98.42 ± 0.23 [b]
QZZ	0 TA	1492.33 ± 6.39 [a]	253.38 ± 14.45 [a]	0.17 ± 0.01 [a]	27.22 ± 1.41 [a]	99.35 ± 0.18 [a]
	0.1% TA	1412.67 ± 28.90 [a]	308.20 ± 10.97 [b]	0.22 ± 0.01 [b]	33.60 ± 1.31 [b]	99.05 ± 0.04 [a]
	0.3% TA	1498.67 ± 37.68 [a]	309.83 ± 5.51 [b]	0.21 ± 0.01 [b]	33.32 ± 0.64 [b]	99.09 ± 0.13 [a]
Two-Factor ANOVA—*p*						
Factor1 (flour)		<0.05	<0.05	<0.05	<0.05	<0.05
Factor2 (TA concentration)		>0.05	<0.05	<0.05	<0.05	<0.05
Factor 1 × factor 2		<0.05	>0.05	>0.05	>0.05	>0.05

Lower-case letters indicate the statistical results of a gas cell characteristic (corresponding to each of the columns) between the control and tannic acid treatments ($p < 0.05$, Tukey's HSD test).

TA supplementation increased the total area of cells, average area of cells, and porosity, consistent with the improved bread volume and specific volume of the TA-added breads. According to a previous report, the addition of barley β-glucan to the bread could increase the total and average area of cells, while the specific volume of the bread became larger [43]. By contrast, the addition of 10–14% soluble oat fiber decreased the specific volume of the bread by 50% compared to the control bread, and the total area of cells and porosity were decreased as well [45].

The total number of cells in a slice of bread serves as an indication of the number of carbon dioxide bubbles captured during proofing [46]. TA addition did not significantly affect the total number of cells compared to the control, indicating that TA addition may not affect the ability to capture carbon dioxide bubbles during proofing. Previous studies showed that bread containing 4–12% of apricot kernel flour does not allow gas cells to expand, resulting in smaller gas chambers in the bread and thus a compact crumb, consistent with the fact that apricot kernel flour is less elastic than gluten proteins [44]. Unlike the undesirable effects brought by apricot kernel flour, TA addition indeed enhanced the dough property to allow CO_2 to expand properly, resulting in a greater porosity.

3.6. Effect of TA Supplementation on the Bread Sensory Attributes

In the sensory evaluation experiments, the six sensory attributes were scored by the semi-trained panelists (Table 5). A two-way ANOVA analysis showed significant differences ($p < 0.05$) in crumb characteristics and taste between breads. The XL bread without TA addition received the highest score in the categories of crumb characteristics and taste, which might be associated with the highest specific volume of non-TA XL bread among all the bread samples. When compared with the non-TA bread samples, only TA-added XL bread samples obtained significantly higher scores for several sensory attributes, while in the other two breads (QZG and QZZ), TA addition did not differ from the control.

The score of textural attributes for XL was significantly increased ($p < 0.05$) at the 0.1% TA addition level compared to the control. When TA was added at 0.3%, the appearance, crumb characteristics, taste, and crust color of XL were significantly increased ($p < 0.05$) compared to the bread without TA addition. This indicated that the TA-added XL breads were more popular. It was found that both appearances decreased when gas cells were smaller and evenly distributed [41]. TA addition led to larger and uneven gas cells in the XL breads, thus producing an improved loaf appearance. Crumb color is an important sensory attribute that influences consumer preferences. For the breads made from all three types of flour, the crumb color scores of the breads added with 0.1% or 0.3% TA did not differ significantly compared to the breads without TA, which was consistent with the colorimeter results (Table 3). Therefore, both results suggest that TA supplementation into breads did not reduce sensory attributes and consumer acceptability.

Table 5. The sensory attributes of the bread samples supplemented with or without TA.

Breads	Samples	Appearance	Crumb Characteristics	Crust Color	Crumb Color	Aroma	Taste
XL	0 TA	7.07 ± 0.18 ^{a}	7.27 ± 0.40 ^{a}	7.00 ± 0.26 ^{a}	7.47 ± 0.24 ^{a}	7.60 ± 0.27 ^{a}	7.53 ± 0.26 ^{a}
	0.1% TA	7.20 ± 0.24 ^{a}	8.13 ± 0.19 ^{b}	7.47 ± 0.22 ^{a}	7.67 ± 0.21 ^{a}	7.80 ± 0.22 ^{a}	8.13 ± 0.19 ^{a}
	0.3% TA	7.80 ± 0.22 ^{b}	8.60 ± 0.19 ^{c}	7.67 ± 0.21 ^{b}	7.80 ± 0.35 ^{a}	7.73 ± 0.25 ^{a}	8.40 ± 0.34 ^{b}
QZG	0 TA	7.27 ± 0.21 ^{a}	7.13 ± 0.24 ^{a}	7.00 ± 0.24 ^{a}	7.33 ± 0.27 ^{a}	7.27 ± 0.32 ^{a}	6.33 ± 0.37 ^{a}
	0.1% TA	7.73 ± 0.21 ^{a}	7.40 ± 0.25 ^{a}	7.33 ± 0.23 ^{a}	7.47 ± 0.24 ^{a}	7.67 ± 0.23 ^{a}	6.63 ± 0.41 ^{a}
	0.3% TA	7.60 ± 0.16 ^{a}	7.60 ± 0.21 ^{a}	7.73 ± 0.23 ^{b}	7.33 ± 0.23 ^{a}	7.40 ± 0.31 ^{a}	7.03 ± 0.23 ^{a}
QZZ	0 TA	7.27 ± 0.28 ^{a}	6.53 ± 0.32 ^{a}	7.27 ± 0.28 ^{a}	7.27 ± 0.18 ^{a}	7.60 ± 0.16 ^{a}	6.27 ± 0.33 ^{a}
	0.1% TA	7.40 ± 0.31 ^{a}	7.13 ± 0.17 ^{a}	7.40 ± 0.24 ^{a}	7.07 ± 0.30 ^{a}	7.13 ± 0.35 ^{a}	6.93 ± 0.28 ^{a}
	0.3% TA	7.53 ± 0.31 ^{a}	7.07 ± 0.27 ^{a}	7.67 ± 0.21 ^{a}	7.20 ± 0.24 ^{a}	6.93 ± 0.45 ^{a}	7.00 ± 0.26 ^{a}
Two-Factor ANOVA—p							
Factor1 (flour)		>0.05	<0.05	>0.05	>0.05	>0.05	<0.05
Factor2 (TA concentration)		>0.05	<0.05	<0.05	>0.05	>0.05	<0.05
Factor 1 × factor 2		>0.05	>0.05	>0.05	>0.05	>0.05	>0.05

Lower-case letters indicate the statistical analysis of a sensory attribute (corresponding to each of the columns) between the control and tannic acid treatments ($p < 0.05$, Tukey's HSD test).

Previous studies have shown that the bread taste changes as the amount of natural ingredients added increases [47–49]. For example, 0.5% or more microencapsulated onion skin extracts made the bread taste obviously bitter compared to those added with low concentrations of the onion skin extracts [48]. The addition of 15% prickly pear peel flour made the bread sour while the addition of 20% changed the bitterness and astringency of the bread [47]. At 4% and higher additions of dandelion root, the bread could be perceived to have an unusual odor and an obvious bitterness [49]. The TA in this study did not change the aroma of the bread as it was added at a mere 0.3%.

4. Conclusions

Our research demonstrates that the addition of TA to commercial high-gluten flours produces breads with greater volume and specific volume. The addition of 0–0.3% TA to commercial flour did not affect the proximate composition of the bread. Importantly, TA addition into bread is likely an effective way to produce breads with high antioxidant activity. The addition of TA to commercial high-gluten bread flour had a lower hardness, gumminess and chewiness. TA supplementation, at least at the concentrations tested herein, did not negatively affect bread color and appearance. Moreover, the sensory characteristics of TA-supplemented bread (made from the high-gluten bread flour) were scored the highest for the appearance, crumb characteristics, taste and crust color in our sensory evaluation. Overall, our study demonstrated that TA represents as an effective bread additive for the development of breads with bioactive compounds and increased antioxidant activity and the quality of the bread and acceptable sensory quality.

Author Contributions: Investigation, Methodology, Data curation, Writing—original draft, Y.G.; Investigation, Methodology, Data curation, Writing—original draft, X.Y.; Validation, Resources, C.P., J.K. and R.W.; Resources, X.L.; Investigation, review and editing, Y.W. (Yuesheng Wang) and M.C.; Investigation, M.L. and Q.W.; Conceptualization, Methodology, Resources, Writing—review and editing, Supervision, G.H.; Conceptualization, Methodology, Resources, Writing—review and editing, Supervision, G.Y. and J.C.; Conceptualization, Investigation, Methodology, Resources, Writing—original draft, Writingvreview and editing, Supervision, Project administration, Funding acquisition, Y.L. and Y.W. (Yaqiong Wang). All authors have read and agreed to the published version of the manuscript.

Funding: This work has been supported by National Genetically Modified New Varieties of Major Projects of China (2016ZX08010004-004), the National Natural Science Foundation of China (32272126, 31771418, 31371614) and the National Natural Science Foundation of Hubei China (2016CFB549). This work has also been supported by the Fundamental Research Funds for Central Universities, HUST to Yin Li (2021XXJS070 and 3004170157), Wuhan Knowledge Innovation Project to Yin Li (2022020801010073), the Innovation Training Project for College Students (HUST: 22YA-32) and the Opening Funding for The Key Laboratory of Molecular Biophysics of Chinese Ministry of Education to Qiong Wang.

Institutional Review Board Statement: Ethical approval (2022S116) has been provided by the Ethics Committee of Huazhong University of Science and Technology (HUST), People's Republic of China.

Data Availability Statement: Data is contained within the article.

Acknowledgments: We thank the Research Core Facilities for Life Science (HUST) MultiSkan for assistance in the determination of the antioxidant activity.

Conflicts of Interest: The authors declare no conflict of interest.

References

1. Sestili, F.; Margiotta, B.; Vaccino, P.; Moscaritolo, S.; Giorgi, D.; Lucretti, S.; Palombieri, S.; Masci, S.; Lafiandra, D. A cross between bread wheat and a 2D(2R) disomic substitution triticale line leads to the formation of a novel disomic addition line and provides information of the role of rye secalins on breadmaking characteristics. *Int. J. Mol. Sci.* **2020**, *21*, 8450. [CrossRef]
2. Shewry, P.R. What is gluten-Why is it special? *Front. Nutr.* **2019**, *6*, 101. [CrossRef] [PubMed]
3. Hemdane, S.; Jacobs, P.J.; Dornez, E.; Verspreet, J.; Delcour, J.A.; Courtin, C.M. Wheat (*Triticum aestivum* L.) bran in bread making: A critical review. *Compr. Rev. Food Sci. Food Saf.* **2016**, *15*, 28–42. [CrossRef]
4. Brouns, F.; van Buul, V.J.; Shewry, P.R. Does wheat make us fat and sick? *J. Cereal Sci.* **2013**, *58*, 209–215. [CrossRef]
5. Fardet, A. New hypotheses for the health-protective mechanisms of whole-grain cereals: What is beyond fibre? *Nutr. Res. Rev.* **2010**, *23*, 65–134. [CrossRef] [PubMed]
6. Hazard, B.; Trafford, K.; Lovegrove, A.; Griffiths, S.; Shewry, P. Strategies to improve wheat for human health. *Nat. Food* **2020**, *1*, 475–480. [CrossRef] [PubMed]
7. Vishwakarma, S.; Genu Dalbhagat, C.; Mandliya, S.; Niwas Mishra, H. Investigation of natural food fortificants for improving various properties of fortified foods: A review. *Food Res. Int.* **2022**, *156*, 111186. [CrossRef]
8. Butt, M.S.; Arshad, M.U.; Alam, M.S.; Nadeem, M.T. Bioavailability and storage stability of vitamin A fortificant (retinyl acetate) in fortified cookies. *Food Res. Int.* **2007**, *40*, 1212–1219. [CrossRef]
9. Liguori, I.; Russo, G.; Curcio, F.; Bulli, G.; Aran, L.; Della-Morte, D.; Gargiulo, G.; Testa, G.; Cacciatore, F.; Bonaduce, D.; et al. Oxidative stress, aging, and diseases. *Clin. Interv. Aging.* **2018**, *13*, 757–772. [CrossRef]
10. Paucean, A.; Muresan, V.; Maria-Man, S.; Chis, M.S.; Muresan, A.E.; Serban, L.R.; Pop, A.; Muste, S. Metabolomics as a tool to elucidate the sensory, nutritional and safety quality of wheat bread-A review. *Int. J. Mol. Sci.* **2021**, *22*, 8945. [CrossRef]
11. Mitelut, A.C.; Popa, E.E.; Popescu, P.A.; Popa, M.E. *Chapter 7—Trends of Innovation in Bread and Bakery Production, in Trends in Wheat and Bread Making*; Galanakis, C.M., Ed.; Academic Press: Cambridge, MA, USA, 2021; pp. 199–226.
12. Melis, S.; Delcour, J.A. Impact of wheat endogenous lipids on the quality of fresh bread: Key terms, concepts, and underlying mechanisms. *Compr. Rev. Food Sci. Food Saf.* **2020**, *19*, 3715–3754. [CrossRef] [PubMed]
13. Mannuramath, M.; Yenagi, N.; Orsat, V. Quality evaluation of little millet *(Panicum miliare)* incorporated functional bread. *J. Food Sci. Technol.* **2015**, *52*, 8357–8363. [CrossRef] [PubMed]
14. Nasir, S.; Allai, F.M.; Gani, M.; Ganaie, S.; Gul, K.; Jabeen, A.; Majeed, D. Physical, textural, rheological, and sensory characteristics of amaranth-based wheat flour bread. *Int. J. Food Sci.* **2020**, *2020*, 8874872. [CrossRef] [PubMed]
15. Tolve, R.; Simonato, B.; Rainero, G.; Bianchi, F.; Rizzi, C.; Cervini, M.; Giuberti, G. Wheat bread fortification by grape pomace powder: Nutritional, technological, antioxidant, and sensory properties. *Foods* **2021**, *10*, 75. [CrossRef]
16. Chlopicka, J.; Pasko, P.; Gorinstein, S.; Jedryas, A.; Zagrodzki, P. Total phenolic and total flavonoid content, antioxidant activity and sensory evaluation of pseudocereal breads. *LWT-Food Sci. Technol.* **2012**, *46*, 548–555. [CrossRef]

17. Han, H.M.; Koh, B.K. Effect of phenolic acids on the rheological properties and proteins of hard wheat flour dough and bread. *J. Sci. Food Agric.* **2011**, *91*, 2495–2499. [CrossRef]
18. Han, H.M.; Koh, B.K. Antioxidant activity of hard wheat flour, dough and bread prepared using various processes with the addition of different phenolic acids. *J. Sci. Food Agric.* **2011**, *91*, 604–608. [CrossRef]
19. Girard, A.L.; Castell-Perez, M.E.; Bean, S.R.; Adrianos, S.L.; Awika, J.M. Effect of condensed tannin profile on wheat flour dough rheology. *J. Agric. Food Chem.* **2016**, *64*, 7348. [CrossRef]
20. Zhang, L.; Cheng, L.; Jiang, L.; Wang, Y.; Yang, G.; He, G. Effects of tannic acid on gluten protein structure, dough properties and bread quality of Chinese wheat. *J. Sci. Food Agric.* **2010**, *90*, 2462–2468. [CrossRef]
21. Wang, Q.; Li, Y.; Sun, F.; Li, X.; Wang, P.; Sun, J.; Zeng, J.; Wang, C.; Hu, W.; Chang, J.; et al. Tannins improve dough mixing properties through affecting physicochemical and structural properties of wheat gluten proteins. *Food Res. Int.* **2015**, *69*, 64–71. [CrossRef]
22. Du, J.; Dang, M.; Khalifa, I.; Du, X.; Xu, Y.; Li, C. Persimmon tannin changes the properties and the morphology of wheat gluten by altering the cross-linking, and the secondary structure in a dose-dependent manner. *Food Res. Int.* **2020**, *137*, 109536. [CrossRef]
23. Sanyal, R.; Darroudi, F.; Parzefall, W.; Nagao, M.; Knasmuller, S. Inhibition of the genotoxic effects of heterocyclic amines in human derived hepatoma cells by dietary bioantimutagens. *Mutagenesis* **1997**, *12*, 297–303. [CrossRef] [PubMed]
24. AOAC. *Official Method of Analysis of AOAC Internation*, 21st ed.; AOAC International: Arlington, VA, USA, 2019.
25. International Association for Cereal Science and Technology. *Standard Methods of the ICC*; The Association: Vienna, Austria, 2003.
26. Supasil, R.; Suttisansanee, U.; Santivarangkna, C.; Tangsuphoom, N.; Khemthong, C.; Chupeerach, C.; On-Nom, N. Improvement of sourdough and bread qualities by fermented water of asian pears and assam tea leaves with co-cultures of lactiplantibacillus plantarum and saccharomyces cerevisiae. *Foods* **2022**, *11*, 2071. [CrossRef] [PubMed]
27. Wang, C.; Zeng, J.; Li, Y.; Hu, W.; Chen, L.; Miao, Y.; Deng, P.; Yuan, C.; Ma, C.; Chen, X.; et al. Enrichment of provitamin A content in wheat (*Triticum aestivum* L.) by introduction of the bacterial carotenoid biosynthetic genes CrtB and CrtI. *J. Exp. Bot.* **2014**, *65*, 2545–2556. [CrossRef] [PubMed]
28. de Erive, M.O.; Wang, T.; He, F.; Chen, G. Development of high-fiber wheat bread using microfluidized corn bran. *Food Chem.* **2020**, *310*, 125921. [CrossRef]
29. Atudorei, D.; Atudorei, O.; Codina, G.G. Dough rheological properties, microstructure and bread quality of wheat-germinated bean composite flour. *Foods* **2021**, *10*, 1542. [CrossRef]
30. Atudorei, D.; Mironeasa, S.; Codina, G.G. Effects of germinated lentil flour on dough rheological behavior and bread quality. *Foods* **2022**, *11*, 2982. [CrossRef]
31. Kazakos, S.; Mantzourani, I.; Plessas, S. Quality characteristics of novel sourdough breads made with functional *Lacticaseibacillus paracasei* SP5 and prebiotic food matrices. *Foods* **2022**, *11*, 3226. [CrossRef]
32. Castellari, M.P.; Simsek, S.; Ohm, J.-B.; Perry, R.; Poffenbarger, H.J.; Phillips, T.D.; Jacobsen, K.L.; Van Sanford, D.A. Genetic variation and heritability of sensory and artisan bread traits in a set of SRW wheat breeding lines. *Foods* **2023**, *12*, 2617. [CrossRef]
33. Pořízka, J.; Slavíková, Z.; Bidmonová, K.; Vymětalová, M.; Diviš, P. Physiochemical and sensory properties of bread fortified with wheat bran and whey protein isolates. *Foods* **2023**, *12*, 2635. [CrossRef]
34. Guo, X.; Shi, L.; Yang, S.; Yang, R.; Dai, X.; Zhang, T.; Liu, R.; Chang, M.; Jin, Q.; Wang, X. Effect of sea-buckthorn pulp and flaxseed residues on quality and shelf life of bread. *Food Funct.* **2019**, *10*, 4220–4230. [CrossRef]
35. Sivam, A.S.; Sun-Waterhouse, D.; Waterhouse, G.I.N.; Quek, S.; Perera, C.O. Physicochemical properties of bread dough and finished bread with added pectin fiber and phenolic antioxidants. *J. Food Sci.* **2011**, *76*, 97–107. [CrossRef] [PubMed]
36. Czubaszek, A.; Czaja, A.; Sokół-Łętowska, A.; Kolniak-Ostek, J.; Kucharska, A.Z. Changes in antioxidant properties and amounts of bioactive compounds during simulated in vitro digestion of wheat bread enriched with plant extracts. *Molecules* **2021**, *26*, 6292. [CrossRef] [PubMed]
37. Oloniyo, R.O.; Omoba, O.S.; Awolu, O.O.; Olagunju, A.I. Orange-fleshed sweet potatoes composite bread: A good carrier of beta (beta)-carotene and antioxidant properties. *J. Food Biochem.* **2021**, *45*, e13423. [CrossRef] [PubMed]
38. Bhardwaj, A.; Sharma, N.; Sharma, V.; Alam, T.; Sahu, J.K.; Hamid, H. Assessing the consumer acceptance and storability of chitosan and beeswax coated cellulose packaging for whole wheat bread. *Food Control.* **2022**, *133*, 108682. [CrossRef]
39. Perri, G.; Coda, R.; Rizzello, C.G.; Celano, G.; Ampollini, M.; Gobbetti, M.; De Angelis, M.; Calasso, M. Sourdough fermentation of whole and sprouted lentil flours: In situ formation of dextran and effects on the nutritional, texture and sensory characteristics of white bread. *Food Chem.* **2021**, *355*, 129638. [CrossRef]
40. Salmenkallio-Marttila, M.; Roininen, K.; Autio, K.; Lahteenmaki, L. Effects of gluten and transglutaminase on microstructure, sensory characteristics and instrumental texture of oat bread. *Agric. Food Sci.* **2004**, *13*, 138–150. [CrossRef]
41. Thiranusornkij, L.; Thamnarathip, P.; Chandrachai, A.; Kuakpetoon, D.; Adisakwattana, S. Comparative studies on physicochemical properties, starch hydrolysis, predicted glycemic index of Hom Mali rice and Riceberry rice flour and their applications in bread. *Food Chem.* **2019**, *283*, 224–231. [CrossRef]
42. Crowley, P.; Grau, H.; Arendt, E. Influence of additives and mixing time on crumb grain characteristics of wheat bread. *Cereal Chem.* **2000**, *77*, 370–375. [CrossRef]
43. Skendi, A.; Biliaderis, C.G.; Papageorgiou, M.; Izydorczyk, M.S. Effects of two barley β-glucan isolates on wheat flour dough and bread properties. *Food Chem.* **2010**, *119*, 1159–1167. [CrossRef]

44. Dhen, N.; Ben Rejeb, I.; Boukhris, H.; Damergi, C.; Gargouri, M. Physicochemical and sensory properties of wheat- Apricot kernels composite bread. *LWT-Food Sci. Technol.* **2018**, *95*, 262–267. [CrossRef]
45. de Erive, M.O.; He, F.; Wang, T.; Chen, G. Development of β-glucan enriched wheat bread using soluble oat fiber. *J. Cereal Sci.* **2020**, *95*, 103051. [CrossRef]
46. Sullivan, P.; O'Flaherty, J.; Brunton, N.; Arendt, E.; Gallagher, E. The utilisation of barley middlings to add value and health benefits to white breads. *J. Food Eng.* **2011**, *105*, 493–502. [CrossRef]
47. Parafati, L.; Restuccia, C.; Palmeri, R.; Fallico, B.; Arena, E. Characterization of prickly pear peel flour as a bioactive and functional ingredient in bread preparation. *Foods* **2020**, *9*, 1189. [CrossRef]
48. Czaja, A.; Czubaszek, A.; Wyspiańska, D.; Sokół-Łętowska, A.; Kucharska, A.Z. Quality of wheat bread enriched with onion extract and polyphenols content and antioxidant activity changes during bread storage. *Int. J. Food Sci. Technol.* **2020**, *55*, 1725–1734. [CrossRef]
49. Cacak-Pietrzak, G.; Dziki, D.; Gawlik-Dziki, U.; Sułek, A.; Kalisz, S.; Sujka, K. Effect of the addition of dried dandelion roots (*Taraxacum officinale* F. H. Wigg.) on wheat dough and bread properties. *Molecules* **2021**, *26*, 7564. [CrossRef]

foods

Article

Effects of Roasting on the Quality of *Moringa oleifera* Leaf Powder and Loaf Volume of *Moringa oleifera*-Supplemented Bread

Takako Koriyama [1,*], Mika Saikawa [1], Yuria Kurosu [1], Michiyo Kumagai [2] and Takahiro Hosoya [1]

[1] Faculty of Food and Nutritional Science, Toyo University, 1-1-1 Izumino, Itakura-machi, Ora-gun, Gunma 374-0193, Japan; mikasaikawa@gmail.com (M.S.); s3c102200029@toyo.jp (Y.K.); hosoya@toyo.jp (T.H.)
[2] Tokyo Seiei College, 1-4-6 Nishishinkoiwa, Katsushika-ku, Tokyo 124-8530, Japan; kumagai-m@tsc-05.ac.jp
* Correspondence: koriyama@toyo.jp; Tel.: +81-276-82-9157

Abstract: Although a decrease in bread volume on adding nutrient-rich *Moringa oleifera* leaf powder (MLP) is known, to our knowledge, improving the swelling of MLP-added bread has not been attempted. This study aimed to investigate the effects of MLP and roasted MLP (RMLP) on bread quality. Bread was supplemented with MLP and RMLP treated at varying temperatures and times; the baked bread was then biochemically evaluated relative to the control. The specific volume of MLP-supplemented bread was 2.4 cm^3/g, which increased to >4.0 cm^3/g on using MLP roasted at 130 °C for ≥20 min, demonstrating remarkable swelling. The specific volume of bread supplemented with MLP roasted at 170 °C for 20 min was 4.6 cm^3/g, similar to that of the control. Additionally, MLP interfered with carbon dioxide production in bread, thus decreasing the abundance of yeast cells; however, RMLP had no such effect and allowed normal fermentation. Scanning electron microscopy revealed gluten formation independent of MLP roasting. Thus, MLP-containing breads generally exhibit suppressed fermentation and expansion due to the bactericidal properties of raw MLP, but these effects are alleviated by heat treatment. These findings highlight the importance of heat treatment in mitigating the effects of MLP on bread fermentation and swelling.

Keywords: *Moringa oleifera* leaf powder; *Moringa oleifera*-supplemented bread; loaf volume; roasting; antioxidant activity

1. Introduction

Moringa (*Moringa oleifera*), a plant belonging to the Moringaceae family native to northwestern India, is widely distributed in Africa, South America, and Asia [1,2]. *M. oleifera* is also called the 'tree of life' [3,4], as its leaves, seeds, flowers, roots, stems, and sap are utilized for their various health-promoting effects [5–7]. In particular, the leaves are rich in protein and minerals, with nine essential amino acids [5,6,8,9], and contain appreciable amounts of phenols and flavonoids, which are used for their antioxidant properties [10,11]. When used as part of a balanced diet, these nutrients scavenge free radicals and may exert immunosuppressive effects [12]. While whole leaves have limited applicability as food ingredients, *M. oleifera* leaf powder (MLP), i.e., ground moringa leaves, is relatively feasible for addition to food. MLP is also rich in proteins and mineral dietary fibers, which is often lacking, and also contains ingredients exhibiting antioxidant activity [11–13]. Consequently, the fortification of nutritionally limited staple foods with MLP can compensate for vitamin, protein, and mineral deficiencies [13,14].

Wheat flour bread, a staple that is consumed worldwide as part of the daily diet, is rich in carbohydrates but is low in protein and functional ingredients. As bread is consumed by people of all age groups, increasing its nutrient content would be beneficial for the population. Given the improved living standards and increased public health awareness, various studies have investigated the nutritional enhancement of bread using various plant materials, such as green tea powder [15], celery powder [16], kenaf leaf powder [17], and vegetable leaf powder [18]. The addition of MLP as a dietary supplement to wheat bread and its nutritional benefits have been explored, and most studies recommend adding MLP in the range of 2–2.5% (*w*/*w*) [14,19], mainly because higher MLP content adversely affects bread swelling and palatability. Govender and Siwela [13] reported that bread containing 5% MLP could contribute to addressing malnutrition associated with protein deficiency; unfortunately, the loaf volume of such bread was reduced compared with that of unsupplemented bread. Thus, enhancing the nutritional content of MLP-supplemented bread while preserving loaf volume is a desirable goal. However, to our knowledge, no previous studies have attempted to improve the swelling of bread with added MLP.

Loaf volume is the most important property of bread and provides a quantitative measure of bread baking quality. During the process of bread swelling, carbon dioxide is typically trapped in small air pockets in the dough. During baking, the trapped carbon dioxide causes the air pockets to expand and the starch to gelatinize, transforming the dough into an elastic bread crumb. The dough should remain puffed up when baked, and starch is part of its skeleton [20]. Sengev et al. [4] proposed that the reduction in height and volume of MLP-supplemented bread is caused by an associated reduction in gluten content. Sengev et al. [14] stipulated that the reduction in bread height and volume was the result of the antimicrobial effect of *M. oleifera* leaves on the leavening activity of yeast during dough fermentation. However, the exact reason for the decrease in bread volume upon MLP addition remains unknown.

We hypothesized that if the reduction in bread volume was caused by the chemical components of MLP, the volume and height of bread baked after the removal of these components would remain the same as those of the unsupplemented control. To test this, we inactivated MLP using dry heat treatment and investigated its effect on bread properties, antioxidant capacity, and volume. Under various conditions, dry-roasted MLP samples were added to bread, and their effect on swelling during baking and palatability was observed. Finally, the role of gluten formation in the suppression of swelling owing to the addition of MLP was evaluated.

2. Materials and Methods

2.1. Roasted Sample Preparation

MLP was purchased from SunRise Ltd. (Okinawa, Japan). The tested roasting temperatures were 110, 130, 150, 170, and 190 °C, and the tested heating times were 10, 20, and 30 min, all processed in an electric oven (NE-BS604, Panasonic, Osaka, Japan). MLP roasting at 170 °C for 30 min and 190 °C for 10 min resulted in burned samples with no material for analysis. Herein, MLP subjected to roasting treatment is referred to as "roasted MLP" (RMLP). The samples were stored in airtight containers at 4 °C and shielded from light until further experiments.

2.2. Bread Baking

For MLP-supplemented bread, 5% of the flour weight was substituted with RMLP. The bread was molded and baked using a standard program of the automatic bakery machine (SD-BMT1001; Panasonic, Japan). Briefly, wheat flour (237.5 g), MLP or RMLP (12.5 g), deionized water (180 mL), butter (10 g), salt (5 g), sugar (17 g), non-fat dry milk (6 g), and dry yeast (2.8 g) were added and mixed in the bread maker. MLP was added in powder form to the bread dough just before kneading. The control bread without MLP and RMLP was also prepared at the same time. All ingredients were purchased from a local retail market. The production process required approximately 4 h and included kneading, fermentation, and baking. After baking, the bread was left standing at 25 °C for 90 min, and its physical properties were measured. Samples for analysis were wrapped in a plastic film, placed in a polyethylene bag, and stored at −30 °C until use.

2.3. Proximate Analysis

The general nutritional contents of the wheat flour and MLP were analyzed. Moisture, crude protein, crude fat, ash, and crude fiber contents were determined according to the standard methods described by the AOAC [21]. Moisture content was measured using the atmospheric pressure drying method (135 °C). Crude protein content was determined using the Kjeldahl method; the nitrogen-to-protein conversion factor was 6.25. Crude fat content was determined using chloroform–methanol extraction. Dietary fiber contents were determined using the Prosky method. Carbohydrate content was calculated using Equation (1) [22]:

$$Total\ carbohydrate\ (g/100\ g) = 100 - [moisture\ (\%) + crude\ protein\ (\%) + crude\ fat\ (\%) + ash\ (\%)] \quad (1)$$

Energy values were calculated using the Atwater system. Mineral components were prepared by the dry-ash method. Sodium, potassium, calcium, magnesium, and iron levels were determined using an atomic absorption spectrophotometer (AA-700; Shimadzu Co., Ltd., Kyoto, Japan). The sample (2.0 g) was mixed with 50 mL of 1.0% hydrochloric acid and stirred for 1 h at 20 °C. The suspension was centrifuged at 3000 rpm for 15 min. Before measurement, the supernatant was filtered through a 0.45 µm nylon filter. Strontium chloride (0.1%) was added to the samples as an interference suppression agent for calcium and magnesium ions.

2.4. Water Absorption Capacity

The water absorption capacities of MLP and RMLP samples were measured using the solvent retention capacity profiles [23]. The sample (5.0 ± 0.05 g) was weighed and transferred to a 50 mL centrifuge tube. Distilled water (25 mL) was added to the centrifuge tube and quickly mixed with a vortex mixer for 5 s to homogenize the entire powder with water. The powder was then allowed to solvate and swell for 20 min with shaking for 5, 10, 15, and 20 min. Following this, centrifugation was performed at 1000× g for 15 min (Hitachi Koki CR22GIII., Tokyo, Japan). Subsequently, the supernatant was discarded, and the centrifuge tube was placed inverted on a paper towel and allowed to stand for at least 10 min to remove water before weighing. The weight of the plastic 50 mL centrifuge tube was subtracted from the weight of the empty centrifuge tube, and the resulting value was used as the gel weight. The water retention capacity (%) was calculated using Equation (2). The experiment was performed in triplicate.

$$Water\text{-}absorption\ capacity\ (\%) = \{(gel\ wt/flour\ wt) - 1 \times \{86/(100 - \%\ flour\ moisture)\} \times 100 \quad (2)$$

2.5. Color Measurements

Color intensity was measured using a spectrophotometer (CM-700d, KONICA MINOLTA, Osaka, Japan) using L^*, a^*, and b^* values according to the CIE color scale. L^* represents the brightness from 0 (black) to 100 (white). The other two coordinates represent redness ($+a^*$) to greenness ($-a^*$) and yellowness ($+b^*$) to blueness ($-b^*$), respectively. The color difference (ΔE) caused by roasting was calculated using Equation (3):

$$\Delta E = \left[(L^*RMLP - L^*MLP)^2 + (a^*RMLP - a^*MLP)^2 + (b^*RMLP - b^*MLP)^2\right]^{1/2} \quad (3)$$

All experiments were performed in triplicate.

2.6. Determination of Physical Properties of Bread

After cooling to 25 °C for 90 min, the bread weight, loaf volume, and crumb texture were measured. Loaf volume was measured using a 3D scanner (Ein Scan SP, Japan 3D

Printer, Co., Ltd. Tokyo, Japan), with the turntable step set to 8 and the mesh level set to high. The specific volume of bread was calculated using Equation (4):

$$SV = V/m \tag{4}$$

where SV is the specific volume of the bread (cm^3/g); V is the volume of the bread (cm^3); and m is the mass of the bread (g). All measurements were performed at least in triplicate.

Textural profile analysis was performed using a texture analyzer (TA XT Plus, Stable Micro Systems, Ltd., Godalming, UK). The bread crumb was cut into 2 cm × 2 cm × 2 cm cubes. Each sample was compressed twice until a 70% strain was achieved at a probe speed of 1 mm/s. A cylindrical probe with a 20 mm diameter was used. Crumb hardness (peak force 1), elasticity (length of the base of area 2/length of the base of area 1), and chewiness (hardness × cohesiveness × springiness) were determined as the characteristic parameters of the crumb texture. Six replicates were analyzed for each loaf sample.

2.7. Extraction of MLP

Briefly, 1 mL of 80% ethanol was added to 10 mg of MLP or RMLP. The suspension was thoroughly mixed and sonicated for 5 min. After centrifugation (12,054× g, 5 min) (CT15RE, HIMAC, Hitachi, Japan), the supernatant was used for further analysis.

2.8. DPPH Radical Scavenging Activity

The antioxidant activity was evaluated via a DPPH radical scavenging assay. For the assay, 10 mg/mL of MLP extract was stepwise diluted in 80% ethanol. Next, 100 µL of the stepwise-diluted solution was transferred to a 96-well microplate, and 100 µL of 0.4 mM DPPH radical solution was added. The reaction was allowed to proceed for 30 min at room temperature. Sample absorbance at 520 nm was measured using a microplate reader (Synegy-HTX, Bio Tek, Shoreline, WA, USA) to determine the percentage of DPPH radicals remaining, from which the half-maximal inhibitory concentration (IC_{50}, mg/mL) value was calculated.

2.9. Determination of Total Polyphenol Content

Total polyphenol content was evaluated using the Folin–Ciocalteu method. Gallic acid used as a standard was dissolved in 80% ethanol, and a series of stepwise dilutions from 2 to 1000 g/mL were prepared. Then, 60 µL of water, 10 µL of stepwise-diluted gallic acid sample, and 15 µL of Folin–Ciocalteu phenol reagent were added to a 96-well plate and incubated for 5 min at room temperature. Next, 75 µL of 2% sodium carbonate was added. After 30 min at room temperature in the dark, sample absorbance at 750 nm was measured using a microplate reader (Synegy-HTX, Bio Tek, USA). A calibration curve was then constructed based on the relationship between absorbance and concentration.

The total polyphenol content of MLP extract was determined in the same manner, with 10 µL of 10 mg/mL MLP extract assayed. The total polyphenol content of MLP extract was calculated as gallic acid equivalents (GAE, mg/mL) based on sample absorbance and the calibration curve for gallic acid.

2.10. Quantitative Analysis of the Antioxidant Quercetin-3-Glucoside (Q3G)

High-resolution (HR) electrospray ionization (ESI)–mass spectroscopy (MS) was performed using an Acquity system (Waters, Milford, MA, USA) equipped with a mass spectrometer (Synapt-G2, Waters, USA). Analysis of Q3G was performed using linear gradient elution with solution A (0.1% formic acid in water) and solution B (CH_3CN containing 0.1% formic acid), delivered at a flow rate of 0.4 mL/min, with a linear gradient from 2% to 100% B over 10 min. The injection volume was 5 µL of sample solution. Data were acquired in the negative MS scanning mode. The mass tolerance of Q3G was within 10 ppm. The Q3G standard was stepwise diluted in ethanol, and the Q3G area under the curve at m/z 463.0877 and Rt 4.62 min was analyzed by LC/MS. A calibration curve was generated from the relationship between the area value and Q3G concentration. The MLP extract was analyzed by LC/MS under the same conditions, the area value of Q3G was calculated, and the final quantification was performed using the calibration curve.

2.11. Fermentation Tests

The fermentation of bread dough was evaluated by measuring the dough volume expansion ratio. The expansion of the dough is attributed to the production of carbon dioxide within the dough during fermentation [24]. The dough was kneaded in an automatic home bakery (SD-BMT1001; Panasonic, Japan) for 20 min using half the amount of ingredients used for bread baking (2.2). The kneaded dough was placed in a bowl and wrapped lightly using plastic wrap, then fermented in an incubator (convection oven, MOV-212F, SANYO, Osaka, Japan) maintained at 38 °C. The dough volume was measured using a 3D scanner (Ein Scan SP; Japan 3D Printer Co., Ltd., Tokyo, Japan) 0, 10, 30, 40, 50, 60, 90, and 120 min after the start of fermentation. The volume expansion ratio was calculated using Equation (5):

$$Dough\ volume\ expansion\ ratio = (V_T - V_0)/V_0 \quad (5)$$

where V_0 refers to the volume at the beginning of fermentation and V_T indicates the volume after fermentation at T min. All measurements were performed at least in triplicate.

An alcoholic fermentation test was conducted using dry yeast to determine the amount of carbon dioxide produced. The yeast solution was prepared by adding 3 g of dry yeast (Nisshin Super Camellia dry yeast, Nisshin Flour Milling Inc., Tokyo, Japan) to 0.3 g of MLP or 0.3 g of RMLP, mixing well, refrigerating at 4 °C for 2 h, and mixing with 100 mL of distilled water. For the assay, 20 mL of yeast solution and 20 mL of 10% sucrose solution were placed in a Kühne fermentation tube, and the tube opening was covered with non-fat cotton. The tube was placed in a thermostatic chamber (SDN-B, TAITEC, Saitama, Japan) set at 40 °C, and the amount of carbon dioxide produced was read after 20 min. A dry yeast solution prepared without the addition of MLP was used as the control.

2.12. Determination of Yeast Viability

Yeast viability was determined using a Cell Counting Kit-8 (WST-8) (Dojindo Laboratories, Kumamoto, Japan). For the assay, 0.30 g of dry yeast, 0.030 g of MLP in 0.30 g of dry yeast, and 0.030 g of RMLP in 0.30 g of dry yeast were mixed and refrigerated at 4 °C for 2 h. Then, 10 mL of purified water was added to each sample and mixed thoroughly. After incubating at room temperature for 30 min, 20 µL of WST-8 reagent was added to 500 µL of each sample, and the mixtures were incubated for 1.5 h at 40 °C. Each solution was centrifuged (12,054× g for 5 min), and the absorbance of the supernatant was measured (Synegy-HTX, Bio Tek, USA) at 450 nm.

2.13. Scanning Electron Microscopy

Gluten networks in the dough were observed using a scanning electron microscope (JSM-6610LA, JEOL Ltd., Akishima, Japan). The samples were freeze-dried and coated with gold using an ion spatter (E-1010; Hitachi, Tokyo, Japan) for preprocessing. An accelerating voltage of 5 kV was applied.

2.14. Sensory Evaluation

The bread crumb was divided into 2 × 2 × 2 cm cubes. A pilot study was conducted before the main study, involving nine participants. Based on the results of the pilot study, the tested roasting conditions were reduced to five for the sensory test, considering the differences between the samples and the ease of comparison: untreated MLP and RMLP obtained by roasting for 20 min at 110 °C, 130 °C, 150 °C, or 170 °C. The sensory evaluation panel consisted of 33 untrained students and faculty (20–65 years old), consisting of 9 males and 24 females, all of whom consumed bread at least once per week. The taste, aroma, appearance, texture, and overall acceptability of the bread were evaluated according to a nine-point hedonic scale (1, extremely dislike; 5, neither like nor dislike; and 9, extremely like). This study was approved by the Toyo University Ethics Committee (approval number TU2022-010).

2.15. Statistical Analysis

Data are presented as the mean ± SD. Significant differences in values were analyzed using one-way analysis of variance (ANOVA) in the SPSS statistical software (ver. 27.0, IBM, Armonk, NY, USA). Tukey's and Bonferroni's methods were used for multiple comparisons, with statistical significance set at $p < 0.05$. All measurements were performed at least in triplicate.

3. Results and Discussion

3.1. General Nutritional Composition of MLP

Table 1 shows the results of the proximate analysis of the wheat flour and MLP. The moisture contents of the flour and MLP were 14.3% and 3.7%, respectively, which were significantly different. Hence, the general nutrient ingredient content was expressed as % dry weight. The protein and lipid contents of MLP were higher than those of wheat flour, and MLP contained a large amount of various minerals. Moreover, the dietary fiber content of MLP was 10 times that of wheat flour. MLP was compared with the foods listed in The Standard Tables of Food Composition, Japan (2015) (7th Edition) [25] for specific nutrient contents. The potassium content of MLP was 6.9 times that of bananas; calcium content 19.1 times that of milk; and iron content 4.4 times that of spinach. This is the first time that the nutritional value of MLP from *M. oleifera* grown in Japan has been reported. The nutritional value of Japanese *M. oleifera* leaves was superior to that of other nutritious vegetables and fruits.

Table 1. Nutritional composition of wheat flour and MLP on a dry weight basis (dwb).

Ingredient	Wheat Flour			MLP		
Water content (%)	14.3	±	0.35	3.7	±	0.49
Protein (% dwb)	13.5	±	0.39	27.7	±	8.22
Carbohydrate (% dwb)	84.6	±	1.06	53.2	±	7.95
Dietary fiber (total, % dwb)	3.2	±	0.04	29.1	±	4.34
Dietary fiber (water soluble, % dwb)	1.4	±	0.02	5.1	±	0.75
Dietary fiber (insoluble, % dwb)	1.8	±	0.02	24.0	±	3.59
Crude fat (% dwb)	2.6	±	0.77	8.8	±	1.13
Ash (% dwb)	0.5	±	0.08	10.8	±	0.06
Energy (kcal/100 g dwb)	420	±	9.80	371	±	36.4
Na (mg/100 g dwb)	0	±	0.00	127	±	0.65
K (mg/100 g dwb)	102	±	2.39	2570	±	26.4
Ca (mg/100 g dwb)	20	±	0.46	2024	±	225
Mg (mg/100 g dwb)	26	±	0.62	470	±	74.1
Fe (mg/100 g dwb)	1	±	0.09	10	±	0.54

3.2. Physicochemical Characteristics of RMLP Obtained under Various Roasting Conditions

As shown in Table 2, the moisture content of unroasted MLP was approximately one-third that of wheat flour. The moisture content decreased as the MLP roasting temperature increased and became one-tenth that of wheat flour after roasting at 170 °C for 20 min. The water absorption capacity of unroasted MLP was very high (515%), approximately 4.3 times that of wheat flour (119%). Although this parameter decreased with roasting, it was more than twice that of wheat flour, even for RMLP obtained by roasting at 170 °C for 20 min (244%). The relatively high water absorption of MLP can be explained by its protein, ash, and dietary fiber contents, which were higher than those of wheat flour. The decrease in water absorption after roasting was presumably due to changes in the MLP components and interactions between the components upon heating.

Table 2. Effect of roasting treatment on MLP properties and bread quality.

		Wheat Flour	MLP											
				110 °C			130 °C			150 °C			170 °C	
			Unroasted	10 min	20 min	30 min	10 min	20 min	30 min	10 min	20 min	30 min	10 min	20 min
Flour														
Moisture content (%)		14.5 ± 0.01 ^{a}	4.9 ± 0.31 ^{b}	3.7 ± 0.11 ^{c}	2.7 ± 0.25 ^{d}	1.8 ± 0.26 ^{e}	2.1 ± 0.23 ^{de}	1.8 ± 0.25 ^{e}	1.6 ± 0.41 ^{e}	1.9 ± 0.30 ^{de}	1.9 ± 0.23 ^{e}	1.8 ± 0.20 ^{e}	2.1 ± 0.46 ^{de}	1.5 ± 0.50 ^{e}
Water absobing capacity (%)		119 ± 6.77 ^{g}	515 ± 17.2 ^{ab}	464 ± 22.1 ^{ab}	465 ± 18.6 ^{ab}	461 ± 22.7 ^{ab}	506 ± 16.1 ^{a}	405 ± 20.3 ^{bcd}	341 ± 15.1 ^{cde}	411 ± 24.2 ^{bc}	314 ± 6.65 ^{ef}	323 ± 3.84 ^{def}	342 ± 22.9 ^{cde}	244 ± 13.1 ^{f}
Color	L*	90.1 ± 0.01 ^{a}	57.9 ± 0.68 ^{b}	56.4 ± 0.06 ^{c}	54.9 ± 0.05 ^{d}	53.9 ± 0.05 ^{e}	53.1 ± 0.06 ^{e}	48.7 ± 0.03 ^{fg}	48.3 ± 0.26 ^{g}	49.5 ± 0.05 ^{f}	39.4 ± 0.19 ^{h}	38.1 ± 0.06 ^{i}	40.0 ± 0.06 ^{h}	32.8 ± 0.20 ^{j}
	a*	−0.8 ± 0.10 ^{f}	−12.7 ± 0.13 ^{m}	−13.2 ± 0.11 ^{l}	−10.6 ± 0.07 ^{k}	−7.3 ± 0.06 ^{i}	−9.9 ± 0.16 ^{j}	−2.1 ± 0.08 ^{g}	0.3 ± 0.04 ^{e}	−2.4 ± 0.05 ^{h}	4.3 ± 0.09 ^{c}	4.7 ± 0.01 ^{b}	3.0 ± 0.04 ^{d}	5.5 ± 0.07 ^{a}
	b*	9.72 ± 0.05 ^{h}	37.2 ± 1.01 ^{b}	38.4 ± 0.08 ^{a}	37.1 ± 0.21 ^{b}	35.2 ± 0.07 ^{c}	36.6 ± 0.14 ^{b}	32.3 ± 0.06 ^{d}	31.4 ± 0.21 ^{d}	34.6 ± 0.03 ^{e}	26.9 ± 0.07 ^{ef}	25.5 ± 0.03 ^{f}	27.1 ± 0.03 ^{e}	22.3 ± 0.19 ^{g}
	ΔE	—	—	1.8 ± 0.07	3.2 ± 0.03	6.7 ± 0.09	6.4 ± 0.17	15.5 ± 0.06	17.7 ± 0.25	13.9 ± 0.07	27.6 ± 0.21	29.1 ± 0.05	26.3 ± 0.05	28.7 ± 7.91
DPPH radical scavenging activity (IC_{50}, mg/mL)		N/A	1.7 ± 0.07 ^{b}	1.6 ± 0.05 ^{bc}	1.7 ± 0.08 ^{b}	1.4 ± 0.07 ^{de}	1.6 ± 0.05 ^{bcd}	1.4 ± 0.03 ^{cde}	1.4 ± 0.06 ^{cde}	1.3 ± 0.06 ^{e}	1.6 ± 0.04 ^{bcde}	1.7 ± 0.05 ^{b}	1.4 ± 0.05 ^{e}	2.3 ± 0.07 ^{a}
Total phenol content (μg GAE/mL)		N/A	149 ± 2.75 ^{d}	152 ± 9.96 ^{d}	149 ± 7.77 ^{d}	173 ± 6.66 ^{bcd}	159 ± 6.06 ^{cd}	178 ± 5.01 ^{bc}	180 ± 21.18 ^{abc}	187 ± 5.39 ^{ab}	174 ± 0.13 ^{bcd}	161 ± 4.92 ^{bcd}	204 ± 9.29 ^{a}	123 ± 6.71 ^{e}
Q3G content (mg/g)		N/A	10.3 ± 1.11 ^{abcd}	9.3 ± 0.96 ^{cd}	9.7 ± 1.11 ^{bcd}	10.6 ± 0.85 ^{abcd}	10.4 ± 0.57 ^{abcd}	11.5 ± 0.57 ^{abc}	12.0 ± 0.58 ^{ab}	12.3 ± 0.87 ^{a}	10.3 ± 0.83 ^{abcd}	8.7 ± 1.11 ^{d}	12.4 ± 1.21 ^{a}	1.9 ± 0.47 ^{e}
Bread (1)														
Moisture content (%)		38.5 ± 0.24 ^{a}	37.1 ± 0.88 ^{abcd}	37.4 ± 0.42 ^{abcd}	36.6 ± 1.39 ^{cd}	36.2 ± 0.38 ^{d}	37.1 ± 1.09 ^{abcd}	37.1 ± 1.19 ^{abcd}	37.8 ± 0.17 ^{abc}	38.2 ± 0.27 ^{ab}	37.1 ± 0.50 ^{abcd}	37.5 ± 0.21 ^{abcd}	36.7 ± 0.93 ^{bcd}	37.4 ± 0.30 ^{abcd}
Loaf volume (cm^3)		1844 ± 40.2 ^{a}	990 ± 41.3 ^{j}	1002 ± 42.1 ^{i}	1250 ± 43.2 ^{g}	1389 ± 45.4 ^{f}	1118 ± 44.9 ^{h}	1624 ± 56.6 ^{d}	1703 ± 52.1 ^{c}	1545 ± 47.8 ^{e}	1818 ± 58.7 ^{b}	1763 ± 42.6 ^{bc}	1723 ± 52.7 ^{c}	1856 ± 59.9 ^{a}
Specific volume (cm^3/g)		4.6 ± 0.12 ^{a}	2.4 ± 0.1 ^{f}	2.5 ± 0.02 ^{f}	3.1 ± 0.11 ^{e}	3.4 ± 0.03 ^{d}	2.8 ± 0.08 ^{ef}	4.0 ± 0.02 ^{c}	4.2 ± 0.07 ^{bc}	3.8 ± 0.12 ^{c}	4.4 ± 0.03 ^{b}	4.4 ± 0.03 ^{b}	4.3 ± 0.05 ^{b}	4.6 ± 0.08 ^{a}
Hardness (N)		3.5 ± 0.91 ^{bef}	26.8 ± 3.47 ^{a}	18.2 ± 3.56 ^{b}	14.5 ± 5.13 ^{c}	7.2 ± 1.9 ^{de}	7.5 ± 1.51 ^{d}	3.1 ± 0.55 ^{def}	3.4 ± 0.68 ^{def}	3.6 ± 0.47 ^{def}	3.2 ± 0.79 ^{def}	2.8 ± 1.04 ^{def}	2.5 ± 0.32 ^{ef}	2.2 ± 0.22 ^{f}
Elasticity		1.2 ± 0.25 ^{ab}	0.9 ± 0.02 ^{b}	0.9 ± 0.05 ^{b}	0.9 ± 0.02 ^{ab}	1.0 ± 0.01 ^{ab}	1.0 ± 0.01 ^{ab}	1.0 ± 0.04 ^{a}	1.0 ± 0.01 ^{ab}	1.0 ± 0.01 ^{ab}	1.1 ± 0.12 ^{ab}	1.0 ± 0.02 ^{ab}	1.0 ± 0.03 ^{ab}	1.1 ± 0.03 ^{ab}
Chewiness		3.2 ± 0.88 ^{c}	11.9 ± 2.33 ^{a}	9.9 ± 2.06 ^{a}	7.8 ± 2.63 ^{b}	4.6 ± 0.91 ^{bc}	4.6 ± 1.09 ^{bc}	2.7 ± 1.76 ^{c}	2.6 ± 0.52 ^{c}	2.1 ± 0.34 ^{c}	2.4 ± 0.36 ^{c}	2.6 ± 0.62 ^{c}	2.3 ± 0.71 ^{c}	1.9 ± 0.64 ^{c}
VRC (2) %		0.4 ± 0.03 ^{a}	0.2 ± 0.01 ^{g}	0.3 ± 0.06 ^{fg}	0.2 ± 0.04 ^{ef}	0.2 ± 0.0 ^{cde}	0.3 ± 0.03 ^{bc}	0.3 ± 0.02 ^{b}	0.3 ± 0.02 ^{b}	0.3 ± 0.01 ^{ab}	0.3 ± 0.01 ^{bcd}	0.3 ± 0.01 ^{bcd}	0.3 ± 0.02 ^{bc}	0.3 ± 0.01 ^{ab}
Crumb color	L*	79.1 ± 0.76 ^{a}	42.6 ± 1.39 ^{cd}	41.7 ± 0.25 ^{de}	45.0 ± 0.64 ^{bcd}	46.0 ± 0.77 ^{bc}	43.6 ± 1.32 ^{cd}	46.0 ± 0.99 ^{bc}	47.0 ± 0.67 ^{b}	46.1 ± 0.09 ^{bc}	42.9 ± 2.18 ^{de}	40.9 ± 0.81 ^{e}	41.5 ± 0.83 ^{de}	40.8 ± 1.39 ^{de}
	a*	−0.9 ± 0.27 ^{de}	−1.9 ± 0.85 ^{e}	−0.9 ± 0.34 ^{e}	−1.0 ± 0.39 ^{e}	−1.0 ± 0.11 ^{e}	−0.3 ± 0.15 ^{de}	1.5 ± 0.74 ^{c}	1.9 ± 0.11 ^{bc}	0.9 ± 0.65 ^{cd}	4.5 ± 0.48 ^{a}	4.6 ± 0.34 ^{ab}	3.1 ± 0.26 ^{ab}	5.07 ± 0.65 ^{a}
	b*	17.5 ± 0.32 ^{f}	32.9 ± 0.72 ^{a}	32.3 ± 1.26 ^{ab}	32.9 ± 0.33 ^{ab}	31.5 ± 0.78 ^{ab}	30.4 ± 1.02 ^{bc}	25.1 ± 2.49 ^{d}	24.5 ± 1.13 ^{d}	28.2 ± 1.21 ^{c}	20.7 ± 0.99 ^{e}	23.2 ± 0.36 ^{de}	23.2 ± 0.36 ^{de}	15.5 ± 0.46 ^{f}
	ΔE	—	—	1.7 ± 0.77	3.5 ± 0.58	4.8 ± 0.94	4.3 ± 1.34	10.1 ± 2.11	11.1 ± 1.18	8.3 ± 0.91	14.4 ± 0.77	12.1 ± 0.43	26.3 ± 0.05	19.3 ± 1.53
Crust color	L*	56.4 ± 2.27 ^{a}	35.7 ± 0.33 ^{c}	37.1 ± 4.22 ^{c}	36.8 ± 3.53 ^{c}	36.0 ± 3.07 ^{c}	36.0 ± 3.05 ^{c}	35.8 ± 1.68 ^{c}	40.8 ± 0.98 ^{bc}	37.0 ± 3.18 ^{bc}	36.8 ± 2.11 ^{c}	38.6 ± 1.41 ^{bc}	42.3 ± 2.35 ^{b}	39.5 ± 1.94 ^{bc}
	a*	14.1 ± 1.06 ^{a}	7.1 ± 0.58 ^{def}	5.1 ± 1.23 ^{def}	4.4 ± 1.99 ^{f}	4.5 ± 1.19 ^{ef}	6.3 ± 0.89 ^{cdef}	7.1 ± 0.58 ^{bc}	6.6 ± 1.26 ^{cdef}	7.0 ± 0.7 ^{bcd}	8.7 ± 1.54 ^{b}	6.7 ± 1.17 ^{bcde}	7.7 ± 0.95 ^{bc}	6.9 ± 0.35 ^{bcde}
	b*	32.7 ± 1.08 ^{a}	21.5 ± 0.67 ^{bc}	17.1 ± 2.79 ^{d}	18.0 ± 2.84 ^{cd}	20.1 ± 1.84 ^{bcd}	19.9 ± 0.54 ^{bcd}	21.4 ± 2.04 ^{bc}	23.9 ± 0.85 ^{b}	22.7 ± 2.21 ^{b}	20.1 ± 1.86 ^{bcd}	20.0 ± 2.41 ^{bcd}	23.2 ± 1.62 ^{b}	18.2 ± 0.93 ^{cd}
	ΔE	—	—	6.5 ± 0.58	6.3 ± 1.574	4.3 ± 1.54	2.5 ± 1.98	2.1 ± 1.65	5.4 ± 1.28	3.0 ± 2.65	4.0 ± 1.02	4.2 ± 1.16	6.8 ± 2.21	5.3 ± 1.33

(1) Bread, bread containing each of the above powders. In all cases, the added MLP or RMLP substituted 5% of the wheat flour weight. (2) VRC, volume recovery coefficient. Values with different letters in the same row are significantly different (Bonferroni's test, $p < 0.05$; $n = 5$).

Raw MLP was dark green in color; thus, it had a lower L^* value and more negative a^* value than wheat flour. This was expected because of the high chlorophyll content and natural green color of *M. oleifera* leaves [19,26]. The L^* values of RMLP were lower than those of MLP, and almost all a^* values of RMLP were higher than those of MLP; the difference was more pronounced with an increasing roasting time and temperature. Because L^* indicates lightness and a^* indicates redness, roasting resulted in a darker and redder (i.e., browning) color of MLP. Color degradation occurs when chlorophyll pigments are rapidly converted into pheophytin (a green-grey pigment) during the roasting process. These phenomena were particularly noticeable when roasting at 150 °C for 20 min or longer. In addition, the decrease in L^* value with increasing a^* and b^* values during roasting is reportedly caused by the formation of melanoidin, a brown pigment, a product of the Maillard reaction [27]. Therefore, the color change of MLP upon roasting can be attributed to the non-enzymatic browning associated with Maillard reaction.

3.3. Antioxidant Activity of RMLP Obtained under Various Roasting Conditions

Neither DPPH radical scavenging activity nor polyphenols were detected in the control sample (pure wheat flour) (Table 2). In unroasted MLP, the DPPH radical scavenging activity was 1.7 IC_{50} mg/mL, and the phenol content was 149 μg GAE/mL, indicating high antioxidant activity. For RMLP obtained at various roasting temperatures, the IC_{50} values ranged from 1.3 to 2.3 mg/mL, and the total phenol content ranged from 149 to 204 μg GAE/mL. The DPPH radical scavenging activity and polyphenol contents of all RMLPs, except for those obtained by roasting at 170 °C for 20 min, were comparable to or higher than those of unroasted MLP.

We also analyzed the Q3G content of MLP (Table 2). Q3G, a major antioxidant found in Moringa leaves [28], has been validated as a significant contributor to the DPPH radical scavenging activity within MLP through activity-guided fractionation. The Q3G content of MLP prepared under various conditions was analyzed using LC/MS. It was comparable to that of bread prepared using MLP, except for the bread prepared using RMLP by roasting at 170 °C for 20 min, likely due to the Q3G decomposition under these conditions.

M. oleifera leaves, both fresh and dried, are an excellent source of antioxidants [19]; however, few studies have investigated the antioxidant activity of roasted MLP. Melanoidin, a brown pigment produced in the late stages of the Maillard reaction, is highly reductive and exhibits strong antioxidant properties [29]. According to previous studies, a positive correlation between color changes related to non-enzymatic browning and total phenol content is observed in many foods [27,30,31]. In this study, a significantly strong positive correlation ($r = 0.946$) was observed between DPPH radical scavenging activity and total phenol content for RMLP obtained by roasting for up to 10 min at 150 °C. These results suggest that the phenolic content increased with the browning of MLP due to roasting. Furthermore, we propose that roasting at 150° C for at least 20 min leads to a simultaneous increase in the number and decomposition of antioxidant substances, and their canceling effect is reflected in the apparent antioxidant activity.

3.4. Effect of MLP Roasting Conditions on Bread Loaf Volume

The results of the investigation of the effects of MLP roasting on bread characteristics are shown in Figure 1 and Table 2. The loaf volume and specific volume of bread supplemented with MLP were approximately half those of the no-MLP control, with poor crust formation and caving-like indentations observed at the loaf top. In addition, the hardness of the MLP-supplemented bread was significantly higher than that of the control bread, and the MLP-supplemented bread was less elastic than the control. These observations were consistent with previous findings for wheat bread enriched with 5% MLP [14,19]. In contrast, for bread supplemented with RMLP, the higher the roasting temperature or the longer the roasting time, the significantly higher the specific volume of bread and significantly lower the hardness. A particularly remarkable improvement in swelling was observed with RMLP obtained by roasting at 170 °C for 20 min, with the specific volume of the supplemented bread equal to that of the control and the hardness lower than that of the control. These observations suggested that MLP roasting effectively improves the loaf volume of RMLP-supplemented bread and that higher roasting temperatures improve the swelling of the RMLP-supplemented bread more effectively.

Figure 1. Improvement in bread loaf volume by the addition of RMLP prepared under various roasting conditions. In all samples, MLP or RMLP added to the bread substituted 5% of the wheat flour weight.

As shown in Table 2, the moisture content of bread was approximately 36.2–38.5%, regardless of the loaf volume, which agrees with the findings of other studies [4,19,32]. This indicates that sufficient moisture evaporation occurs during baking, regardless of the loaf volume. The lightness (L^*) of the crumb and crust decreased significantly with increasing MLP roasting temperature or time. The redness (a^*) of the crumb increased with MLP roasting temperature. The bread color exhibited the same tendency as the powder. Breads with added RMLP obtained by roasting at high temperatures appeared brown.

In general, the volume of bread depends on the distribution and size of air bubbles, which in turn depend on the extensibility of the thin film of gluten. Hence, gluten formation is essential for well-swollen bread. Gluten develops as wheat proteins absorb water, undergo hydration, and interact with one another. Accordingly, one of the reasons for the suppression of swelling in MLP-supplemented bread is the high water absorbency of MLP: wheat proteins are deprived of water that they would normally absorb to form dough gluten, resulting in poorly swollen bread. Nonetheless, although RMLP achieved through roasting at 170 °C for 20 min exhibited lower water absorption than MLP, it still surpassed that of the no-MLP control by more than twofold (Table 2). Hence, additional factors may contribute to the suppression of swelling.

3.5. Effect of RMLP on Bread Dough Fermentation

Various fermentation tests were conducted using RMLP obtained by roasting at 170 °C for 20 min, i.e., the roasting condition with the best bread swelling relative to MLP. The effects of roasting on dough volume expansion curves are shown in Figure 2. The control and RMLP-supplemented doughs swelled immediately after the start of fermentation, and the expansion ratios increased with fermentation time. The dough volume expansion ratios of the control and RMLP-supplemented doughs after 120 min of fermentation were similar.

In contrast, the MLP (unroasted)-supplemented dough tended to swell significantly less than the other doughs, and its volume was approximately half that of RMLP-supplemented dough after 120 min of fermentation. This trend was consistent with the difference in the specific volumes of *Moringa*-supplemented bread after baking.

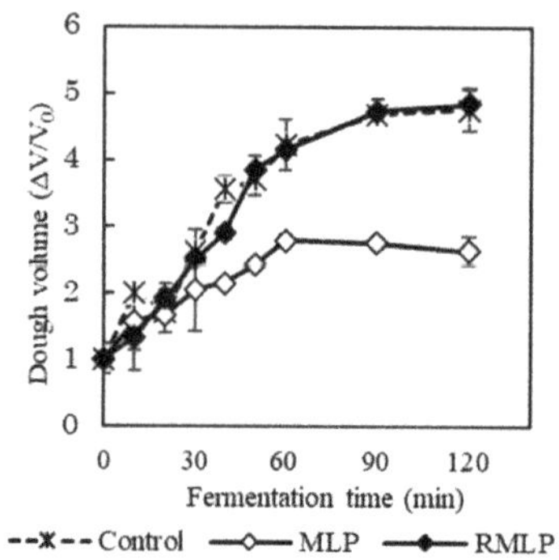

Figure 2. Effect of MLP roasting on the fermentation curve of dough. RMLP was roasted at 170 °C for 20 min. Each bread dough was kneaded in an automatic bakery for 20 min and fermented at 38 °C for 120 min. Dough volume was measured using a 3D scanner. Control, 100% wheat flour bread dough; MLP, bread dough with 5% unroasted MLP; RMLP, bread dough with 5% roasted MLP. Data are expressed as the mean (n = 3).

Generally, bread swells the most during fermentation and baking [24]. In particular, during fermentation, yeast produces carbon dioxide gas, which creates bubbles and expands the dough. The gluten film spreads around the bubbles formed by the carbon dioxide gas and traps the gas inside the dough so that it does not escape. As the amount of carbon dioxide gas produced increases, the gluten film is pushed outwards from the inside and becomes more flexible [33].

However, in this study, MLP-supplemented dough did not notably swell at the time of fermentation. Therefore, we investigated if poor swelling during fermentation was caused by poor dough formation or poor carbon dioxide production. We investigated alcoholic fermentation and viable yeast counts during dry yeast fermentation. As shown in Figure 3, the amount of carbon dioxide produced by yeast mixed with unroasted MLP was significantly lower than that in the no-MLP control (approximately half that of the control). However, the amount of carbon dioxide produced by yeast mixed with RMLP was higher than that by yeast mixed with MLP and was not significantly different from that in the control. As shown in Figure 4, viable yeast counts were also significantly lower in the presence of unroasted MLP than in the control, whereas the presence of RMLP did not significantly affect the viable yeast counts. The trends in carbon dioxide production and viable yeast counts are consistent.

These findings indicate that MLP kills yeast and inhibits alcoholic fermentation, but these effects are reversed by MLP roasting. This is in agreement with the strong antimicrobial properties of MLP. Aida et al. [34] described that the addition of dry MLP to labneh suppressed the growth of yeast and mold. Similarly, Abd El-Fat et al. [35] reported a lack of yeast or mold growth in cream cheese containing an MLP extract after four weeks of storage. Collectively, the reduction in the height and volume of bread supplemented with MLP is caused by the antimicrobial effect of *M. oleifera* leaves on yeast during fermentation, thereby causing the yeast to become inactive and disabling the complete formation of the gas cells. Of note, although untreated MLP inhibited yeast fermentation, the inhibition was easily ameliorated by MLP heat treatment.

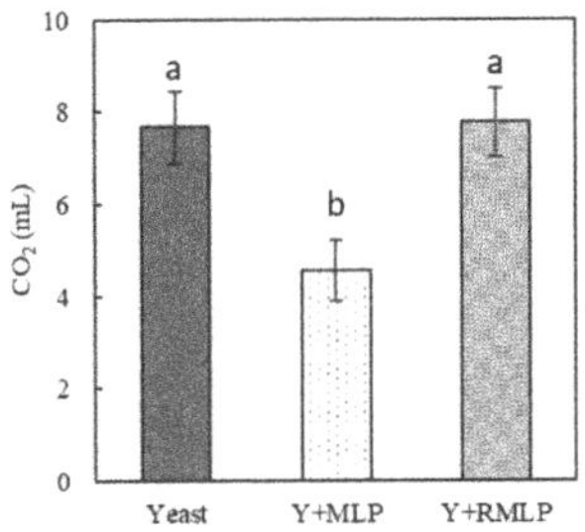

Figure 3. Amount of carbon dioxide generated by alcoholic fermentation using dry yeast. The following yeast solutions were tested: yeast, dry yeast only; Y + MLP, 10% MLP added to dry yeast; Y + RMLP, 10% RMLP added to dry yeast. Significant differences are indicated by different lowercase letters (Tukey's test, $p < 0.01$; $n = 3$).

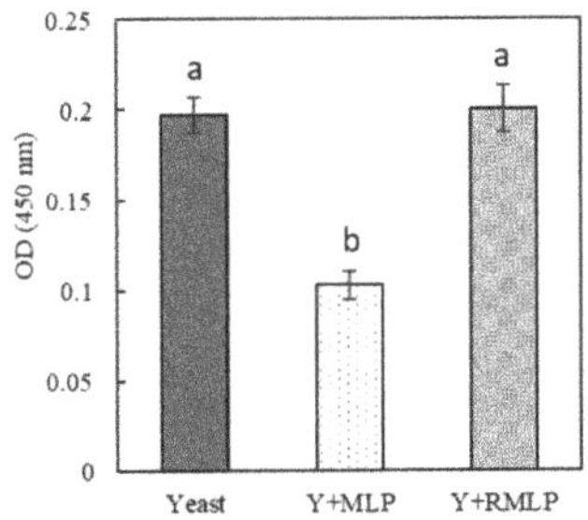

Figure 4. Effect of MLP roasting on dry yeast viability. The following yeast solutions were tested: yeast, dry yeast only; Y + MLP, 10% MLP added to dry yeast; Y + RMLP, 10% RMLP added to dry yeast. Significant differences are indicated by different lowercase letters (Tukey's test, $p < 0.01$; $n = 3$).

3.6. *Effect of RMLP on Gluten Formation and Bread Structure*

The dough contains a continuous system of gluten proteins with embedded starch, with the gluten proteins creating a fibrous web-like structure (Figure 5A,C). Scanning electron microscopy (SEM) analysis revealed mesh-like gluten formed by globular and fibrous proteins in the control, MLP-supplemented, and RMLP-supplemented dough. However, variations in the network of starch granules and the protein matrix were observed. In particular, the control and RMLP-supplemented dough contained small and large voids with traces of air bubbles in the gluten, whereas the MLP-supplemented dough had a generally compact gluten network structure, with no large voids. No major differences were observed in air bubbles in the different types of bread after baking. These observations indicate that gluten formed normally regardless of MLP supplementation, although the bread containing MLP did not fully expand during baking.

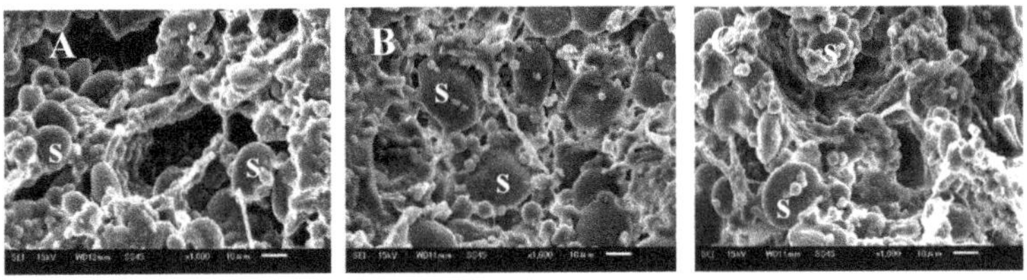

Figure 5. Scanning electron micrographs of different dough and bread samples: (**A**) control dough (no MLP), (**B**) 5% MLP-containing dough, and (**C**) 5% RMLP-containing dough. RMLP was obtained by roasting at 170 °C for 20 min. S: starch.

3.7. Sensory Evaluation

As shown in Table 3, all MLP-supplemented breads, irrespective of whether they were enriched with MLP or RMLP, were rated 5.7 or above, with no items rated as unfavorable. The different bread samples greatly differed in appearance and texture. High desirability ratings for both were noted for RMLP obtained by roasting at 130 °C and 170 °C, respectively. Particularly, the bread texture obtained higher acceptability ratings for bread made using RMLP obtained by roasting at a high temperature. This indicates that the greatest improvement in desirability related to roasting is associated with texture. The taste of bread supplemented with RMLP obtained by roasting at 130 °C and 170 °C rated higher than that of bread supplemented with unroasted MLP. Among all the tested breads, the bread supplemented with RMLP obtained by roasting at 130 °C for 20 min was the most favored in terms of taste, aroma, appearance, and overall quality. In previous studies, a tendency to not like the green color of Moringa and negative evaluations of its aroma were reported [4]. However, it is thought that Japanese people do not have a sense of objection to the green color and unique scent of Moringa and rather tend to prefer it. This could be because of the Japanese food culture of using green powders such as green tea, matcha, and yomogi. Furthermore, green tea is high in catechins and has high antioxidant activity; therefore, it is thought that Japanese people associate green food that smells of leaves with healthy and nutritious food.

Table 3. Sensory evaluation of desirability of bread containing RMLP prepared under different roasting conditions.

MLP		Taste	Aroma	Appearance	Texture	Overall
Unroasted		6.7 ± 1.6	6.3 ± 1.8	7.1 ± 1.4 [a]	6.5 ± 1.8 [bc]	6.8 ± 1.4
Roasted	110 °C	6.3 ± 2.0	6.6 ± 1.7	6.9 ± 1.5 [ab]	6.3 ± 2.1 [c]	6.3 ± 1.9
	130 °C	7.3 ± 1.5	6.6 ± 1.8	7.2 ± 1.5 [a]	7.5 ± 1.3 [ab]	7.0 ± 1.5
	150 °C	6.3 ± 1.9	5.7 ± 2.0	6.0 ± 1.8 [b]	7.5 ± 1.6 [ab]	6.4 ± 2.0
	170 °C	7.1 ± 1.6	5.8 ± 2.1	6.1 ± 2.0 [ab]	7.9 ± 1.0 [a]	6.8 ± 1.7

Values with different letters in the same row are significantly different (Tukey's test, $p < 0.05$; $n = 33$).

Our findings indicate that supplementing bread with RMLP not only increased the volume but also enhanced palatability. In particular, MLP roasting at 130 °C for 20 min, which made the RMLP-supplemented bread the most palatable, improved bread swelling and increased antioxidant capacity while maintaining Moringa's unique green color. Roasting MLP at 170 °C for 20 min is recommended for consumers who do not like the green color and strong aroma of Moringa.

4. Conclusions

The loaf volume and specific volume of bread supplemented with MLP were reduced by approximately half compared to those of the MLP-free control. However, RMLP-supplemented bread roasted at 130 °C for at least 20 min exhibited improved swelling. Notably, a remarkable enhancement in swelling was observed with RMLP obtained by roasting at 170 °C for 20 min, showing a similar volume to that of the control. After supplementation with MLP and roasting at 170 °C for 20 min, carbon dioxide production was hindered, leading to the killing of approximately half of the yeast cells. In contrast, yeast cells supplemented with RMLP were unaffected, fermenting normally and producing carbon dioxide. SEM observation revealed that gluten formation in the dough occurred irrespective of MLP roasting. These findings imply that the bactericidal component in raw MLP restrained yeast fermentation and impeded bread swelling, and roasting deactivated this component, thereby enabling normal fermentation. These findings underscore the importance of heat treatment in mitigating the effects of MLP on bread fermentation and swelling. In the future, it will be necessary to examine in detail the changes in components caused by heating.

Author Contributions: Conceptualization, supervision, and writing—review and editing, T.K. and T.H.; formal analysis and methodology, M.S., Y.K. and M.K. All authors have read and agreed to the published version of the manuscript.

Funding: This research received no external funding.

Institutional Review Board Statement: This study was approved by the Toyo University Ethics Committee (approval number TU2022-010).

Informed Consent Statement: Informed consent was obtained from all subjects involved in the study.

Data Availability Statement: The data used to support the findings of this study can be made available by the corresponding author upon request.

Acknowledgments: The authors are grateful to the Iida Industrial Technology Testing Laboratory, Japan, for providing SEM images.

Conflicts of Interest: The authors declare no conflict of interest.

References

1. Pandey, A.; Pradheep, K.; Gupta, R.; Nayar, E.R.; Bhandari, D.C. 'Drumstick tree' (*Moringa oleifera* Lam.): A multipurpose potential species in India. *Genet. Resour. Crop Evol.* **2011**, *58*, 453–460. [CrossRef]
2. Verdcourt, B. A synopsis of the Moringaceae. *Kew Bull.* **1985**, *40*, 1–23. [CrossRef]
3. Fuglie, L.J. *The Miracle Tree: Moringa oleifera, Natural Nutrition for the Tropics*; Church World Service: Dakar, Senegal, 1999.
4. Sengev, I.A.; Abu, J.O.; Gernah, D.I. Effect of *Moringa oleifera* leaf powder supplementation on some quality characteristics of wheat bread. *Food Nutr. Sci.* **2013**, *4*, 270–275.
5. Foidl, N.; Makkar, H.P.S.; Becker, K. The potential of *Moringa oleifera* for agricultural and industrial uses. In *The Miracle Tree: The Multiple Attributes of Moringa*; Fuglie, L.J., Ed.; Church World Service: Dakar, Senegal, 2001; pp. 47–67.
6. Price, M.L. The Moringa Tree. *ECHO Tech. Note* **1985**, 1–19. Available online: https://allaboutmoringa.info/wp-content/uploads/2020/11/Ebook-The-Moringa-Tree-echo.pdf (accessed on 1 January 2020).
7. Ramachandran, C.; Peter, K.V.; Gopalakrishnan, P.K. Drumstick (*Moringa oleifera*): A multipurpose Indian vegetable. *Econ. Bot.* **1980**, *34*, 276–283. [CrossRef]
8. Leone, A.; Fiorillo, G.; Criscuoli, F.; Ravasenghi, S.; Santagostini, L.; Fico, G.; Spadafranca, A.; Battezzati, A.; Schiraldi, A.; Pozzi, F.; et al. Nutritional characterization and phenolic profiling of *Moringa oleifera* leaves grown in Chad, Sahrawi Refugee Camps, and Haiti. *Int. J. Mol. Sci.* **2015**, *16*, 18923–18937. [CrossRef]
9. Palada, M.C. Moringa (*Moringa oleifera* Lam.): A versatile tree crop with horticultural potential in the subtropical United States. *Hortic. Sci.* **1996**, *31*, 794–797. [CrossRef]
10. Saini, R.K.; Shetty, N.P.; Giridhar, P. Carotenoid content in vegetative and reproductive parts of commercially grown *Moringa oleifera* Lam. cultivars from India by LC–APCI–MS. *Eur. Food Res. Technol.* **2014**, *238*, 971–978. [CrossRef]
11. Wang, Y.; Gao, Y.; Ding, H.; Liu, S.; Han, X.; Gui, J.; Liu, D. Subcritical ethanol extraction of flavonoids from *Moringa oleifera* leaf and evaluation of antioxidant activity. *Food Chem.* **2017**, *218*, 152–158. [CrossRef] [PubMed]
12. Saini, R.K.; Shetty, N.P.; Prakash, M.; Giridhar, P. Effect of dehydration methods on retention of carotenoids, tocopherols, ascorbic acid and antioxidant activity in *Moringa oleifera* leaves and preparation of a RTE product. *J. Food Sci. Technol.* **2014**, *51*, 2176–2182. [CrossRef]
13. Govender, L.; Siwela, M. The Effect of *Moringa oleifera* Leaf Powder on the Physical Quality, Nutritional Composition and Consumer Acceptability of White and Brown Breads. *Foods* **2020**, *9*, 1910. [CrossRef] [PubMed]
14. Sengev, I.A.; Agbanyi, M.M.; Sule, S. Effect of dry shredded *Moringa oleifera* leaves and vitamin C on the physicochemical properties of the dough and bread. *J. Curr. Res. Food Sci.* **2021**, *2*, 35–39.
15. Ning, J.; Hou, G.G.; Sun, J.; Wan, X.; Dubat, A. Effect of green tea powder on the quality attributes and antioxidant activity of whole-wheat flour pan bread. *L.W.T. Food Sci. Technol.* **2017**, *79*, 342–348. [CrossRef]
16. Wang, N.; Xu, Y.; Chao, H.; Zhang, M.; Zhou, Y.; Wang, M. Effects of celery powder on wheat dough properties and textural, antioxidant and starch digestibility properties of bread. *J. Food Sci. Technol.* **2020**, *57*, 1710–1718. [CrossRef]
17. Lim, P.Y.; Sim, Y.Y.; Nyam, K.L. Influence of kenaf (*Hibiscus cannabinus* L.) leaves powder on the physico-chemical, antioxidant and sensorial properties of wheat bread. *J. Food Meas. Charact.* **2020**, *14*, 2425–2432. [CrossRef]
18. Odunlade, T.V.; Famuwagun, A.A.; Taiwo, K.A.; Gbadamosi, S.O.; Oyedele, D.J.; Adebooye, O.C. Chemical composition and quality characteristics of wheat bread supplemented with leafy vegetable powders. *J. Food Qual.* **2017**, *2017*, 9536716. [CrossRef]
19. Bourekoua, H.; Różyło, R.; Gawlik-Dziki, U.; Benatallah, L.; Zidoune, M.N.; Dziki, D. Evaluation of physical, sensorial, and antioxidant properties of gluten-free bread enriched with *Moringa oleifera* leaf powder. *Eur. Food Res. Technol.* **2018**, *244*, 189–195. [CrossRef]
20. Gan, Z.; Ellis, P.R.; Schofield, J.D. Gas cell stabilisation and gas retention in wheat bread dough. *J. Cereal Sci.* **1995**, *21*, 215–230. [CrossRef]

21. AOAC. *Official Methods of Analysis of the Association of Official Analytical Chemists*, 20th ed.; AOAC International: Rockville, MD, USA, 2016.
22. Monro, J.; Burlingame, B. Carbohydrates and related food components: INFOODS tagnames, meanings, and uses. *J. Food Compos. Anal.* **1996**, *9*, 100–118. [CrossRef]
23. American Association of Cereal Chemists. *Approved Methods of the AACC*, 10th ed.; Methods 56-11; The Association: St. Paul, MN, USA, 2000.
24. Autio, K.; Laurikainen, T. Relationships between flour/dough microstructure and dough handling and baking properties. *Trends Food Sci. Technol.* **1997**, *8*, 181–185. [CrossRef]
25. MEXT. *Standard Tables of Food Composition in Japan—2015—(Seventh Revised Version)*; Ministry of Education, Culture, Sports, Science and Technology (MEXT): Tokyo, Japan, 2015.
26. Karim, O.; Kayode, R.; Oyeyinka, S.; Oyeyinka, A. Physicochemical properties of stiff dough 'amala' prepared from plantain (*Musa paradisca*) flour and moringa (*Moringa oleifera*) leaf powder. *Food Health Dis.* **2015**, *4*, 48–58.
27. Manzocco, L.; Calligaris, S.; Mastrocola, D.; Mocoli, M.C.; Lerici, C.R. Review of non-enzymatic browning and antioxidant capacity in proceeded foods. *Trends Food Sci. Technol.* **2001**, *11*, 340–346. [CrossRef]
28. Maiyo, F.C.; Moodley, R.; Singh, M. Cytotoxicity, antioxidant and apoptosis studies of quercetin-3-*o* glucoside and 4-(β-D-glucopyranosyl-1→4-α-L-rhamnopyranosyloxy)-benzyl isothiocyanate from *Moringa oleifera*. *Anticancer. Agents Med. Chem.* **2016**, *16*, 648–656. [CrossRef] [PubMed]
29. Hayase, F.; Hirashima, S.; Okamoto, G.; Kato, H. Scavenging of active oxygens by melanoidins. *Agric. Biol. Chem.* **1989**, *53*, 3383–3385.
30. Jogihalli, P.; Singh, L.; Kumar, K.; Sharanagat, V.S. Physico-functional and antioxidant properties of sand-roasted chickpea (*Cicer arietinum*). *Food Chem.* **2017**, *237*, 1124–1132. [CrossRef] [PubMed]
31. Koriyama, T.; Teranaka, K.; Tsuchida, M.; Kasai, M. Effects of storage and roasting condition on the antioxidant activity of soybeans with different colors of seed coat. *Foods* **2022**, *12*, 92. [CrossRef] [PubMed]
32. Kokoh, A.; Elleingand, E.; Koffi, E. Physico-chemical and sensory properties of breads produced from wheat and fermented yam composite flour fortified with moringa leaves powder. *J. Food Nutr. Res.* **2019**, *7*, 772–777. [CrossRef]
33. Grenier, D.; Rondeau-Mouro, C.; Dedey, K.B.; Morel, M.H.; Lucas, T. Gas Cell Opening in Bread Dough During Baking. *Trends Food Sci. Technol.* **2021**, *109*, 482–498. [CrossRef]
34. Aida, S.S.; Wafaa, M.S.; Hassanein, A.M.; El-Ghandour, H.M.A. Enhancement of nutritional and biological values of labneh by adding dry leaves of *Moringa oleifera* as innovative dairy products. *World Appl. Sci. J.* **2013**, *22*, 1594–1602.
35. El-Fataah Mohamed, F.A.; Hassan Salama, H.; El-Sayed, S.M.; El-Sayed, H.S.; Abdel-Hady Zahran, H. Utilization of natural antimicrobial and antioxidant of *Moringa oleifera* leaves extract in manufacture of cream cheese. *J. Biol. Sci.* **2018**, *18*, 92–106. [CrossRef]

foods

Article

Effect of Gluten Composition in Low-Allergy O-Free Wheat Flour on Cookie-Making Performance Compared with Flours with Different Gluten Strengths

Nayeon Baek [1], Yujin Moon [2], Jeongeon Kim [2] and Meera Kweon [2,3,*]

1 Department of Nutrition Education, Pusan National University, Busan 46241, Republic of Korea; qorsk100@pusan.ac.kr
2 Department of Food Science and Nutrition, Pusan National University, Busan 46241, Republic of Korea; amoebacul@pusan.ac.kr (Y.M.); wjddjs8@pusan.ac.kr (J.K.)
3 Kimchi Research Institute, Pusan National University, Busan 46241, Republic of Korea
* Correspondence: meera.kweon@pusan.ac.kr; Tel.: +82-51-510-2716

Abstract: The increasing demand for allergen-free and reduced-allergen foods has led to an investigation into the potential use of O-free wheat, a low-allergy wheat cultivar, in cookie production. This study focused on assessing the gluten composition of O-free flour and comparing its suitability for cookie making in comparison to flours with varying gluten strengths. Several analyses were conducted, including gluten composition, solvent retention capacity (SRC), thermal and pasting properties, dough-mixing characteristics, and cookie-making performance. The gluten composition of O-free flour by SDS-PAGE confirmed the absence of ω-gliadins and the reduced levels of low-molecular-weight glutenins and γ-gliadins. The SRC values of O-free flour fell between the flours with weak and medium gluten strengths. While thermal and pasting properties showed significant differences in sucrose solution but not across flour types, indicating similar starch structures, mixograms displayed distinct variations influenced by both sucrose solution and flour type, highlighting the importance of gluten quality and composition. Cookies made with O-free flour demonstrated similarities to those produced with weak gluten flour, known for their favorable cookie characteristics. This study emphasizes the significant influence of flour gluten composition on cookie-making performance and advocates for the adoption of O-free flour in the development of allergy-friendly cookies.

Keywords: low-allergy wheat; SDS-PAGE; thermal and pasting property; mixogram; cookie

1. Introduction

Food allergies are hypersensitivity reactions caused by an abnormal immune response to specific foods [1]. The prevalence of food allergies is increasing worldwide, including domestically [2]. Cereals containing gluten, such as wheat, are frequently associated with allergic reactions in Korean adults. Wheat allergies pose a particular challenge as they can lead to wheat-dependent exercise-induced anaphylaxis (WDEIA), which can be fatal [3]. To address this, the Rural Development Administration in Korea has recently developed a specialty wheat variety called O-free, which does not contain the main allergens such as ω-5-gliadins, making wheat consumption safe for individuals with wheat allergies [4]. Previous research has investigated the performance of O-free flour in bread and noodle production, broadening its potential applications [5–7]. However, there is a lack of in-depth studies on the performance of O-free flour in cookie making.

In cookie production, sucrose, a primary ingredient, suppresses gluten development during mixing and delays starch gelatinization and pasting during baking, resulting in cookies with large diameters and small stack heights [8–11]. Numerous studies have examined the effects of wheat flour protein content and composition on cookie quality [12–17].

Protein content negatively affects cookie quality, reducing diameter and increasing hardness [14,16]. Gluten, a component of wheat protein, negatively affects cookie diameter by limiting the dough spread time [13] and preventing collapse through gluten entanglement and/or cross-linking during baking [12]. Moreover, the deletion of *Glu-D1*-encoded high-molecular-weight glutenin subunits resulted in poor gluten quality but improved cookie quality [15]; subunits 2 + 12 among the high-molecular-weight glutenin subunits encoded by the *Glu-D1* locus positively influenced cookie quality [16].

Sulfur-containing amino acid residues in proteins play a crucial role in the formation of disulfide bonds through free thiols during gluten development. Gliadins and glutenins, excluding ω-gliadin, contain these sulfur amino acid residues, which impact gluten development and strength, ultimately leading to reduced cookie spread [18,19]. However, previous study by Uthayakumaran et al. [20] has reported the effects of gliadin alone on cookie quality, such as increased diameter and thickness, softer texture, and enhanced dough extensibility of the γ-gliadin fraction, which was expected to improve cookie quality. Barak et al. [21] also demonstrated a significant positive correlation between the cookie spread factor and the gliadin/glutenin ratio. Therefore, it is valuable to confirm the effect of gluten composition variation in O-free flour on cookie-making performance.

When comparing O-free wheat to its two parent lines, Keumkang and Olgeuru, a deficiency of ω-5-gliadin was identified through RNA sequencing analysis, revealing a simple deletion in the chromosome. This finding was further corroborated by the similarity analysis using the expressed sequence tags (ESTs) for gliadins and glutenins. The wheat also exhibited a reduced low-molecular-weight glutenin subunit [4]. However, there has been few reports on the gluten protein composition of O-free flour in comparison to regular wheat flour using SDS-PAGE gel electrophoresis. It is valuable to confirm the gluten protein composition of O-free flour in relation to its performance in making cookies.

Therefore, this study aims to analyze the quality characteristics of low-allergy O-free wheat flour by comparing its gluten protein composition, SRC, pasting, thermal properties, and dough-mixing properties with commercial flours of varying gluten strengths. It also evaluates the quality of cookies made with O-free flour.

2. Materials and Methods

2.1. Materials

The flour used in this study was milled from a low-allergy wheat cultivar, O-free, supplied by the National Institute of Crop Science (Wanju, Jeollabukdo, Republic of Korea). For a comparative study, three commercial flours with varied gluten strengths (strong, medium, and weak) were purchased from a local market: the flour with strong gluten strength (CJ Jeiljedang, Yangsan, Republic of Korea), the flour with medium gluten strength (referred to as all-purpose flour), and the flour with weak gluten strength (Samyang Corp., Asan, Republic of Korea).

2.2. Analysis of Moisture and Protein Content of Flour Samples

The moisture and ash contents of the wheat flour samples were measured according to Method 44-15.02 and 08-01,01, respectively [22]. For moisture content analysis, approximately 3 g of each flour was placed on pre-weighed aluminum containers and dried in a convection oven (FO 600-M, Jeio Tech Co., Ltd., Daejeon, Republic of Korea) at 130 °C for 1 h. The containers were then cooled in a desiccator for 30 min, and then the containers with the dried samples were reweighed. To assess ash content, approximately 3 g of each sample was weighed in a crucible. The crucibles were placed in a furnace at 550 °C (DMF-4.5.T, SciFinetech, Seoul, Republic of Korea) for 15 h. After cooling, the crucibles containing the samples were reweighed to determine the weight of the remaining inorganic materials.

The protein contents of the wheat flour samples were measured according to Method 46-30.01 [22]. The flour samples were combusted using an Elementar instrument (rapid N exceed, Elementar Analysensysteme GmbH, Langelselbold, Germany), and the nitrogen content was calculated. The crude protein content was determined in duplicate, as it exhibited good reproducibility, and it was calculated by multiplying the nitrogen content by the typical protein factor of wheat (5.7).

2.3. Extraction of Glutein Components and SDS-PAGE Gel Electrophoresis of Flours

Following the method outlined by Lookhart and Bean [23], albumin and globulins were extracted from 500 mg of the defatted sample using 5 mL of 0.15 M NaCl for 2 h. Subsequently, the mixture was centrifuged at 20,000× *g* for 15 min. Gliadin was then extracted from the remaining pellet by employing 5 mL of 70% ethanol for a duration of 12 h, followed by centrifugation under identical conditions. Glutenin, on the other hand, was extracted from the pellet after gliadin extraction, using a solution of 50% 1-propanol and 1% DTT for two hours, followed by centrifugation.

For the electrophoresis of the gliadin and glutenin extracts, a 4× Laemmli sample buffer containing 0.4 M DTT was combined with the protein extract in a 2:1 (v/v) ratio. A running buffer consisting of 10× Tris/Glycine/SDS Buffer was diluted tenfold and employed. Each extract (10 μL) was loaded onto 12% TGX Stain-Free Precast Gels, and electrophoresis was conducted at 70 V for 10 min, followed by 100 V for 90 min. Upon completion of electrophoresis, the staining solution (comprising 40% methanol, 10% acetic acid, and 0.1% Coomassie Brilliant Blue G-250) was applied for 30 min, followed by the destaining solution (40% methanol and 10% acetic acid) for 1 h. The extraction and electrophoresis experiments were conducted in triplicate.

2.4. Analysis of Solvent Retention Capacity of Flours in Four Solutions

To evaluate flour quality, the SRC was determined according to Method 56-11.02 [22]. Four 50 mL conical tubes were weighed, and 5 g of each flour (5 g) was placed in each tube. Distilled water, 5% (w/w) lactic acid, 5% (w/w) sodium carbonate, and 50% (w/w) sucrose solutions were prepared, and 25 g of each solution was added to each tube containing the flours. The tubes were then shaken for 20 min at 5 min intervals to disperse and hydrate the flour and centrifuged at 1000× *g* for 15 min (LaboGene 1248, Gyrozen Inc., Daejeon, Republic of Korea). After centrifugation, the supernatant in each tube was discarded, and the pellet weight was measured to determine SRC percentages using the AACCI method [22].

2.5. Analysis of Thermal Characteristics of Flour Using Differential Scanning Calorimetry

Differential scanning calorimeter was used to analyze the gelatinization properties of starch in the flour with water, according to the method described by Kweon et al. [24]. Water or pre-dissolved sucrose solution (50% w/w) and flour were mixed in a 1:1 (w/w) ratio, and approximately 40 mg of the mixture was placed in an empty aluminum pan and sealed. A pan containing the sample was placed in a DSC instrument (DSC 6000, Penkin Elmer Co., Waltham, MA, USA) and heated at a rate of 10 °C/min from 10 to 140 °C. An empty aluminum pan was used as a reference, and the Pyris Software (ver. 11, Perkin Elmer Co., Ltd.) was used to calculate gelatinization temperatures and enthalpies.

2.6. Determination of Flour Pasting Property Using Rapid Visco-Analyzer

The pasting properties of starch in the flours were investigated using a RVA (RVA4, Newport Scientific, Sydney, Australia) based on the method described by Kweon et al. [24]. The RVA canister was filled with 25 mL of water or the pre-dissolved sucrose solution (50% w/w), and 3.5 g of the flour was added and thoroughly mixed using a plastic paddle. The RVA was run according to standard 1 profile (50–95 °C, 12.2 °C/min heating; 95 °C, 2.5 min maintained; 95–50 °C, 12.2 °C/min cooling; 50 °C, 2 min maintained). Pasting parameters such as peak, breakdown, final, and setback viscosities; peak time; and peak temperature were calculated using RVA software (Thermocline for Windows ver. 2.5).

2.7. Analysis of Dough-Mixing Property of Flour using Mixograph

A mixograph test was conducted using a 10 g Mixograph machine (National Manufacturing Inc., Lincoln, NE, USA). Each flour sample (10 g) was placed in a mixograph bowl, and 6.0–7.0 g of water or 6.5 g of pre-dissolved sucrose solution (50% *w*/*w*) was added. The Mixograph machine was run for 10 min, and the peak height and time of the graph were measured. Each sample was tested in duplicate.

2.8. Preparation of Cookies

Cookies were prepared following Method 10-53.01 [22], and the ingredients and formula are presented in Table 1. Shortening (90.0 g) and granulated sucrose (94.5 g) were weighed separately, placed in a bowl (N50; Hobart, Troy, OH, USA), and mixed for 3 min at speed 1 (low speed). Then, the corn syrup (3.4 g), water (49.5 g), and ammonia bicarbonate (1.1 g) were combined, placed in a bowl, and mixed at speed 1 for 1 min, followed by mixing at speed 2 (medium speed) for 1 min. Subsequently, skim milk powder (2.3 g), refined sea salt (2.8 g), and sodium bicarbonate (2.3 g) were thoroughly combined with flour (225.0 g), placed in a bowl, and mixed for 2 min at speed 1. After mixing, the dough was divided into four 60 g masses and placed on an aluminum baking pan. The dough was flattened using a rolling pin, and shaped using a circular cutter (6 cm in diameter). The dough was baked in an oven (Phantom M301 Combi; Samjung, Gyeonggi, Republic of Korea) at 215 °C for 12 min. After cooling the cookies for approximately 1 h at 25 °C, they were sealed in an aluminum foil bag, and their quality was assessed after 24 h.

Table 1. Ingredients and formula for AACC Method 10-53.01 cookie baking.

Ingredient	Standard Weight (g)
Flour	225.0 [1]
Shortening	90.0
Sucrose	94.5
Corn syrup	3.4
Sodium bicarbonate	2.3
Non-fat dry milk	2.3
Salt	2.8
Ammonium bicarbonate	1.1
Water	49.5

[1] Method 10-53 assumes 13% flour water content.

2.9. Assessment of Cookie Quality

The length and width of each cookie were measured, and the average value was calculated using the measurements of the four cookies. Cookie height was measured using a caliper, arranging four cookies side by side vertically, and the average value was calculated from four measurements taken in different orders.

The top surface color of the cookies was measured by L* (lightness), a* (redness), and b* (yellowness) values using a chromometer (CR-20, Minolta, Co., Ltd., Tokyo, Japan), and the average value was obtained from four measurements.

Cookie firmness was determined using a Texture Analyzer (CT3, Brookfield, Middleboro, MA, USA). was measured using a Texture Analyzer. The measurement conditions were as follows: Mode, measure force in compression; pre-test speed, 1.0 mm/s; test speed, 0.5 mm/s; probe, 2 mm cylinder probe; penetration distance, 10 mm. The measurement was repeated three times, and the average cookie firmness was calculated from these three measurements.

2.10. Statistical Analysis

All experiments, except for moisture, protein and ash content, were conducted in triplicate unless otherwise specified. All data were obtained through multiple measurements at least three times. The difference between the mean values of the samples was analyzed using analysis of variance with Tukey's HSD test at a significance level of $p < 0.05$, using SPSS 22.0 (SPSS Inc., Armonk, New York, NY, USA) program.

3. Results and Discussion

3.1. Physicochemical Properties of Flour and SDS-PAGE gel Electrophoresis of Gluten Protein

The physicochemical properties of the flour samples are shown in Table 2. The moisture content of the flour samples was 13.08–14.0%, which is consistent with the typical range of flour moisture (12–14%) [25], indicating that the water added in the tempering process during the milling of the flour samples was not significantly different [24]. The ash content of the flour samples ranged from 0.37 to 0.43%; the ash content of the flour with weak gluten strength (W) was significantly lower than that of the other flour samples ($p < 0.05$), suggesting the lowest extraction rate during milling [26].

Table 2. Moisture, ash, and protein contents, and SRC values of the flour with different gluten strengths and O-free flour.

Flour (1)	Moisture (%)	Ash (%)	Protein (3) (%)
S	13.72 ± 0.03 c(2)	0.42 ± 0.00 b	13.98 ± 0.02 d
M	13.50 ± 0.04 b	0.43 ± 0.01 b	10.56 ± 0.02 b
W	14.01 ± 0.01 d	0.37 ± 0.00 a	8.50 ± 0.02 a
O	13.10 ± 0.00 a	0.43 ± 0.00 b	11.49 ± 0.01 c

(1) S, M, and W, flours with strong, medium, and weak gluten strength, respectively; O, O-free flour. (2) Results are expressed as the mean ± SD. Values with the same letter within the same column are not significantly different ($p < 0.05$), according to Tukey's HSD test. (3) Protein indicates the protein content adjusted for the moisture content of 14%.

The protein content of the flour samples ranged from 8.50% to 13.98%. The flour with strong gluten strength (S) had the highest protein content, while the flour with weak gluten strength (W) had the lowest protein content. This is consistent with the relationship between flour protein and gluten strength, where flours milled from hard wheat (strong gluten) have higher protein content, and flours milled from soft wheat (weak gluten) have lower protein content [27]. The protein content of O-free flour (O) was 11.5%, which was lower than the flour with strong gluten strength but higher than the flours with medium and weak gluten strengths. This is in line with a previous study by Park et al. [28]. Kang et al. [16] and Ozturk et al. [29] reported that the flour protein content affected the cookie quality. When cookies are made with high-protein flour, he elasticity of the cookie dough increases, and the cookie spread decreases during baking, resulting in cookies with small diameters and large heights [21]. Therefore, flours with low protein content and weak gluten strength, such as the W flour sample, were suggested to be the more suitable for cookie production.

The SDS-PAGE gel electrophoresis of glutenin and gliadin extracted from the flour samples was presented in Figure 1. In the glutenin and gliadin extract, SDS-PAGE gel electrophoresis of O-free flour exhibited significant differences compared to the other three flours. O-free flour displayed five distinct bands in the molecular weight range between 75 and 150 kDa, representing high-molecular-weight glutenin subunits (HWG-GS) [19,30,31], in contrast to the other flours which showed only three bands. Notably, glutenin extract exhibited more distinct bands in this molecular weight region compared to the gliadin extract. The intensity of these three bands in both extracts decreased in the order of S > M > W, reflecting the reduced amount of HWG-GS in the flour samples.

Importantly, no clear bands in the molecular weight range of 50–75 kDa were observed in the gluten extract, but approximately three bands appeared in the gliadin extract of S, M, and W flours, except for O-free flour. These results indicated the absence of ω-gliadins in O-free flour, which is consistent with findings from ω-gliadin-free wheat reported by Waga and Skoczowski [32]. The band intensity decreased in the order of S > M > W, indicating reduced amounts of ω-gliadins.

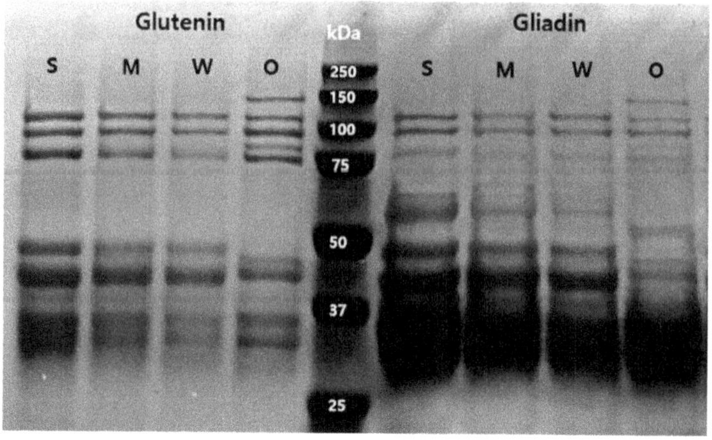

Figure 1. Sodium dodecyl sulfate-polyacrylamide gel electrophoresis of glutenin and gliadin extract from the flour with different gluten strengths and O-free flour: S, flour with strong gluten strength; M, flour with medium gluten strength; W, flour with weak gluten strength; O, O-free flour.

Additionally, the bands in O-free flour in the range of MW 37–50 kDa appeared significantly different from those of the other flours. Gliadin extract exhibited much thicker bands than glutenin extract, which could be attributed to differences in γ-gliadins and low-molecular-weight glutenin subunits (LMW-GS) [19]. In particular, the band near 50 kDa in both extracts was absent in O-free flour, indicating a reduction or absence of γ-gliadins and LMW-GS components. Meanwhile, the other flours exhibited a decrease in intensity similar to the bands of HWG-GS.

These results clearly demonstrate variations in the amount of glutenin and gliadin in the flours. Overall, the SDS-PAGE gel electrophoresis results confirm the differences in the gluten protein composition of O-free flour compared to flours with varying protein contents and gluten strengths [4]. Therefore, it can be assumed that the gluten protein composition and protein content of O-free flour may impact its performance in cookie making.

3.2. Solvent Retention Capacity of Flours

Figure 2 shows the solvent retention capacity (SRC) results of the flour samples in four different solutions: water, lactic acid, sodium carbonate, and sucrose. The SRC values indicate significant differences among the flour samples, following the order of flour with strong gluten strength > flour with medium gluten strength > O-free flour > flour with weak gluten strength ($p < 0.05$). This result was consistent with a study on the SRC of O-free flour conducted by Moon et al. [5].

Water SRC values for the flours with strong, medium, and weak gluten strengths were 72.9%, 64.8%, and 51.4%, respectively, while the O-free flour had a water SRC value of 55.4%. The flour with strong gluten strength had the highest water SRC value, indicating a potentially high amount of damaged starch generated during the milling of hard wheat. Water SRC values are associated with water absorption [33], suggesting that flours with weak gluten strength and O-free flour require less water to form machinable dough for making cookies.

Lactic acid SRC is related to flour gluten strength [33]. The SRC values of the flours with strong, medium, and weak gluten strengths were 165.3%, 125.6%, and 91.4%, respectively, while the O-free flour had a lactic acid SRC value of 120.4%. These values indicate significant differences depending on the gluten strength. Notably, the O-free flour, which had higher protein content, exhibited a lower lactic acid SRC value compared to the flour with medium gluten strength. This can be attributed to the reduced low-molecular-weight glutenin protein content observed in O-free flour, as demonstrated in SDS-PAGE gel electrophoresis (Figure 1). This reduction in low-molecular-weight glutenin proteins has an impact on lactic acid SRC and overall gluten quality [4]. Among the flour samples, the flour with weak gluten strength was identified as the most suitable for cookie production, consistent with previous studies on cookie flour [29].

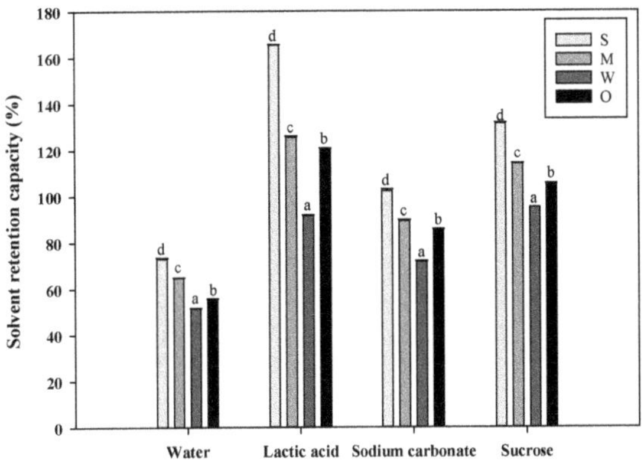

Figure 2. Solvent retention capacity (SRC) of the flour with different gluten strengths and O-free flour: S, flour with strong gluten strength; M, flour with medium gluten strength; W, flour with weak gluten strength; O, O-free flour. The same letters indicated above the bars are not significantly different at $p = 0.05$, according to Tukey's HSD test.

Sodium carbonate SRC is associated with damaged starch in wheat flour [33]. The SRC values of the flours with strong, medium, and weak gluten strengths were 102.2%, 89.2%, and 71.6%, respectively, while the O-free flour had a sodium carbonate SRC value of 85.6%. The flour with strong gluten strength showed a significantly high sodium carbonate SRC value due to the high kernel hardness of hard wheat, resulting in a high flour extraction yield and damaged starch during milling. Conversely, the flour with weak gluten strength exhibited a significantly low sodium carbonate SRC value due to the low kernel hardness of soft wheat, leading to a low extraction yield and damaged starch during milling [34].

Sucrose SRC is associated with the contribution of arabinoxylans in wheat flour [33]. The SRC values of the flours with strong, medium, and weak gluten strengths were 131.4%, 114.0%, and 95.1%, respectively, while the O-free flour had a sucrose SRC value of 104.9%. The results indicated a significantly higher arabinoxylan content in the flour with strong gluten strength compared to the flour with medium and weak gluten strengths and O-free flour. Arabinoxylans play a significant role in water absorption compared to damaged starch or gluten proteins [33]. Consequently, the poor cookie-making performance of the flour with strong gluten strength can be attributed to the challenge of moisture evaporation during baking, while the cookie-making performance of O-free flour may be superior to that of flours with strong and medium gluten strengths.

Gaines [35] reported that ideal cookies could be made with specific SRC values of wheat flour: water SRC $\leq$ 51%, lactic acid SRC $\geq$ 87%, sodium carbonate SRC $\leq$ 64%, and sucrose SRC $\leq$ 89%. Therefore, flours with low water, sodium carbonate, and sucrose SRC values allow for quick water evaporation during baking, making them favorable for cookie production [8,36]. In this context, the flour with weak gluten strength and the O-free flour, which exhibited relatively low SRC values in water, sodium carbonate, and sucrose solutions, are suitable for cookie production. Furthermore, the O-free flour is suitable for producing low-allergy cookies based on its solvent retention capacity.

3.3. Gelatinization Properties of O-Free Flour in Water and Sucrose Solution Measured Using Differential Scanning Calorimetry (DSC)

The DSC results for the flour samples in both water and pre-dissolved 50% (w/w) sucrose solution are presented in Table 3. The pre-dissolved sucrose solution was employed to investigate the gelatinization of starch in the flour samples under conditions that mimic a sugar concentration found in cookie formulations. The gelatinization peak temperature (Tpeak) of the flours in water ranged from 68.2 to 70.8 °C, which was significantly lower than the Tpeak observed in the sucrose solution (ranging from 96.9 to 101.6 °C). This indicates a notable delay in the gelatinization of starch induced by sucrose solution ($p < 0.05$). This result is consistent with previous studies on the effect of sucrose solution on starch gelatinization [24,37–39].

Table 3. Thermal characteristics of the flour samples in water and 50% (*w*/*w*) pre-dissolved sucrose solution.

Solvent [(1)]	Flour [(2)]	Tonset (°C)	Tpeak (°C)	Tend (°C)	Heat of Transition (ΔQ, J/g)
Water	S	60.8 ± 0.2 [ab(3)]	70.8 ± 0.1 [a]	90.6 ± 0.1 [a]	3.0 ± 0.1 [a]
	M	58.9 ± 0.2 [a]	69.2 ± 0.5 [a]	91.0 ± 0.6 [a]	3.5 ± 0.1 [ab]
	W	61.5 ± 0.1 [b]	68.2 ± 0.0 [a]	86.0 ± 0.3 [a]	3.7 ± 0.0 [bc]
	O	64.1 ± 0.3 [c]	69.8 ± 0.3 [a]	87.9 ± 1.3 [a]	3.5 ± 0.1 [ab]
Sucrose	S	88.5 ± 0.2 [de]	101.6 ± 0.7 [c]	120.4 ± 1.5 [b]	3.4 ± 0.1 [ab]
	M	86.3 ± 0.8 [d]	98.3 ± 1.1 [bc]	119.1 ± 1.5 [b]	3.9 ± 0.1 [bc]
	W	87.5 ± 0.2 [d]	96.9 ± 0.3 [b]	118.9 ± 0.2 [b]	4.2 ± 0.0 [c]
	O	89.8 ± 0.6 [e]	98.2 ± 1.2 [bc]	117.0 ± 3.2 [b]	3.9 ± 0.2 [bc]

[(1)] Flour: solvent ratio = 1:1 (*w*/*w*). [(2)] S, M, and W, flours with strong, medium, and weak gluten strength, respectively; O, O-free flour. [(3)] Results are expressed as the mean ± SD. Values with the same letter within the same column are not significantly different ($p < 0.05$) according to Tukey's HSD test.

Both the gelatinization peak and end temperatures of the flours, including those with strong, medium, and weak gluten strengths, as well as O-free flour, in water, exhibited values of 70.8, 69.2, 68.2, and 69.8 °C, respectively. Similarly, in the sucrose solution, these temperatures displayed a comparable trend across all flour samples. The gelatinization of starch in flour during the baking of cookies can significantly impact various attributes of the cookies, such as size, color, moisture content, and texture. Sucrose has been shown to delay the network formation of starch in flour during baking, leading to cookies with larger diameters and smaller heights [11]. These results suggest that the influence of flours with varying gluten strengths, as well as O-free flour, on cookie attributes related to starch gelatinization is likely to be similar.

3.4. Pasting Properties of O-Free Flour in Water and Sucrose Solution Measured Using Rapid Visco-Analyzer (RVA)

The RVA results for the flour samples in both water and pre-dissolved 50% (*w*/*w*) sucrose solutions are presented in Table 4. As previously utilized in the DSC analysis, the pre-dissolved sucrose solution, which simulated the cookie formulation, was employed to evaluate starch pasting in the flour samples within the RVA analysis. The pasting pattern of flours in water differed from that of flours in sucrose solution because the starch granules in water swelled quickly during heating to achieve the highest pasting viscosity. The swollen starch granules were then broken by continuous shearing, thereby reducing the viscosity. The viscosity increased again during cooling, indicating gel formation. In contrast, the pasting pattern of flours in sucrose solution showed a slow increase in viscosity because the starch granules did not swell quickly during heating. Furthermore, no noticeable peak was observed because the less swollen granules were not broken even after continuous shearing [24,38,40]. Notably, the pasting temperature of the flours in water (67.3–87.5 °C) was lower than that in the sucrose solution (90.7–92.8 °C).

The higher peak viscosity was observed in water, ranging from 2454 to 3059 cP, compared to sucrose solution, ranging from 547 to 1742 cP. This indicates greater swelling of starch granules in water [41]. The breakdown viscosity, which reflects the breakdown of swollen starch granules and amylose release during continuous shearing, was higher in water (761–1221 cP) compared to sucrose solution (64–214 cP) [10,36]. The final viscosities of the flours in water (2752–3519 cP) were higher than those in sucrose solution (673–1987 cP).

The setback viscosity, which indicates the binding of starch molecules and the formation of a gel network during cooling [42], was greater in water (1157–1644 cP) than in sucrose solution (189–543 cP), suggesting that starch aging and gel formation occurred more quickly in water. Further, the peak time to reach the highest viscosity was shorter in water (6.1–6.2 min) than in sucrose solution (6.9–7.0 min).

Table 4. Pasting characteristics of the flour samples in water and 50% (*w/w*) pre-dissolved sucrose solution.

Solvent [(1)]	Flour [(2)]	Peak Viscosity (cP)	Breakdown Viscosity (cP)	Final Viscosity (cP)	Setback Viscosity (cP)	Peak Time (min)	Pasting Temp (°C)
Water	S	3059 ± 18 [g(3)]	1221 ± 19 [g]	3482 ± 1 [g]	1644 ± 3 [f]	6.2 ± 0.0 [a]	67.4 ± 0.2 [a]
	M	2454 ± 3 [e]	859 ± 1 [e]	2752 ± 1 [e]	1157 ± 1 [d]	6.1 ± 0.1 [a]	67.3 ± 0.4 [a]
	W	2732 ± 21 [f]	761 ± 8 [d]	3519 ± 21 [g]	1548 ± 8 [e]	6.1 ± 0.0 [a]	81.9 ± 0.4 [b]
	O	2804 ± 13 [f]	1022 ± 40 [f]	3354 ± 21 [f]	1572 ± 31 [e]	6.2 ± 0.1 [a]	87.5 ± 0.5 [c]
Sucrose	S	1299 ± 23 [c]	103 ± 4 [ab]	1399 ± 23 [b]	203 ± 4 [a]	7.0 ± 0.0 [b]	91.5 ± 0.5 [d]
	M	1742 ± 7 [d]	214 ± 4 [c]	1987 ± 15 [d]	459 ± 4 [b]	7.0 ± 0.1 [b]	90.7 ± 0.7 [d]
	W	1177 ± 13 [b]	187 ± 6 [bc]	1533 ± 9 [c]	543 ± 2 [c]	7.0 ± 0.0 [b]	92.0 ± 0.1 [d]
	O	547 ± 0 [a]	64 ± 2 [a]	673 ± 4 [a]	189 ± 6 [a]	6.9 ± 0.1 [b]	92.8 ± 0.0 [d]

[(1)] Flour: solvent ratio = 7:50 (*w/v*). [(2)] S, M, and W, flours with strong, medium, and weak gluten strength, respectively; O, O-free flour. [(3)] Results are displayed as the mean ± SD. According to Tukey's HSD test, values with the same letter within the same column did not differ significantly ($p < 0.05$).

Although there were variations in pasting viscosity values among the four flours when mixed with water, their pasting patterns displayed similarities. Additionally, similar trends were observed in the pasting properties of the flours when subjected to a sucrose solution. These results suggest that the pasting properties of O-free flour did not significantly differ from those of the other three flours, indicating that starch's impact on cookie-making performance is minimal. The pasting viscosity values of O-free flour in a sucrose solution were relatively low, which presents a challenge when attempting to explain the results. Factors such as arabinoxylans, proteins, lipids, and particle size, in addition to starch in the wheat flour, may influence pasting viscosities.

3.5. Dough-Mixing Properties of Flour Samples

Mixograms of the flours in water and pre-dissolved 50% (*w/w*) sucrose solutions are depicted in Figure 3. The peak center height of the curve is a parameter indicating dough development and strength, which are associated with protein content and quality [43,44]. The flour with strong gluten strength exhibited a distinct peak at around 4 min, and the width of the dough-mixing band was maintained for approximately 10 min. The flour with medium gluten strength showed a broader peak, but the width of the dough-mixing band was still maintained. In contrast, the flour with weak gluten strength did not exhibit a peak. These mixograms are typical of flours with different gluten strengths [45]. The O-free flour displayed a distinct peak at approximately 2 min and 30 s, similar to that of the flour with strong gluten strength. However, the dough strength of the O-free flour weakened during continuous mixing and the width of the dough-mixing band narrowed, resembling the flour with weak gluten strength. Notably, the gluten network formed quickly and reached its maximum strength, but it collapsed and rapidly weakened when kneaded for a certain period.

Edwards et al. [46] and Khatkar et al. [47] analyzed the dough patterns of different flours with varying protein contents and quality. The wheat flour with strong gluten strength maintained a broad bandwidth even after reaching its maximum strength, while the wheat flour with weak gluten strength narrowed the bandwidth after reaching its maximum strength. These findings confirmed the effects of wheat flour quality and gluten protein content on dough properties. Furthermore, ω-gliadins were found to have the least positive effects on peak dough resistance in mixing characteristics and loaf volume [48].

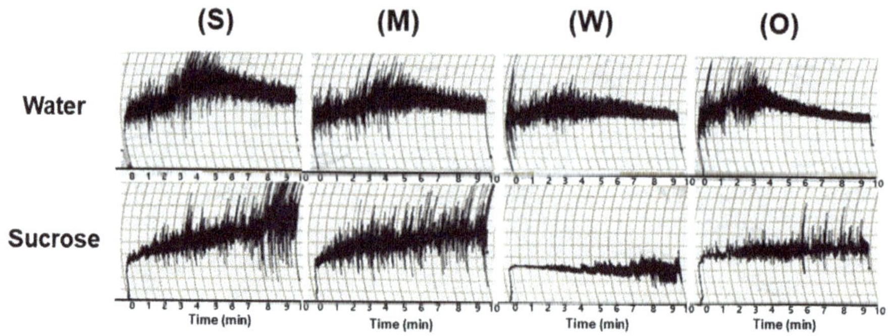

Figure 3. Mixograms of the flour with different gluten strengths and O-free flour in water and pre-dissolved 50% (*w*/*w*) sucrose solution. S, flour with strong gluten strength; M, flour with medium gluten strength; W, flours with weak gluten strength; O, O-free flour.

Comparing the mixograms of flour in water to those of the flours in sucrose solutions, we observe that the latter did not display any distinct peaks. This absence of peaks indicates a delay in gluten development. Additionally, the flours with strong and medium gluten strengths exhibited broader bandwidths when contrasted with the flours with weak gluten strength and O-free flour.

These results suggest that gluten development occurred more rapidly in the flours with strong and medium gluten strengths than in those with weak gluten strength. The accelerated development of gluten during cookie dough mixing resulted in reduced cookie spread and structural collapse, leading to smaller cookie diameters and greater cookie heights. Consequently, this variance in gluten development could potentially result in lower-quality cookies when using flours with stronger gluten characteristics. The quantity of glutenins and gliadins in the flour with weak gluten strength, along with the distinct gluten protein composition of O-free flour as illustrated in Figure 1, are crucial factors that likely influence their dough-mixing properties.

3.6. Cookie-Making Performance of Flour Samples

The top and side views of the cookies made using the various flour samples are depicted in Figure 4. The characteristics of the dough and cookies are shown in Table 5.

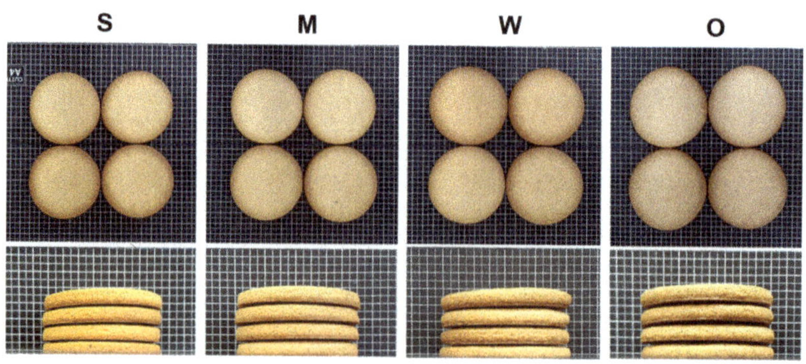

Figure 4. Top and side views of the cookies prepared using the flours with different gluten strengths and O-free flour: S, flour with strong gluten strength; M, flour with medium gluten strength; W, flour with weak gluten strength; O, O-free flour.

The width and length of the cookies followed the order: O-free flour ≥ flour with weak gluten strength > flour with medium gluten strength ≥ flour with strong gluten strength. This observation aligns with the predictions based on the solvent retention capacity (SRC) results of the flour samples (Table 2). Despite O-free flour having a higher protein content (Table 2), its gluten strength is relatively lower than that of flour with medium gluten strength, resulting in larger cookies. Typically, gluten imparts a viscoelastic property to dough by forming disulfide bonds between gliadin and glutenin [49,50]. O-free flour lacks ω-5 gliadin and has fewer low-molecular-weight glutenin subunits compared to the other three flours. In particular, ω-gliadin lacks sulfur and thiol groups, which are not involved in gluten development; however, low-molecular-weight glutenins contribute to gluten development [17,49]. Therefore, cookies prepared using O-free flour exhibited similar characteristics to those made with flour of weak gluten strength, indicating the suitability of O-free flour for producing high-quality cookies. The top surface colors of the cookies made with the different flour samples are also presented in Table 4. The cookies made with O-free flour did not significantly differ from those made with flour of weak gluten strength. A slight difference in b* of O-free cookies may be attributed to variations in protein composition and amino acids, leading to distinct color development through the Maillard reaction.

Table 5. Geometry, top surface color and firmness of the cookies prepared using the flours with different gluten strengths and O-free flour.

Flour [(1)]	Cookie Geometry (cm)			Top Surface Color			Cookie Firmness (N)
	Width	Length	Height	L*	a*	b*	
S	7.0 ± 0.1 [a(2)]	7.0 ± 0.6 [a]	0.94 ± 0.01 [a]	65.1 ± 0.1 [b]	12.5 ± 0.3 [ab]	32.8 ± 0.2 [b]	12.9 ± 0.1 [a]
M	7.1 ± 0.8 [a]	7.2 ± 0.3 [a]	1.06 ± 0.01 [b]	65.8 ± 0.2 [b]	12.4 ± 0.3 [a]	32.6 ± 0.3 [b]	14.3 ± 0.2 [b]
W	7.4 ± 0.6 [b]	7.5 ± 0.4 [b]	1.06 ± 0.01 [b]	62.2 ± 0.4 [a]	13.4 ± 0.2 [b]	32.4 ± 0.5 [b]	17.7 ± 2.0 [d]
O	7.5 ± 0.6 [b]	7.5 ± 1.0 [b]	1.16 ± 0.01 [c]	62.4 ± 0.1 [a]	13.1 ± 0.1 [ab]	31.0 ± 0.1 [a]	16.5 ± 1.3 [c]

[(1)] S, M, and W, flours with strong, medium, and weak gluten strength, respectively; O, O-free flour. [(2)] Results are expressed as the mean ± SD. Values with the same letter within the same column are not significantly different ($p < 0.05$), according to Tukey's HSD test.

The cookie firmness followed the order: flour with weak gluten strength > O-free flour > flour with medium gluten strength > flour with strong gluten strength. Generally, cookies made from wheat flour with low protein content are known to exhibit high firmness and excellent crispness due to high moisture baked out [33]. This suggests that O-free flour can be used to create crispy cookies.

4. Conclusions

This study explores the potential of using O-free wheat, a low-allergy wheat cultivar, in cookie production to meet the demand for allergen-free and reduced-allergen foods. The research highlights the unique gluten composition of O-free flour, showcasing its absence of ω-gliadins and reduced levels of low-molecular-weight glutenins and γ-gliadins. The flour's solvent retention capacity (SRC) places it between weak and medium gluten flours, emphasizing its distinctive quality. While thermal and pasting properties show similarities, mixograms reveal variations, indicating the critical role of gluten quality and composition in cookie dough characteristics. Cookies made with O-free flour closely resemble those produced with weak gluten flour, known for their appealing cookie attributes. In conclusion, this study supports using O-free flour for allergy-friendly cookies, providing a safe and tasty option for consumers with dietary needs.

Author Contributions: Conceptualization, N.B. and M.K.; data curation, N.B. and J.K.; formal analysis, N.B. and Y.M.; investigation, M.K.; funding acquisition, M.K.; methodology, N.B., Y.M. and J.K.; project administration, M.K.; resources, M.K.; supervision, M.K.; validation, N.B., Y.M. and J.K.; visualization, M.K.; writing—original draft preparation, N.B., Y.M. and J.K.; writing—review and editing, M.K. All authors have read and agreed to the published version of the manuscript.

Funding: This research was supported by the Cooperative Research Program of the Agriculture Science and Technology Department (Project No. PJ016031) funded by the Rural Development Administration (Republic of Korea).

Data Availability Statement: The data presented in this study are available upon request from the corresponding author.

Acknowledgments: We thank Chon-Sik Kang and Kyounghoon Kim for gifting the sample of O-free wheat flour.

Conflicts of Interest: The authors declare no conflict of interest.

References

1. Satitsuksanoa, P.; Jansen, K.; Głobińska, A.; Van de Veen, W.; Akdis, M. Regulatory immune mechanisms in tolerance to food allergy. *Front. Immunol.* **2018**, *9*, 2939. [CrossRef]
2. Tang, M.L.; Mullins, R.J. Food allergy: Is prevalence increasing? *Intern. Med. J.* **2017**, *47*, 256–261. [CrossRef]
3. Kang, C.S.; Yun, Y.M.; Choi, C.H.; Son, J.H.; Lee, J.Y.; Jeong, H.Y.; Kim, Y.J.; Cheong, Y.K.; Park, T.I. Gluten intolerance and reduced wheat-dependent exercise-induced anaphylaxis (WDEIA) wheat 'O-free'. In Proceedings of the 2019 Annual Meeting of the Korean Society of Crop Science, Jinju, Republic of Korea, 18–19 April 2019.
4. Lee, Y.G.; Choi, S.C.; Kang, Y.; Kang, C.S.; Kim, C. Genomics approach to identify the cause of the missing omega-5 gliadin protein in O-free wheat. *Plant Breed. Biotech.* **2018**, *6*, 413–425. [CrossRef]
5. Moon, Y.J.; Cho, H.J.; Baek, N.Y.; Xia, T.Y.; Kweon, M. Quality of low allergy wheat 'O-free' flour and its bread- and noodle-making performance. *Korean J. Food Cook. Sci.* **2021**, *37*, 492–500.
6. Xia, T.; Kim, K.; Kweon, M. Quality of low-allergy wheat ('O-free') flour and optimization of its bread- and noodle-making performance. *Foods* **2022**, *11*, 3399. [CrossRef] [PubMed]
7. Cho, H.J.; Kim, H.; Moon, Y.; Kim, J.; Kweon, M. Effects of water amount and mixing time on fresh noodle-making performance of low-allergy wheat 'O-free' flour. *Korean J. Food Cook. Sci.* **2022**, *38*, 367–375.
8. Slade, L.; Levine, H. Structure-function relationships of cookie and cracker ingredients. In *The Science of Cookie and Cracker Production*; Faridi, H., Ed.; Springer: New York, NY, USA, 1994; pp. 23–141.
9. Pareyt, B.; Delcour, J.A. The role of wheat flour constituents, sugar, and fat in low moisture cereal-based products: A review on sugar-snap cookies. *Crit. Rev. Food Sci. Nutr.* **2008**, *48*, 824–839. [CrossRef] [PubMed]
10. Kweon, M.; Slade, L.; Levine, H.; Martin, R.; Souza, E. Exploration of functionality of sugar functionality in sugar-snap and wire-cut cookie baking: Implications for potential sucrose replacement or reduction. *Cereal Chem.* **2009**, *88*, 537–552. [CrossRef]
11. Kweon, M.; Slade, L.; Levine, H.; Gannon, D. Cookie- versus cracker-baking-What's the difference? Flour functionality requirements explored by SRC and Alveography. *Crit. Rev. Food Sci. Nutr.* **2014**, *54*, 115–138. [CrossRef] [PubMed]
12. Pareyt, B.; Brijs, K.; Delcour, J.A. Sugar-snap cookie dough settting: The impact of sucrose on gluten functionality. *J. Agric. Food Chem.* **2009**, *57*, 7814–7818. [CrossRef]
13. Pareyt, B.; Bruneel, C.; Biijs, K.; Goesaert, H.; Delcour, J.A. Flour sodium dodecyl sulfate (SDS)-extractable protein level as a cookie flour quality indicator. *J. Agric. Food Chem.* **2010**, *58*, 353–360. [CrossRef] [PubMed]
14. Moiraghi, M.; Vanzetti, L.; Bainotti, C.; Helguera, M.; León, A.; Pérez, G. Relationship between soft wheat flour physicochemical composition and cookie-making performance. *Cereal Chem.* **2011**, *88*, 130–136. [CrossRef]
15. Zhang, X.; Zhang, B.; Wu, H.; Lu, C.; Lű, G.; Liu, D.; Li, M.; Jiang, W.; Song, G.; Gao, D. Effect of high-molecular-weight glutenin subunit deletion on soft wheat quality properties and sugar-snap cookie quality estimated through near-isogenic lines. *J. Integr. Agric.* **2018**, *17*, 1066–1073. [CrossRef]
16. Ma, F.; Kim, J.; Baik, B.K. Influences of high-molecular-weight glutenin subunits and rye translocations on dough-mixing properties and sugar-snap cookie-baking quality of soft winter wheat. *J. Sci. Food Agric.* **2020**, *100*, 3850–3856. [CrossRef] [PubMed]
17. Kang, C.S.; Jung, J.U.; Baik, B.K.; Park, C.S. Relationship between physicochemical characteristics of flour and sugar-snap cookie quality in Korean wheat cultivar. *Int. Food Res. J.* **2014**, *21*, 617–624.
18. Veraverbeke, W.S.; Delcour, J.A. Wheat protein composition and properties of wheat glutenin in relation to breadmaking functionality. *Crit. Rev. Food Sci. Nutr.* **2002**, *42*, 179–208. [CrossRef] [PubMed]
19. Wieser, H.; Koehler, P.; Scherf, K. Chemistry of wheat glutenin proteins: Qualitative composition. *Cereal Chem.* **2022**, *100*, 23–35. [CrossRef]
20. Uthayakumaran, S.; Tomoskozi, S.; Tatham, A.S.; Savage, A.W.J.; Gianibelli, M.C.; Stoddard, F.L.; Bekes, F. Effects of gliadin fractions on functional properties of wheat dough depending on molecular size and hydrophobicity. *Cereal Chem.* **2001**, *78*, 138–141. [CrossRef]
21. Barak, S.; Mudgil, D.; Khatkar, B.S. Effect of composition of gluten proteins and dough rheological properties on the cookie-making quality. *Br. Food J.* **2013**, *115*, 564–574. [CrossRef]
22. AACC. *Approved Method of Analysis*, 11th ed.; Method 10-53.01, 44-15.02, 46-30.01, 56-11.02; AACC International Inc.: St. Paul, MN, USA, 2010.
23. Lookhart, G.; Bean, S. Separation and characterization of wheat protein fractions by high-performance capillary electrophoresis. *Cereal Chem.* **1995**, *72*, 527–532.
24. Kweon, M.; Slade, L.; Levine, H. Potential sugar reduction in cookies formulated with sucrose alternatives. *Cereal Chem.* **2016**, *93*, 576–583. [CrossRef]

25. Van Hung, P.; Maeda, T.; Morita, N. Waxy and high-amylose wheat starches and flours—Characteristics, functionality and application. *Trends Food Sci. Technol.* **2006**, *17*, 448–456. [CrossRef]
26. Kweon, M.; Martin, R.; Souza, E. Effect of tempering conditions on milling performance and flour functionality. *Cereal Chem.* **2009**, *86*, 12–17. [CrossRef]
27. Choi, C.H.; Park, J.H.; Mo, Y.J.; Kim, K.M.; Kim, K.H.; Kang, C.S.; Park, T.I.; Cho, S.W.; Park, C.S. Effect of heterozygous genotypes at the Pinb-D1 loci on physicochemical properties of flour and end-use quality. *Korean J. Breed. Sci.* **2020**, *52*, 131–139. [CrossRef]
28. Park, J.H.; Yoon, Y.M.; Son, J.H.; Choi, C.H.; Kim, K.H.; Kim, K.M.; Cheong, Y.K.; Kang, C.S.; Yang, J. Enhancement of cookie quality by microwave treatment of allergy reaction-reduced "Ofree" wheat flour. *Korean J. Crop Sci.* **2019**, *64*, 366–372.
29. Ozturk, S.; Kahraman, K.; Tiftik, B.; Koksel, H. Predicting the cookie quality of flours by using Mixolab®. *Eur. Food Res. Technol.* **2008**, *227*, 1549–1554. [CrossRef]
30. Dupont, F.M.; Samoil, V.; Chan, R. Extraction of up to 95% of wheat (*Triticum aestivum*) flour protein using warm sodium dodecyl sulfate (SDS) without reduction or sonication. *J. Agric. Food Chem.* **2008**, *56*, 7431–7438. [CrossRef] [PubMed]
31. Dahesh, M.; Banc, A.; Duri, A.; Morel, A.H.; Ramos, L. Polymeric assembly of gluten proteins in an aqueous ethanol solvent. *J. Phys. Chem. B* **2014**, *118*, 11065–11076. [CrossRef]
32. Waga, J.; Skoczowski, A. Development and characteristics of ω-gliadin-free wheat genotypes. *Euphytica* **2014**, *195*, 105–116. [CrossRef]
33. Kweon, M.; Slade, L.; Levine, H. Solvent retention capacity (SRC) testing of wheat flour: Principles and value in predicting flour functionality in different wheat-based food processes in wheat breeding—A review. *Cereal Chem.* **2011**, *88*, 537–552. [CrossRef]
34. Jung, S.H.; Jeon, S.; Young, M.; Moon, Y.J.; Hong, Y.E.; Kweon, M. Quality of cookies formulated with Jeju magma seawater. *Korean J. Food Cook. Sci.* **2017**, *33*, 292–299. [CrossRef]
35. Gaines, C.S. Collaborative study of methods for solvent retention capacity profiles (AACC Method 56-11). *Cereal Foods World* **2000**, *45*, 303–306.
36. Faridi, H.; Gaines, C.; Finney, P. Soft wheat quality in production of cookies and crackers. In *Wheat: Production, Properties and Quality*; Bushuk, W., Rasper, V.F., Eds.; Springer: New York, NY, USA, 1994; pp. 154–168.
37. Jung, E.J.; Moon, Y.; Kweon, M. Processing suitability of healthful carbohydrates for potential sucrose replacement to produce muffins with staling retardation. *LWT-Food Sci. Technol.* **2020**, *131*, 109565.
38. Allan, C.; Rajwa, B.; Mauer, L.J. Effects of sugars and sugar alcohols on the gelatinization temperature of wheat starch. *Food Hydrocoll.* **2018**, *84*, 593–607. [CrossRef]
39. Renzetti, S.; van den Hoek, I.A.; van der Sman, R.G. Mechanisms controlling wheat starch gelatinization and pasting behaviour in presence of sugars and sugar replacers: Role of hydrogen bonding and plasticizer molar volume. *Food Hydrocoll.* **2021**, *119*, 106880. [CrossRef]
40. Young, M.; Jeon, S.; Kweon, M. Study on applicability of allulose as a sucrose replacer in cookie making. *J. East Asian Soc. Diet. Life* **2016**, *26*, 450–456. [CrossRef]
41. Suhendro, E.L.; Kunetz, C.F.; McDonough, C.M.; Rooney, L.W.; Waniska, R.D. Cooking characteristics and quality of noodles from food sorghum. *Cereal Chem.* **2000**, *77*, 96–100. [CrossRef]
42. Batey, I.L. Interpretation of RVA curves. In *The RVA Handbook*; Crosbie, G.B., Ross, A.S., Eds.; AACC International Inc.: St. Paul, MN, USA, 2007; pp. 19–30.
43. Isaak, C.; Sapirstein, H.; Wu, Y.; Graf, R. Effects of water absorption and salt on discrimination of wheat gluten strength assessed by dough mixing and protein composition. *J. Cereal Sci.* **2019**, *89*, 102752. [CrossRef]
44. Kovacs, M.I.P.; Fu, B.X.; Woods, S.M.; Khan, K. Thermal stability of wheat gluten protein: Its effect on dough properties and noodle texture. *J. Cereal Sci.* **2004**, *39*, 9–19. [CrossRef]
45. Finnie, S.; Atwell, W.A. Composition of commercial flour. In *Wheat Flour*; Finnie, S., Atwell, W.A., Eds.; AACC International Inc.: St. Paul, MN, USA, 2016; pp. 31–42.
46. Edwards, N.M.; Gianibelli, M.C.; McCaig, T.N.; Clarke, J.M.; Ames, N.P.; Larroque, O.R.; Dexter, J.E. Relationships between dough strength, polymeric protein quantity and composition for diverse durum wheat genotypes. *J. Cereal Sci.* **2007**, *45*, 140–149. [CrossRef]
47. Khatkar, B.S.; Bell, A.E.; Schofield, J.D. A comparative study of the inter-relationships between mixograph parameters and bread-making qualities of wheat flours and glutens. *J. Sci. Food Agric.* **1996**, *72*, 71–85. [CrossRef]
48. Khatkar, B.S.; Fido, R.J.; Tatham, A.S.; Schofield, J.D. Functional properties of wheat gliadins. I Effects on mixing characteristics and bread making quality. *J. Cereal Sci.* **1996**, *35*, 299–306. [CrossRef]
49. Day, L.; Augustin, M.A.; Batey, I.L.; Wrigley, C.W. Wheat-gluten uses and industry needs. *Trends Food Sci. Technol.* **2006**, *17*, 82–90. [CrossRef]
50. Liu, R.; Sun, W.; Zhang, Y.; Huang, Z.; Hu, H.; Zhao, M.; Li, W. Development of a novel model dough based on mechanically activated cassava starch and gluten protein: Application in bread. *Food Chem.* **2019**, *300*, 125196. [CrossRef]

foods

MDPI

Article

Development of Oleogel-Based Fat Replacer and Its Application in Pan Bread Making

Sung-Huo Kim [1,2], Yeon-Ji Jo [2,3], Sung Ho Lee [4] and Sung-Hoon Park [1,2,*]

1 Department of Food & Nutrition, Gangneung-Wonju National University, Gangneung 25457, Republic of Korea; rlatjdgn0616@gwnu.ac.kr
2 Haeram Institute of Bakery Science, Gangneung-Wonju National University, Gangneung 25457, Republic of Korea
3 Department of Marin Bio Food Science, Gangneung-Wonju National University, Gangneung 25457, Republic of Korea; joyeonji@gwnu.ac.kr
4 SPC Group Research Institute of Food and Biotechnology, Seoul National University, Seoul 08826, Republic of Korea; sungholee@spc.co.kr
* Correspondence: sungpark@gwnu.ac.kr; Tel.: +82-33-640-2332; Fax: +82-33-640-2331

Abstract: In recent years, the bakery industry has been exploring alternative fats to replace traditional solid fats. Shortening, a common baking ingredient, is produced through the hydrogenation of vegetable oils, resulting in high levels of saturated and trans fatty acids, despite its vegetable oil origin. The excessive consumption of these fats has been associated with negative health effects, including dyslipidemia and cardiovascular issues. Oleogels, incorporating hydroxypropyl methylcellulose (HPMC), xanthan gum (XG), and olive oil, were utilized to replace shortening in the production of white pan bread. The substitution of shortening with oleogel in the white pan bread preparation demonstrated potential reductions in saturated fat, trans fat, and the ratio of saturated fat to unsaturated fatty acids. Specifically, with the complete substitution of shortening with oleogel, saturated fatty acids decreased by 52.46% and trans fatty acids by 75.72%, with unsaturated fatty acids increasing by 57.18%. Our findings revealed no significant difference in volume between bread made with shortening and bread with up to 50% shortening substitution. Moreover, when compared to bread made with shortening and 50% oleogel substitution, no adverse effects on the quality characteristics of volume and expansion properties were observed, and the retrogradation rate was delayed. This study suggests that incorporating oleogels, formed with hydrocolloids such as HPMC and XG, to replace shortening in bread, in conjunction with traditional solid fats, provides positive effects on the quality and nutritional aspects of the bread compared to using oleogel alone. Through this study, we demonstrate the use of oleogels as a healthier alternative to shortening, without reducing the bread's quality, thus offering a practical solution to reduce unhealthy fats in bakery products.

Keywords: oleogels; shortening; white pan bread; bakery products; hydrocolloids

1. Introduction

Numerous ingredients and additives play crucial roles in the baking process. Among them, fats such as butter, shortening, or margarine are particularly important elements that significantly influence the texture, structure, freshness, and flavor of the finished product. Shortening, a semisolid, plastic fat, is composed entirely of fats derived from refined vegetable oils. It can be produced by hydrogenation and also interesterification. Its primary functions include aiding air entrapment, providing a smooth texture, and contributing to a rich flavor [1].

In yeast-leavened products, a tough texture is commonly observed due to the cohesion of starch particles and gluten proteins. Shortening plays a pivotal role in this context by disrupting the protein and starch structures, lubricating the gluten, and ultimately resulting in soft bakery products [2].

However, the shortening and margarine commonly used in the baking process contain high levels of saturated fats and trans fats, which can have negative effects on human health. The excessive consumption of trans fatty acids, in particular, is reported to increase low-density lipoprotein cholesterol levels in the body, negatively affecting cholesterol metabolism and posing a risk of diabetes and cardiovascular diseases [3–5]. Moreover, the use of vegetable oils in baking, while rich in unsaturated fatty acids, can lead to reduced product processability and storage stability [6].

In recent years, there has been a concerted effort to explore alternatives that allow for a reduction in fat content while preserving product texture and quality without significant changes in chemical composition [7]. One notable avenue in these efforts involves research on using oleogels as a substitute for shortening. Oleogels, characterized as thermally reversible, three-dimensional structures exhibiting viscoelastic behavior, trap organic liquids using low-molecular-weight and high-molecular-weight oleogelators. In comparison to liquid oils, they provide enhanced oxidation stability, improved texture, spreadability, increased processing suitability, and extended storage stability, making them a promising choice for fat replacement in the food industry [8–10].

Hydrocolloids, water-soluble polysaccharides with various functional properties, are widely used in food technology. A recent innovative method has been proposed to produce oleogels using hydroxypropyl methylcellulose (HPMC) for applications in baking products [5]. Hydrocolloids induce structural changes in the major components of flour during both the processing and storage stages of the baking process, notably enhancing storage stability and quality [11–14].

HPMC, a versatile polymer, reduces interfacial tension in both hydrophilic and hydrophobic liquid interfaces [15,16]. Notably, cellulose derivatives including HPMC, obtained by adding methyl and hydroxypropyl groups to cellulose, exhibit uniform properties compared to the variable nature of hydrocolloids from natural sources. Additionally, HPMC retains partial hydrophilicity of cellulose despite the presence of hydrophobic groups [12,17].

In baking, cellulose derivatives like HPMC are reported to enhance moisture absorption and impede protein binding and the aggregation of starch and gluten, thereby improving product quality by reducing retrogradation, increasing product volume, and providing a soft crumb texture [18–22].

Xanthan gum (XG) is another hydrocolloid commonly used as a thickening agent, stabilizer, and, in baking, as a dough enhancer and quality improver in gluten-free bakery products, within the food industry [23–25]. The inclusion of gums is known to augment the viscosity of starch and affect starch gelatinization and retrogradation [26]. Moreover, previous research has confirmed that using XG and gelatin in the formation of oleogels contributes to the stabilization of emulsions through protein–polysaccharide interactions at the oil–water interface [27]. In other studies, the combination of XG and HPMC in oleogel formation was observed to result in the formation of stable emulsions during oleogel formation and demonstrated a degree of stability allowing for moisture removal during the drying process [28].

In this study, oleogels composed of olive oil, HPMC, and XG were formulated to assess their potential and applicability as shortening substitutes in bread production. It is anticipated that the findings will contribute to supporting better food choices for consumers and fostering advancements in the realm of alternative ingredients in the baking industry.

2. Materials and Methods

2.1. Materials

For the preparation of oleogels, the following ingredients were used: olive oil (extra virgin, Beksul CJ Co. Ltd., Incheon, Republic of Korea), HPMC (AN6, ES Food, Seongnam-si, Republic of Korea), and XG (80MESH, ES Food, Seongnam-si, Republic of Korea). In the preparation of white pan bread, the ingredients included 100 g of wheat flour (Mildawon Co. Inc., Gongju, Republic of Korea), 6 g of refined sugar (CJ CO. Inc., Seoul, Republic of

Korea), 6 g of shortening (Samyangsa Co. Inc., Incheon, Republic of Korea), 1.5 g of salt, and 6 g of yeast (Saf-instant gold, Lesaffre Pte Ltd., Marcq-en-Baroeul, France).

2.2. Preparation of Oleogel

The oleogels were prepared with modifications to the methods described previously [29,30]. Initially, 2 g of HPMC was dispersed in 76.8 g of cold water and mixed using a food processor (Vitamix E310, Vita-Mix, Cleveland, OH, USA). Subsequently, 1.2 g of XG was added to the resulting aqueous solution and mixed for 5 min. Following this, 120 g of oil was gradually added and homogenized in the processor for 5 min. The resulting emulsions were spread on a Teflon tray using a silicone pastry bag and a plastic nozzle. Drying of the emulsions was achieved using forced convection air in a dry oven (WOF-W155, DAIHAN Scientific Co., Ltd., Wonju, Republic of Korea) at 80 °C for 4 h. This duration was determined as the minimum time required to reach a constant dry weight, maintaining a moisture content of 1.75% ± 0.51%, under the specified conditions. Finally, the dried products were ground in a grinder for 4 s to produce the oleogels.

2.3. Preparation of White Pan Bread with Oleogel

White pan bread was prepared using a slightly modified AACC (American Association of Cereal Chemists) standard straight dough method [31]. The ingredients were initially mixed at low speed for 3 min and then at high speed for 7 min using a SK-20 Mixer (SKMixer Co., LTD. Saitama, Japan). The mixed dough underwent a fermentation period of 1 h at room temperature (25 °C) with a humidity of 75–80%, followed by dividing the dough into 380 g portions, setting, shaping, and proofing at 38 °C with humidity ranging between 85% and 95% for 50 min. Subsequently, the bread was baked at 175 °C for 35 min. After baking, the finished product was cooled to an internal temperature of either 25 °C or 35 °C and then packaged in a polyethylene bag. The experimental groups were created by substituting 0% (S100), 50% (S50/O50), and 100% (O100) oleogel instead of shortening. All experiments were conducted in triplicate.

2.4. Proximate Analysis of Oleogel Bread

The moisture content of oleogel bread was analyzed using an infrared moisture determination balance (FD-720, KETT Electric Laboratory, Tokyo, Japan). For each sample, 3 g was placed on an 8 cm diameter aluminum foil dish, and measurements were taken at 105 °C for 8 min. The determination of ash, crude protein, and crude fat content in the oleogel bread samples followed the methods recommended by AOAC International [32]. General component analysis of the oleogel bread involved determining the carbohydrate content, calculated by subtracting the combined amounts of moisture, crude protein, crude fat, and ash from 100 g of the sample. The results of the general component analysis in food are typically expressed as percentages [33].

2.5. Baking Loss Measurement

The baking loss was calculated using the following equation, which represents the ratio of the weight difference between the dough before and after baking:

$$\text{Baking Loss }(\%) = \frac{\text{Before Baking(g)} - \text{After Baking (g)}}{\text{Before Baking (g)}} \times 100 \quad (1)$$

2.6. Specific Volume of Oleogel Bread

The specific volume of white pan bread was measured using a laser scanning system (Volscan Profiler-VSP600, Stable Micro System Ltd., Surrey, UK) following the AACC-approved method (10.00–16.01). Initially, samples of white pan bread were prepared for measurement.

2.7. Crumb Structure Analysis of Oleogel Bread

The crumb structure analysis of the bread followed the method outlined in [34]. Sliced bread, with a thickness of 18 mm, underwent scanning using an image scanner (Samsung SL-C1454FW, Republic of Korea), and subsequent analysis was performed using ImageJ analysis software version 1.49 (NIH, Bethesda, MD, USA). The color image was converted into an 8-bit grayscale binary image, and the image resolution was set to 300 dpi. A field of view measuring 30 × 30 mm from the center of the slice was selected and segmented using a threshold to determine metrics such as cell density (cells/cm^2), average cell area (ratio of pore area to the number of cells), and porosity (ratio of pore area to 3 cm^2).

2.8. Color Analysis of Oleogel Bread

The crust and crumb colors of the white pan bread were assessed using a chromameter (CR-400 KONICA MINOLTA, Tokyo, JAPAN). Color values were quantified in terms of L* (lightness/darkness), a* (redness/greenness), and b* (yellowness/blueness). The browning index was employed to illustrate the color alteration during the baking process [35].

2.9. Texture Profile Analysis of Oleogel Bread

Texture profile analysis (TPA) was performed on 18 mm thick slices of bread using a CTX Texture Analyzer (AMETEK Brookfield, MA, USA). The TPA utilized a cylindrical probe with a diameter of 50 mm, a crosshead speed of 60 mm/min, and a strain of 50%. Measurements were taken on day 1, 2, and 4. The texture parameters determined were hardness (N), springiness (mm), cohesiveness, and chewiness (J).

2.10. Fatty Acid Composition Analysis of Oleogel Bread

The fatty acid composition of bread prepared using oleogel was determined following the AOAC 996.06 method [36], with triundecanoin (C11:0) serving as an internal standard. The procedure involved acid hydrolysis using 8.3 M HCl for fat extraction. The homogenized samples underwent acid hydrolysis, followed by fat extraction using organic solvents (diethyl ether and petroleum ether). Subsequently, methylation was carried out by adding 0.5 N NaOH in MeOH and 14% BF3 in MeOH. Fatty acid methyl ester analysis was conducted using GC-FID (Hewlett-Packard 6890, Agilent Technologies, Palo Alto, CA, USA) with an SPTM-2560 capillary column (100 m × 0.25 mm ID × 0.2 μm thickness, Supelco, Bellefonte, PA, USA). Helium served as the carrier gas at a flow rate of 0.75 mL/min, with a split ratio of 200:1. The injector temperature was maintained at 225 °C, and the detector temperature at 285 °C. The oven temperature programming included an initial temperature of 100 °C held for 4 min, followed by a gradual increase at a rate of 3 °C/min up to 240 °C, where it was maintained for 15 min. The obtained data were compared to bread prepared using shortening.

2.11. Statistical Analysis

Statistical analysis was performed using the SPSS statistics program, version 25 (IBM, New York, NY, USA). To identify significant differences ($p < 0.05$), a Tukey's multiple-comparisons test within a one-way analysis of variance was employed. The reported values are presented as mean ± standard deviation, with distinct letters indicating significant differences ($p < 0.05$). The data reported in all of the tables are average values of triplicates.

3. Results and Discussion

3.1. Physicochemical Quality Characteristics of Oleogel Bread

The external structure and physicochemical quality characteristics of the fat-substituted bread are presented in Tables 1 and 2, and Figure 1. Volume measurement is a crucial factor in evaluating bread quality [37]. Specific volume analysis revealed no significant difference between the S100 (5.32) and S50/O50 (5.35), whereas O100 (4.85) exhibited a lower value. Additionally, the baking loss rate analysis showed no significant difference between S100 (15.83) and S50/O50 (15.74), but O100 (13.42) exhibited a notably higher difference with increased replacement. The moisture content showed no significant difference between S100 (33.50) and S50/O50 (33.63), but O100 (35.26) displayed a significantly higher moisture content.

Table 1. Physicochemical quality characteristics of 0% shortening substitution (S100), 50% shortening substitution (S50/O50), and 100% shortening substitution (O100) of white pan bread prepared with shortening and oleogels.

	S100	S50/O50	O100
Baking loss (%)	15.83 ± 0.07 [a]	15.74 ± 0.08 [a]	13.42 ± 0.22 [b]
Specific volume (ml/g)	5.32 ± 0.02 [a]	5.35 ± 0.10 [a]	4.85 ± 0.02 [b]
Cell density (cells/cm^2)	29.52 ± 0.24 [a]	29.04 ± 0.17 [a]	28.18 ± 0.17 [b]
Average cell area (mm^2)	0.54 ± 0.03 [c]	0.66 ± 0.02 [b]	1.08 ± 0.02 [a]
Porosity (%)	16.03 ± 0.78 [c]	19.16 ± 0.38 [b]	30.53 ± 0.50 [a]

All experiments were performed in triplicate. Data are shown as means ± standard deviation. Values with different letters in the same column differ significantly ($p < 0.05$).

Table 2. Proximate analysis of 0% shortening substitution (S100), 50% shortening substitution (S50/O50), and 100% shortening substitution (O100) of white pan bread prepared with shortening and oleogels.

g/100 g	S100	S50/O50	O100
Crude protein	10.08 ± 0.03 [a]	10.09 ± 0.06 [a]	9.92 ± 0.05 [b]
Crude fat	4.75 ± 0.35 [a]	5.14 ± 0.11 [a]	4.81 ± 0.01 [a]
Ash	1.31 ± 0.16 [a]	1.46 ± 0.01 [a]	1.52 ± 0.01 [a]
Moisture content (%)	33.50 ± 0.26 [a]	33.63 ± 0.43 [a]	35.26 ± 0.29 [b]
Carbohydrate	52.10 ± 0.44 [a]	51.70 ± 0.17 [ab]	50.69 ± 0.53 [b]
Calorie	291.70 ± 1.61 [a]	293.30 ± 0.58 [a]	286.70 ± 0.23 [b]

All experiments were performed in triplicate. Data are shown as means ± standard deviation. Values with different letters in the same column differ significantly ($p < 0.05$).

Figure 1. The general shape and appearance of pan bread containing oleogels: (S100) control; (S50/O50) oleogel with 50% shortening substitution; (O100) oleogel with 100% shortening substitution.

These results suggest that substituting with 50% oleogel does not have a major impact on the volume properties and moisture content of the bread. However, O100 results in increased moisture content and decreased volume characteristics. These results may be attributed to the higher content of XG added with increasing levels of oleogel, which, as indicated in [12,38], holds excess moisture and prevents proper oven spring, leading to decreased volume characteristics.

3.2. Crumb Structure of Oleogel Bread

The characteristics of the internal structure, such as the crumb structure of bread with oleogel substitution, are shown in Table 1 and Figure 2. The size and number of cells inside the crumb significantly influence the texture and sensory results of the bread [34,39]. The stability of the gluten–starch matrix plays a crucial role in preventing unevenness and adhesion within expanding gas cells. Ensuring the stability of gas cells is contingent upon the ability of the dough to prevent collapse [40]. The analysis of the crumb structure of the bread reveals a consistent pattern with increasing levels of oleogel replacement. Specifically, higher oleogel replacement levels correspond to an increase in the average cell area (total area/cells) and porosity (total area/cm^2), along with a decrease in cell counts and cell density. The porosity percentages were 0.15% for S100, 0.19% for S50/O50, and 0.30% for O100. This indicates that greater oleogel substitution led to significant alterations in crumb structure, influencing the cell size, density, and porosity.

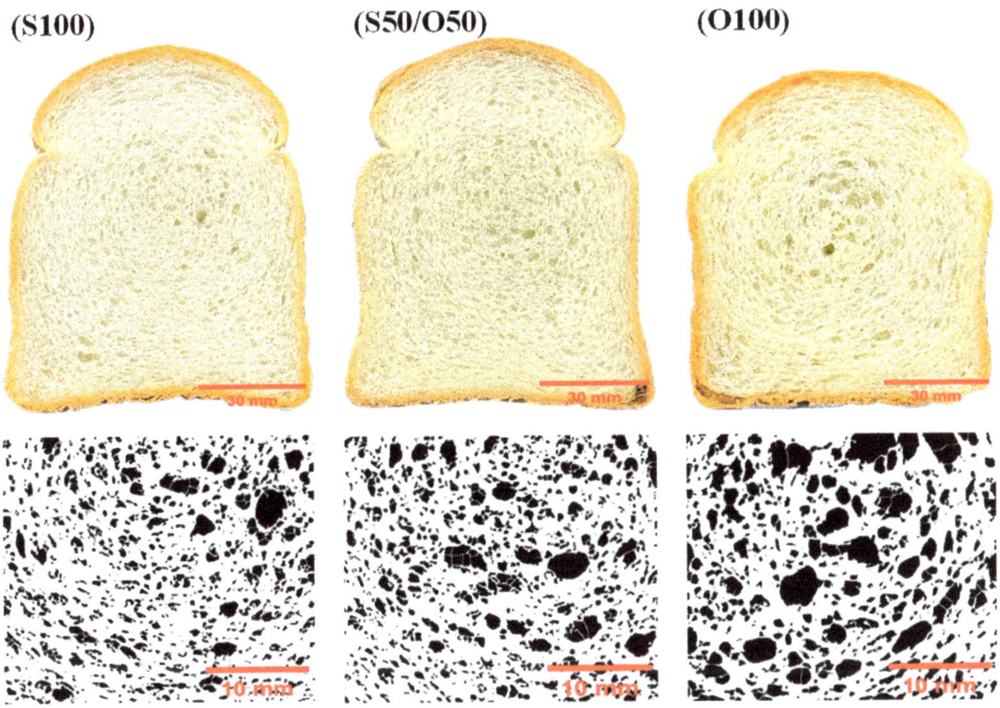

Figure 2. Cross sectional image analysis of crumb structure with different oleogel contents: (S100) control; (S50/O50) oleogel with 50% shortening substitution; (O100) oleogel with 100% shortening substitution.

The observed decrease in volume may be attributed to the substantial amount of XG, which enhances dough elasticity and viscosity, as reported in previous studies [38,41]. The increased dough elasticity and firmness hinder gas cell expansion, resulting in thicker and harder air cell walls. This affects the stability and structure of the gluten–starch matrix in the crumb, leading to fewer mall cells and a more even crumb structure, consistent with findings in the literature [42,43]. Consequently, the addition of XG results in a more open crumb structure, as demonstrated by a previous study [44]. These findings suggest that similar effects on crumb structure are observed even with 100% oleogel replacement in bread production.

3.3. Color Analysis of Oleogel Bread

The color analysis of oleogel bread examined the impact of replacing shortening with oleogel on the bread's color (Table 3). Comparing the S100 group with the S50/O50 and O100 groups, a decrease in the L* value of the crust was observed in the S100. However, visual examination (Figure 1) did not reveal distinct differences in crust color.

Table 3. Color analysis of 0% shortening substitution (S100), 50% shortening substitution (S50/O50), and 100% shortening substitution (O100) of white pan bread prepared with shortening and oleogels.

		S100	S50/O50	O100
Crust	L*	61.70 ± 0.21 b	62.50 ± 0.01 a	57.60 ± 0.30 c
	a*	16.47 ± 0.02 b	16.41 ± 0.05 b	18.61 ± 0.08 a
	b*	30.89 ± 0.07 b	31.63 ± 0.09 a	29.25 ± 0.01 c
	BI (Browning index)	85.99	86.71	91.32
Crumb	L*	83.16 ± 0.06 b	83.83 ± 0.11 a	80.48 ± 0.34 c
	a*	2.53 ± 0.03 a	2.42 ± 0.02 b	2.11 ± 0.01 c
	b*	9.61 ± 0.08 c	10.45 ± 0.04 b	12.67 ± 0.13 a

All experiments were performed in triplicate. Data are shown as means ± standard deviation. Values with different letters in the same column differ significantly ($p < 0.05$).

In the crumb, except for the b* value, the L* and a* values showed similar results between the S100 group and the S50/O50 group. However, an increase in oleogel substitution was associated with a significant increase in the b* value, indicating a stronger yellow color. The O100 group exhibited a significant decrease in L* and a* values, while the b* value showed the highest increase. This trend was also evident in the bread's appearance (Figure 2), where an increase in oleogel replacement was linked to a more pronounced yellow color in the crumb.

The effect of oleogel on bread color was shown by a decrease in the L* value in the crust upon 100% replacement and a significant increase in the b* value in the crumb, indicating a more pronounced yellow color.

These results may be attributed to the color of the added olive oil, which becomes more pronounced with an increase in the replacement amount of oleogel. Moreover, it is worth noting that the fatty acid composition of oleogel bread, differs from that of traditional shortening-based bread. While shortening typically has a white color and is primarily composed of 16:0 saturated fatty acids, oleogels produced from olive oil have a higher content of 18:1 unsaturated fatty acids. This distinction in fatty acid composition may contribute to the observed differences in color parameters.

3.4. *Texture Profile Analysis of Oleogel Bread*

The bread, with shortening replaced by oleogel, was preserved for 4 days, during which texture measurements were conducted. The results are shown in Table 4. A consistent increase in hardness during storage was observed across all groups, irrespective of oleogel substitution. Specifically, the O100 group exhibited the highest hardness measurement (8.43) on day 1, and this trend continued on the fourth day, with O100 recording the highest measurement (14.00). These results stem from the inclusion of XG, as explained by [38] previously, during the production of bread by completely replacing oleogel. This addition of XG leads to increased moisture content, reduced oven spring, and the weakening of the gluten structure. Additionally, solid fats play a vital role in stabilizing bubbles at the air–matrix interface [45]. When shortening is entirely replaced with oleogel, the challenges arise in replicating the mechanism of stabilizing bubbles, a characteristic of solid fats. This could accelerate the reduction in expansion and weakening of the gluten structure, resulting in higher hardness and lower volume.

On the contrary, in the S50/O50 group, retrogradation was delayed the most. In the S100, hardness increased by 99.61% (5.16) on day 1 (5.18) and day 4 (10.34). In contrast, the S50/O50 group exhibited an increase of 77.95% (3.33) on the day 1 (4.40) and day 4 (7.83). These findings indicate that the S50/O50 showed a 35.46% (1.83) delayed rate of hardness increase compared to the S100. Moreover, the substitution of 50% of shortening with oleogel not only led to decreased bread hardness, but also suggested delayed retrogradation. Notably, the S50/O50 group showed lower values of hardness and chewiness, although their specific volume and moisture content values did not differ significantly from those of the S100. These results indicate a positive effect of oleogel on the product without compromising quality when replacing 25–50% of solid fat [5,46]. Additionally, these results indicate that the characteristics of HPMC hydrocolloid, known for reducing bread hardness, manifest when applied up to 50% in the production of bread with oleogel. Substituting oleogel with other baking fats, rather than using oleogel alone, appears to have positive effects on hardness and retrogradation. Although springiness and cohesiveness did not show significant differences with the addition of oleogel, chewiness exhibited notable differences (S100: 34.54, S50/O50: 29.17). Considering that hardness and chewiness are negatively correlated with bread quality, these observations are consistent with previous findings on the effects of bread quality [47–49]. Future studies should explore the mechanisms by which oleogels influence these textural properties to further understand their potential benefits in baking applications.

Table 4. Texture profile analysis of 0% shortening substitution (S100), 50% shortening substitution (S50/O50), and 100% shortening substitution (O100) of white pan bread prepared with shortening and oleogels.

	S100	S50/O50	O100
Hardness, day 1 (N)	5.18 ± 0.03 [b]	4.40 ± 0.02 [c]	8.43 ± 0.04 [a]
Hardness, day 2 (N)	7.82 ± 0.04 [b]	5.86 ± 0.04 [c]	12.23 ± 0.01 [a]
Hardness, day 4 (N)	10.34 ± 0.02 [b]	7.83 ± 0.07 [c]	14.00 ± 0.11 [a]
Springiness, day 1 (mm)	9.64 ± 0.01 [a]	9.71 ± 0.02 [a]	9.64 ± 0.20 [a]
Cohesiveness, day 1	0.67 ± 0.01 [a]	0.68 ± 0.01 [a]	0.65 ± 0.01 [a]
Chewiness, day 1 (J)	34.54 ± 0.45 [b]	29.17 ± 0.65 [c]	53.08 ± 0.49 [a]

All experiments were performed in triplicate. Data are shown as means ± standard deviation. Values with different letters in the same column differ significantly ($p < 0.05$).

3.5. Fatty Acid Composition of Oleogel Bread

The comparison of fatty acid compositions in breads produced by replacing shortening with oleogel is shown in Table 5. The primary fatty acid composition of olive oil consists of 21.9% saturated fatty acids and 78.1% unsaturated fatty acids, with oleic acid (18:1, 68.4%) being the predominant fatty acid [50]. In the O100 group bread, oleic acid (18:1, 61.24) is the predominant component, with saturated fatty acids at 23.80%, unsaturated fatty acids at 75.76%, and trans fatty acids at 0.42%. While these may be slight variations based on olive oil types, the fatty acid composition of this study's olive oil, used in oleogel and bread preparation, did not show significant changes.

For the S100 group bread, saturated fatty acids comprise 50.06%, unsaturated fatty acids account for 48.20%, and trans fatty acids make up 1.73%. In contrast, the O100 group showed a 52.46% decrease in saturated fatty acids (to 23.80%), a 75.72% decrease in trans fatty acids (to 0.42%), and a 57.18% increase in unsaturated fatty acids (to 75.76%) compared to the S100 group. Consequently, when applying oleogel prepared using HPMC, XG, and olive oil as a replacement for shortening in bread production, it is shown that a nutritionally superior bread with a high level of unsaturated fatty acids, along with reduced levels of saturated and trans fatty acids, can be produced without significant alterations to the overall fatty acid composition. These results provide insights into how the type of oil and fatty acid composition used in oleogel preparation influence the fatty acid composition of bread. Furthermore, this understanding can enhance the potential benefits of incorporating oleogels into bakery products.

Table 5. Fatty acid compositions of 0% shortening substitution (S100), 50% shortening substitution (S50/O50), and 100% shortening substitution (O100) of white pan bread prepared with shortening and oleogels.

Fatty Acid (g/100 g)	S100	S50/O50	O100
C4:0	0.38 ± 0.00 [a]	0.39 ± 0.01 [a]	0.42 ± 0.02 [a]
C6:0	0.26 ± 0.01 [a]	0.25 ± 0.01 [a]	0.26 ± 0.01 [a]
C8:0	0.20 ± 0.01 [a]	0.18 ± 0.01 [b]	0.17 ± 0.01 [c]
C10:0	0.41 ± 0.00 [a]	0.38 ± 0.01 [b]	0.37 ± 0.01 [c]
C12:0	0.92 ± 0.01 [a]	0.71 ± 0.01 [b]	0.52 ± 0.01 [c]
C14:0	2.51 ± 0.00 [a]	1.90 ± 0.01 [b]	1.31 ± 0.01 [c]
C16:0	37.66 ± 0.02 [a]	26.68 ± 0.20 [b]	15.78 ± 0.15 [c]
C16:1	0.82 ± 0.01 [c]	0.88 ± 0.01 [b]	0.96 ± 0.01 [a]
C18:0	7.43 ± 0.00 [a]	5.96 ± 0.04 [b]	4.61 ± 0.10 [c]
Trans C18:1	1.35 ± 0.01 [a]	0.83 ± 0.03 [b]	0.33 ± 0.03 [c]
C18:1	33.44 ± 0.02 [c]	47.48 ± 0.29 [b]	61.24 ± 0.15 [a]
Trans C18:2	0.38 ± 0.01 [a]	0.24 ± 0.00 [b]	0.09 ± 0.00 [c]
C18:2	13.25 ± 0.04 [a]	12.86 ± 0.01 [b]	12.42 ± 0.09 [c]
C18:3	0.49 ± 0.01 [c]	0.68 ± 0.01 [b]	0.88 ± 0.01 [a]
C20:0	0.29 ± 0.01 [c]	0.33 ± 0.01 [b]	0.37 ± 0.00 [a]
C20:1	0.21 ± 0.00 [c]	0.23 ± 0.01 [b]	0.26 ± 0.01 [a]
SFA (%)	50.06 ± 0.03 [a]	36.79 ± 0.29 [b]	23.80 ± 0.21 [c]
USFA (%)	48.20 ± 0.05 [c]	62.13 ± 0.30 [b]	75.76 ± 0.25 [a]
TFA (%)	1.73 ± 0.01 [a]	1.07 ± 0.03 [b]	0.42 ± 0.03 [c]
SFA/USFA	1.04 ± 0.00 [a]	0.59 ± 0.01 [b]	0.31 ± 0.01 [c]

All experiments were performed in triplicate. Data are shown as means ± standard deviation. Values with different letters in the same column differ significantly ($p < 0.05$).

4. Conclusions

In this study, oleogel was produced by utilizing the hydrocolloid properties of HPMC and XG, combined with unsaturated fatty acid-rich, plant-based olive oil, to replace shortening in bread formulation. The incorporation of HPMC and XG during oleogel production exhibited intrinsic properties that enhanced crumb softening and dough stability, preserving the hydrocolloid attributes even when used in the production of oleogel. In particular, a 50% replacement revealed reduced hardness and retrogradation compared to the control group. Additionally, substituting shortening with 100% oleogel produced using the developed methodology resulted in a significant reduction in saturated and trans fatty acids, enabling the production of bread rich in unsaturated fatty acids. Substituting up to 50% of shortening with oleogel resulted in high-quality bread with reduced levels of saturated and trans fatty acids, without a significant loss of volume or texture compared to the control group. The adjustment of oleogel content facilitated the control of saturated and trans fatty acid content while enhancing bread quality. Challenges associated with fat replacers and oleogels encompass volume reduction, texture alterations, and sensory impacts. However, through the incorporation of HPMC and XG to create oleogel, the appropriate application in bread formulation led to the production of improved bread quality with controlled levels of saturated and trans fatty acids, thereby enhancing both preference and nutritional aspects.

In conclusion, this study demonstrates that utilizing oleogel produced with HPMC and XG offers the potential to create high-quality bread with a controlled fatty acid composition and improved sensory and nutritional attributes, providing an effective approach to tackle the challenges of fat replacers and oleogels in bread formulation. However, a higher content of unsaturated fatty acids may impact oxidative stability and shelf life, necessitating further research on their effects on preservation and storage conditions.

Author Contributions: Conceptualization, supervision, and writing—review and editing, S.-H.P. and Y.-J.J.; investigation, S.-H.P. and Y.-J.J.; writing—original draft preparation, S.-H.K. and methodology and visualization, S.-H.K., S.H.L., and S.-H.P. All authors have read and agreed to the published version of the manuscript.

Funding: This research received by Basic Science Research Program through the National Research Foundation of Korea (NRF) funded by the Ministry of Education (2022R1F1A1063628).

Institutional Review Board Statement: Not applicable.

Informed Consent Statement: Not applicable.

Data Availability Statement: The original contributions presented in the study are included in the article, further inquiries can be directed to the corresponding author.

Acknowledgments: The authors are grateful to the SPC Research Institute of Food Biotechnology, Republic of Korea, for providing wheat flour and baking ingredients.

Conflicts of Interest: The authors declare no conflicts of interest.

References

1. Ghotra, B.S.; Dyal, S.D.; Narine, S.S. Lipid shortenings: A review. *Food Res. Int.* **2002**, *35*, 1015–1048. [CrossRef]
2. Chrysam, M. Table spreads and shortenings. *Bailey's Ind. Oil Fat Prod.* **1985**, *3*, 41–125.
3. Oomen, C.M.; Ocké, M.C.; Feskens, E.J.; van Erp-Baart, M.-A.J.; Kok, F.J.; Kromhout, D. Association between trans fatty acid intake and 10-year risk of coronary heart disease in the Zutphen Elderly Study: A prospective population-based study. *Lancet* **2001**, *357*, 746–751. [CrossRef]
4. Micha, R.; Mozaffarian, D. Trans fatty acids: Effects on metabolic syndrome, heart disease and diabetes. *Nat. Rev. Endocrinol.* **2009**, *5*, 335–344. [CrossRef] [PubMed]
5. Oh, I.K.; Lee, S. Utilization of foam structured hydroxypropyl methylcellulose for oleogels and their application as a solid fat replacer in muffins. *Food Hydrocoll.* **2018**, *77*, 796–802. [CrossRef]
6. Lim, J.; Inglett, G.E.; Lee, S. Response to consumer demand for reduced-fat foods; multi-functional fat replacers. *Jpn. J. Food Eng.* **2010**, *11*, 147–152. [CrossRef]
7. Co, E.D.; Marangoni, A.G. Organogels: An alternative edible oil-structuring method. *J. Am. Oil Chem. Soc.* **2012**, *89*, 749.
8. Stortz, T.A.; Zetzl, A.K.; Barbut, S.; Cattaruzza, A.; Marangoni, A.G. Edible oleogels in food products to help maximize health benefits and improve nutritional profiles. *Lipid Technol.* **2012**, *24*, 151–154. [CrossRef]
9. Ögütcü, M.; Yilmaz, E. Oleogels of virgin olive oil with carnauba wax and monoglyceride as spreadable products. *Grasas Aceites* **2014**, *65*, e040.
10. Yi, B.; Kim, M.-J.; Lee, S.Y.; Lee, J. Physicochemical properties and oxidative stability of oleogels made of carnauba wax with canola oil or beeswax with grapeseed oil. *Food Sci. Biotechnol.* **2017**, *26*, 79–87. [CrossRef]
11. Appelqvist, I.A.M.; Debet, M.R.M. Starch-biopolymer interactions—A review. *Food Rev. Int.* **1997**, *13*, 163–224. [CrossRef]
12. Guarda, A.; Rosell, C.M.; Benedito, C.; Galotto, M.J. Different hydrocolloids as bread improvers and antistaling agents. *Food Hydrocoll.* **2004**, *18*, 241–247. [CrossRef]
13. Rosell, C.M.; Haros, M.; Escrivá, C.; Benedito de Barber, C. Experimental approach to optimize the use of α-amylases in breadmaking. *J. Agric. Food Chem.* **2001**, *49*, 2973–2977. [CrossRef] [PubMed]
14. Sharadanant, R.; Khan, K. Effect of hydrophilic gums on the quality of frozen dough: II. Bread characteristics. *Cereal Chem.* **2003**, *80*, 773–780. [CrossRef]
15. Li, X.; Al-Assaf, S.; Fang, Y.; Phillips, G.O. Competitive adsorption between sugar beet pectin (SBP) and hydroxypropyl methylcellulose (HPMC) at the oil/water interface. *Carbohydr. Polym.* **2013**, *91*, 573–580. [CrossRef] [PubMed]
16. Schober, T.J.; Bean, S.R.; Boyle, D.L.; Park, S.-H. Improved viscoelastic zein–starch doughs for leavened gluten-free breads: Their rheology and microstructure. *J. Cereal Sci.* **2008**, *48*, 755–767. [CrossRef]
17. Sarkar, N.; Walker, L. Hydration—Dehydration properties of methylcellulose and hydroxypropylmethylcellulose. *Carbohydr. Polym.* **1995**, *27*, 177–185. [CrossRef]
18. Davidou, S.; Le Meste, M.; Debever, E.; Bekaert, D. A contribution to the study of staling of white bread: Effect of water and hydrocolloid. *Food Hydrocoll.* **1996**, *10*, 375–383. [CrossRef]
19. Andersson, H.; Öhgren, C.; Johansson, D.; Kniola, M.; Stading, M. Extensional flow, viscoelasticity and baking performance of gluten-free zein-starch doughs supplemented with hydrocolloids. *Food Hydrocoll.* **2011**, *25*, 1587–1595. [CrossRef]
20. Rosell, C.M.; Rojas, J.A.; De Barber, C.B. Influence of hydrocolloids on dough rheology and bread quality. *Food Hydrocoll.* **2001**, *15*, 75–81. [CrossRef]
21. Rosell, C.M.; Rojas, J.A.; Benedito de Barber, C. Combined effect of different antistaling agents on the pasting properties of wheat flour. *Eur. Food Res. Technol.* **2001**, *212*, 473–476. [CrossRef]
22. Rosell, C.M.; Foegeding, A. Interaction of hydroxypropylmethylcellulose with gluten proteins: Small deformation properties during thermal treatment. *Food Hydrocoll.* **2007**, *21*, 1092–1100. [CrossRef]
23. Kaur, M.; Sandhu, K.S.; Arora, A.; Sharma, A. Gluten free biscuits prepared from buckwheat flour by incorporation of various gums: Physicochemical and sensory properties. *LWT-Food Sci. Technol.* **2015**, *62*, 628–632. [CrossRef]

24. Kim, C.; Yoo, B. Rheological properties of rice starch–xanthan gum mixtures. *J. Food Eng.* **2006**, *75*, 120–128. [CrossRef]
25. Ozkoc, S.O.; Seyhun, N. Effect of gum type and flaxseed concentration on quality of gluten-free breads made from frozen dough baked in infrared-microwave combination oven. *Food Bioprocess Technol.* **2015**, *8*, 2500–2506. [CrossRef]
26. Yoshimura, M.; Takaya, T.; Nishinari, K. Rheological studies on mixtures of corn starch and konjac-glucomannan. *Carbohydr. Polym.* **1998**, *35*, 71–79. [CrossRef]
27. Patel, A.R.; Rajarethinem, P.S.; Cludts, N.; Lewille, B.; De Vos, W.H.; Lesaffer, A.; Dewettinck, K. Biopolymer-Based Structuring of Liquid Oil into Soft Solids and Oleogels Using Water Continuous Emulsions as Templates. *Langmuir* **2015**, *31*, 2065–2073. [CrossRef] [PubMed]
28. Meng, Z.; Qi, K.; Guo, Y.; Wang, Y.; Liu, Y. Effects of thickening agents on the formation and properties of edible oleogels based on hydroxypropyl methyl cellulose. *Food Chem.* **2018**, *246*, 137–149. [CrossRef]
29. Bascuas, S.; Salvador, A.; Hernando, I.; Quiles, A. Designing Hydrocolloid-Based Oleogels with High Physical, Chemical, and Structural Stability. *Front. Sustain. Food Syst.* **2020**, *4*, 111. [CrossRef]
30. Bascuas, S.; Morell, P.; Quiles, A.; Salvador, A.; Hernando, I. Use of Oleogels to Replace Margarine in Steamed and Baked Buns. *Foods* **2021**, *10*, 1781. [CrossRef]
31. AACCI. AACC Methods 10-10.03 (Optimized Straight-Dough Bread-Making Method). In *Approved Methods of American Association of Cereal Chemists*, 10th ed.; AACC: Arnold, MD, USA, 2000.
32. Horwitz, W.; Latimer, G. *Official Methods of Analysis of AOAC International*, 17th ed.; Association of Analytical Chemists International: Gaithersburg, MD, USA, 2000.
33. AACC. *Approved Methods of the AACC*; AACC: Arnold, MD, USA, 2000.
34. Yu, W.; Xu, D.; Li, D.; Guo, L.; Su, X.; Zhang, Y.; Wu, F.; Xu, X. Effect of pigskin-originated gelatin on properties of wheat flour dough and bread. *Food Hydrocoll.* **2019**, *94*, 183–190. [CrossRef]
35. Bal, L.M.; Kar, A.; Satya, S.; Naik, S.N. Kinetics of colour change of bamboo shoot slices during microwave drying. *Int. J. Food Sci. Technol.* **2011**, *46*, 827–833. [CrossRef]
36. AOAC. AOAC Official Method 996.06, Fat (Total, Saturated, and Unsaturated) in Foods. In *Official Methods of Analysis of AOAC International*; Oxford University Press: Oxford, UK, 2001; pp. 20–24.
37. Tronsmo, K.; Faergestad, E.; Schofield, J.; Magnus, E. Wheat protein quality in relation to baking performance evaluated by the Chorleywood bread process and a hearth bread baking test. *J. Cereal Sci.* **2003**, *38*, 205–215. [CrossRef]
38. Jafari, M.; Koocheki, A.; Milani, E. Functional effects of xanthan gum on quality attributes and microstructure of extruded sorghum-wheat composite dough and bread. *LWT* **2018**, *89*, 551–558. [CrossRef]
39. Turbin-Orger, A.; Boller, E.; Chaunier, L.; Chiron, H.; Della Valle, G.; Réguerre, A.-L. Kinetics of bubble growth in wheat flour dough during proofing studied by computed X-ray micro-tomography. *J. Cereal Sci.* **2012**, *56*, 676–683. [CrossRef]
40. Sroan, B.S.; Bean, S.R.; MacRitchie, F. Mechanism of gas cell stabilization in bread making. I. The primary gluten–starch matrix. *J. Cereal Sci.* **2009**, *49*, 32–40. [CrossRef]
41. Ozkoc, S.O.; Sumnu, G.; Sahin, S. The effects of gums on macro and micro-structure of breads baked in different ovens. *Food Hydrocoll.* **2009**, *23*, 2182–2189. [CrossRef]
42. Lazaridou, A.; Duta, D.; Papageorgiou, M.; Belc, N.; Biliaderis, C.G. Effects of hydrocolloids on dough rheology and bread quality parameters in gluten-free formulations. *J. Food Eng.* **2007**, *79*, 1033–1047. [CrossRef]
43. Crockett, R.; Ie, P.; Vodovotz, Y. How do xanthan and hydroxypropyl methylcellulose individually affect the physicochemical properties in a model gluten-free dough? *J. Food Sci.* **2011**, *76*, E274–E282. [CrossRef]
44. Sciarini, L.S.; Ribotta, P.D.; León, A.E.; Pérez, G.T. Incorporation of several additives into gluten free breads: Effect on dough properties and bread quality. *J. Food Eng.* **2012**, *111*, 590–597. [CrossRef]
45. Rodríguez-García, J.; Puig, A.; Salvador, A.; Hernando, I. Optimization of a sponge cake formulation with inulin as fat replacer: Structure, physicochemical, and sensory properties. *J. Food Sci.* **2012**, *77*, C189–C197. [CrossRef] [PubMed]
46. Jung, D.; Oh, I.; Lee, J.; Lee, S. Utilization of butter and oleogel blends in sweet pan bread for saturated fat reduction: Dough rheology and baking performance. *LWT* **2020**, *125*, 109194. [CrossRef]
47. Monthe, O.C.; Grosmaire, L.; Nguimbou, R.M.; Dahdouh, L.; Ricci, J.; Tran, T.; Ndjouenkeu, R. Rheological and textural properties of gluten-free doughs and breads based on fermented cassava, sweet potato and sorghum mixed flours. *LWT* **2019**, *101*, 575–582. [CrossRef]
48. Ma, Y.; Sang, S.; Xu, D.; Jin, Y.; Chen, Y.; Xu, X. The contribution of superheated steam treatment of wheat flour to the cake quality. *LWT* **2021**, *141*, 110958. [CrossRef]
49. Ma, M.; Xie, Y.; Wang, C. Effect of anthocyanin-rich extract from black soybean coat on wheat dough rheology and noodle texture. *J. Food Process. Preserv.* **2021**, *45*, e15007. [CrossRef]
50. Kim, J.; Kim, D.N.; Lee, S.H.; Yoo, S.-H.; Lee, S. Correlation of fatty acid composition of vegetable oils with rheological behaviour and oil uptake. *Food Chem.* **2010**, *118*, 398–402. [CrossRef]

foods

MDPI

Article

Nutrient Composition, Physical Characteristics and Sensory Quality of Spinach-Enriched Wheat Bread

Ritnesh Vishal Prasad [1,2], Sushil Dhital [3], Gary Williamson [2,4] and Elizabeth Barber [2,4,*]

1 School of Chemistry, Faculty of Science, Monash University, Clayton, VIC 3800, Australia; rpra0019@student.monash.edu

2 Department of Nutrition, Dietetics and Food, Monash University, BASE Facility, 264 Ferntree Gully Road, Notting Hill, VIC 3168, Australia; gary.williamson1@monash.edu

3 Bioresource Processing Research Institute of Australia (BioPRIA), Department of Chemical and Biological Engineering, Faculty of Engineering, Monash University, Clayton, VIC 3800, Australia; sushil.dhital@monash.edu

4 Victorian Heart Institute, Monash University, Victorian Heart Hospital, 631 Blackburn Road, Clayton, VIC 3168, Australia

* Correspondence: elizabeth.barber@monash.edu

Abstract: Food innovation that utilises agricultural waste while enhancing nutritional value is important for waste valorisation and consumer health. This study investigated incorporating spinach (*Spinacia oleracea*), as a model leafy agricultural waste, into wheat bread. We analysed the nutrient content, colour, texture, sensory attributes and purchase/consume intention ratings. Adding 10–40% spinach (w/w) yielded loaves with similar heights but significantly different colour and texture ($p < 0.05$) from white bread. Increasing spinach decreased total carbohydrates (including starch) while significantly increasing other nutrients (protein, fibre, iron, magnesium, potassium, zinc, calcium, vitamins A, C, E, folate, niacin, pyridoxine, nitrate/nitrite and polyphenols) ($p < 0.05$). Spinach addition increased bread porosity, linked to higher pasting parameters (peak, trough, breakdown, final and setback viscosity) with reduced pasting time and temperature. Texture analysis resulted in decreased hardness, chewiness, gumminess and firmness while increasing cohesiveness, with maximum resilience at 20% spinach enrichment. Sensory analysis with 21 untrained panellists revealed decreased visual appeal, less preferred taste, odour and overall liking ($p < 0.05$) with increasing spinach, with no significant difference in texture acceptance, but the 20% enrichment had comparable acceptance to white bread. Enriching staple foods like bread with leafy vegetable waste offers a promising approach for increasing daily vegetable intake.

Keywords: food fortification; functional food; sensory attributes; consumer acceptability; nutritional value; rheological properties; natural bioactive compounds

1. Introduction

The valorisation of food waste, particularly through the incorporation of nutrient-rich components into foods to increase functionality and value, has gained significant attention in recent years. This approach not only reduces food waste significantly but also contributes to the development of functional food products to improve the health of consumers for longevity [1].

Inadequate consumption of fruits and vegetables is associated with poorer overall health [2]. Strong and consistent epidemiological evidence shows a lower relative and absolute disease risk with greater survival time when diet quality is improved [3,4]. Plant-based foods are reservoirs of highly bioactive health-protective compounds, among which polyphenols have shown multiple promising effects in improving overall health [5–7].

Spinach (*Spinacia oleracea*) is a green leafy vegetable with highly beneficial nutrients such as fibre, vitamins (i.e., A, C, E, K, B6 and folate) and minerals (i.e., zinc, manganese, iron, calcium, potassium and selenium) [8,9]. Raw spinach is rich in polyphenols, particularly flavonoids (i.e., luteolin, kaempferol, myricetin, quercetin and quercetagetin) [10–13]. Spinach has a global production of 30.1 million tonnes, with China leading at 27.52 million tonnes (91% of global production), followed by the United States, Turkey and Japan. Despite its high nutrient value, 35% of fresh produce is wasted during household consumption due to short shelf stability, while another ~6.5% is lost during packaging and at the retail level [14]. Given the food waste issue and inadequate fruit and vegetable intake [2], this opens an opportunity to rethink food innovation by enriching nutrient-rich ingredients, such as spinach, into commonly consumed household foods, such as bread.

Bread is an easily accessible and digestible staple consumed by most populations. In 2023, the volume of bread production was approximately 185 billion kg, with a predicted market growth of 3.8% in 2025 [15]. Regular consumption of refined white bread low in nutrients and natural bioactive compounds can be associated with a poor diet, yet it is often more generally favoured compared to whole grain or wholemeal bread [16]. To improve the health of the population, bread could be a promising vehicle for vegetable enrichment to increase the consumption of plant bioactive compounds. We therefore assessed the physicochemical characteristics, nutrient composition and sensory properties of bread with added freshly ground spinach. We found that 20% spinach can be effectively added to the dough to produce almost equivalent sensory and physical properties to that of a control bread, with each slice of 40 g containing approximately 16 g of spinach. This manuscript should provide valuable insights and practical guidance on the integration of spinach, modelled as an agricultural waste product, into bread formulations and the assessment of sensory and physical characteristics, highlighting its potential as a newly developed food with enhanced nutritional profiles. This innovative approach not only enriches the bread with essential nutrients and bioactive compounds but also serves as a practical solution to increase vegetable consumption through a commonly consumed staple while addressing the issue of food waste.

2. Materials and Methods

2.1. Bread Formulation

Wallaby Bakers plain flour (Laucke, Strathalbyn, SA, Australia), Wallaby Bakers bread improver (Laucke, Strathalbyn, SA, Australia), instant dried yeast (Lowan Whole Foods, Glendenning, NSW, Australia), 100% gluten flour (Lotus, Kadac Pty Ltd., Braeside, VIC, Australia), salt, sugar, butter and packaged raw baby spinach (90% moisture) were purchased from a local supermarket (Coles Supermarkets Australia Pty Ltd., Hawthorn East, VIC, Australia). The basic formulation was consistent for all bread types except flour, water, gluten and blended spinach (Table 1). Spinach was freshly ground thoroughly at room temperature (25 °C) using a blender (Blixer®® 2; Robot Coupe, Vincennes, France) and added to the mixture at increasing concentrations from 10% to 40% (*w*/*w*) while reducing flour by 2% (*w*/*w*) and water by 8% (*w*/*w*). Additional gluten was added to compensate for the losses of flour ratio. The control white bread consisted of 54.0% (*w*/*w*) flour and 39.9% (*w*/*w*) water without spinach. All measured ingredients were added one at a time into the dough mixer (stand mixer 4.3 L model 5KSM150. KitchenAid, Benton Harbor, MI, USA) and mixed at level 2 for ~15–20 min until the dough was no longer sticking onto the mixing bowl.

The steps involved in the bread-making process used here are illustrated in Figure 1. Once kneaded thoroughly, the dough was covered and rested in a transparent bowl at room temperature (20–24 °C) for 1 h to allow the first fermentation stage. The risen dough was then punched, kneaded briefly to break large gas bubbles, shaped and placed in a bread baking tin (27 cm × 13.5 cm × 10 cm (length × width × depth)), and covered for the second fermentation stage for another hour in a temperature-controlled room of 28 °C to double in size. Once the dough was baked at 180 °C for 25 min in an oven (Rational CombiMaster® Plus CM101, Landsberg am Lech, Germany), the baked bread was left at room temperature to cool down for at least 2 h, after which the height (ends and mid-points) and weight were recorded. This process was repeated in triplicate for each bread for physicochemical and nutrient analysis. The crust of the freshly baked bread samples was removed before sensory analysis.

Table 1. Bread formulations for white bread and spinach-enriched bread.

Ingredients [g (%) (*w*/*w*)]	Control Bread	Spinach Bread (%) (*w*/*w*)			
		10%	20%	30%	40%
1. Bread flour	480 (54.0)	450 (51.6)	440 (49.5)	420 (47.3)	400 (45.0)
2. Warm water	355 (39.9)	283 (31.8)	212 (23.9)	141 (15.9)	70 (7.9)
3. Gluten substitute	0.0	2.3 (0.3)	4.6 (0.5)	6.9 (0.8)	9.2 (1.0)
4. Spinach	0.0	89.7 (10.1)	178.4 (20.1)	267.1 (30.1)	355.8 (40.0)
5. Yeast (*Saccharomyces cerevisiae*)	10 (1.1)	10 (1.1)	10 (1.1)	10 (1.1)	10 (1.1)
6. Bread improver	10 (1.1)	10 (1.1)	10 (1.1)	10 (1.1)	10 (1.1)
7. Sugar, caster	10 (1.1)	10 (1.1)	10 (1.1)	10 (1.1)	10 (1.1)
8. Butter	10 (1.1)	10 (1.1)	10 (1.1)	10 (1.1)	10 (1.1)
9. Salt	5 (0.6)	5 (0.6)	5 (0.6)	5 (0.6)	5 (0.6)
10. Gluten	8.9 (1.0)	8.9 (1.0)	8.9 (1.0)	8.9 (1.0)	8.9 (1.0)
Total weight (g)	888.9	888.9	888.9	888.9	888.9

Ingredients: 1–4 varied in concentration, while ingredients 5–10 were consistent throughout the bread formulation. Values are in weight (g) with the percentage of total weight shown in brackets.

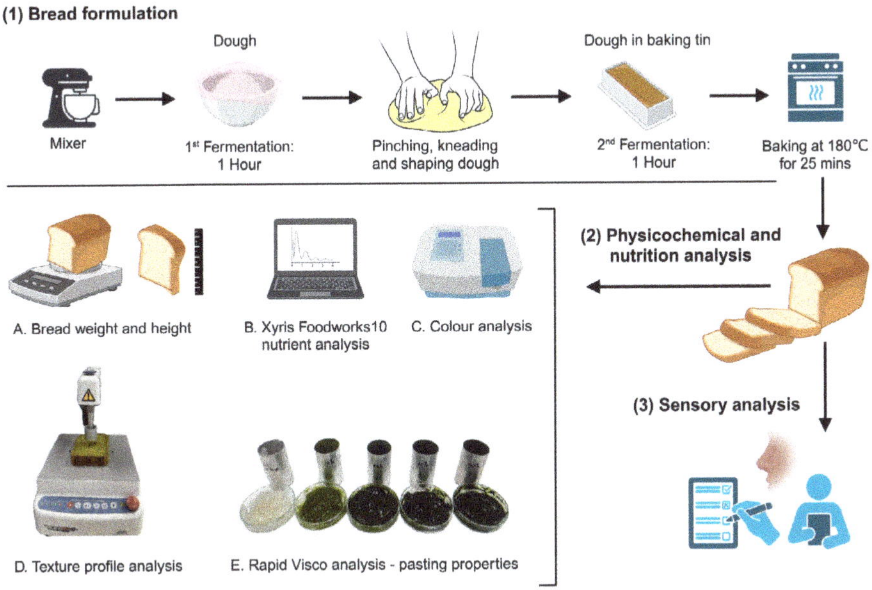

Figure 1. Bread formulation and subsequent analyses. (**1**) Bread formulation process, (**2**) physicochemical and nutritional analysis, (**3**) sensory analysis, $n = 21$ panellists. Created with BioRender.com. "https://app.biorender.com/illustrations/66612687f10e1aa131a6a97f (accessed on 29 June 2024)".

2.2. *Physiochemical and Nutritional Analysis*

2.2.1. Bread Height

The height of the bread was measured at the mid-point using a ruler in triplicate obtained from bread samples baked on different days.

2.2.2. Nutrient Composition Analysis

The nutrient composition of the bread was calculated based on its ingredients' composition using FoodWorks10 Professional software (Xyris Pty Ltd., Spring Hill, QLD, Australia). The weight of ingredients, total yield and weights before and after baking were entered into the software, where the yield was calculated using the following formula:

$$\text{Yield }(\%) = \frac{\text{final baked weight (g)}}{\text{raw dough weight (g)}} \times 100\% \tag{1}$$

Retention factors were applied to spinach and flour to account for the nutrient losses during the baking process. The nitrate (296 mg/100 g) and nitrite (3.8 mg/100 g) content adapted from [17] and the total polyphenol (248.14 mg/100 g), kaempferol (7.86 mg/100 g), quercetin (5.87 mg/100 g) and luteolin (1.11 mg/100 g) content adapted from [11] were manually added to the FoodWorks10 Professional software for data analysis.

2.2.3. Colour Analysis

The colour of baked bread was analysed using ColorFlex EZ (HunterLab, Reston, VA, USA). Samples (25 mm) were placed in HunterLab sample cups and analysed in triplicate. Values of L*, a*, b* and ΔE* were obtained as previously described [18].

2.2.4. Texture Profile Analysis (TPA)

The texture profile of the bread was analysed using a TA.XT Plus texture analyser (Stable Micro Systems, Godalming, England, UK). TPA was performed in accordance with the American Association of Cereals Chemists (AACC) Standard 74-09 [19], specified by the Exponent Connect Software, as reported previously [20]. The crust was removed, and the bread was sliced into equal thicknesses of 25 mm using a ruler and a knife. A 36 mm radius cylindrical flat probe with a 5 kg load cell was used for the double compression test at a 50% compression rate, with a test speed of 1.7 mm/s and a post-test speed of 10 mm/s performed in duplicate. While the firmness (g) and hardness (g) values were provided by the software, the remaining parameters were calculated as follows:

$$\text{Springiness }(\%) = \frac{\text{distance of 2nd compression (mm)}}{\text{distance of 1st compression (mm)}} \times 100\% \tag{2}$$

$$\text{Cohesiveness }(\%) = \frac{\text{area under the 2nd compression curve (g.s)}}{\text{area under the 1st compression curve (g.s)}} \times 100\% \tag{3}$$

$$\text{Resilience }(\%) = \frac{\text{area under 1st compression curve during recovery phase (g.s)}}{\text{area under 1st compression curve during compression phase (g.s)}} \times 100\% \tag{4}$$

$$\text{Chewiness (g)} = \text{hardness (g)} \times \text{cohesiveness} \times \text{springiness} \tag{5}$$

$$\text{Gumminess (g)} = \text{hardness (g)} \times \text{cohesiveness} \tag{6}$$

2.2.5. Pasting Properties

The spinach was first freeze-dried (Labconco Corporation, Kansas City, MO, USA) and was then ground using a mortar and pestle before sieving through a 300 μm sieve to obtain a homogenised and consistent powder. The powder and the flour samples were placed in a drying oven (N-BIOTEK, Inc., Gyeonggi-do, Korea) overnight at 40 °C to equalise moisture content before measuring the pasting characteristics.

The pasting properties of dough were analysed using a Rapid Visco Analyser (RVA) 4500 (Perten Instruments, Stockholm, Sweden). Flour (3.0 g dry basis) and spinach powder of 10%, 20%, 30% and 40% (*w*/*w*) dry weight were added to a canister and topped up with distilled water to obtain a constant weight of 28 g flour–spinach mixture. The slurry was homogenised using an RVA spindle. For RVA measurements, the samples underwent a temperature cycle: 50 °C for 1 min, increased to 95 °C for 3.75 min, held at 95 °C for 2.5 min, cooled to 50 °C for 3.75 min and then maintained at 50 °C for 2 min. After the analysis, peak, trough, breakdown, final viscosity, setback, peak time and the pasting temperature were obtained to determine the pasting profiles, as adapted from [21,22].

2.3. Sensory Evaluation and Food Purchasing Ominr Consumption Rating Test

The sensory evaluation was performed following the ISO 6658:2017 standard [23] using 21 untrained panellists over 18 years old, obtained from the Department of Nutrition, Dietetics and Food, Monash University. The ethics was approved by the Monash University Human Research Ethics Committee (MUHREC #26497). A minimum of 20 panellists was sufficient to provide adequate heterogeneity in consumer preferences, as previously mentioned [24]. The bread crust was removed, and the crumb was cut into bite-size pieces, packed and refrigerated (4 °C) for no more than 2 days for sensory evaluation. The samples were thawed to room temperature for >30 min before serving to the panellists for assessment, in randomised three-digit codes (Figure 2A) to conceal the identities. Sensory analysis was conducted in a room with red lighting (Figure 2B) to mask the colour and appearance of the green hue to minimise bias, modified from [25]. The 9-scale hedonic rating test and 7-scale purchase or consume intent rating tests were used for sensory evaluation. The hedonic scale test consists of 1 as "dislike extremely", 5 as "neither like nor dislike" and 9 as "like extremely". The sensory evaluation was conducted at 0930-1130 (between breakfast and lunch) when the panellists were not too hungry or full. During the evaluation, the panellists tasted one sample at a time, rinsed mouths in between with water, and evaluated taste, appearance, odour, texture and overall liking attributes as described previously [23].

(A) **(B)**

Figure 2. Sensory evaluation of bread. (**A**) Samples presented on a tray with three-digit codes with water. (**B**) Room with red lighting used to conduct sensory evaluation.

The 7-point purchase or consume intent rating test was presented with a set of statements as follows:

Score 7—"I would buy/consume this product at every opportunity I had"
Score 6—"I would buy/consume this product very often"
Score 5—"I like this product and would buy/consume it now and then"
Score 4—"I would buy/consume this product if available but would not go out of my way"
Score 3—"I do not like this product but would buy/consume it on occasion"
Score 2—"I would hardly ever buy/consume this product"
Score 1—"I would buy/consume this product only if forced to"

Greater scores (5–7) indicated high purchasing/consumption intention, a score of 4 indicated neutrality and lower scores (3–1) indicated negative purchasing or consumption perception. The original method [26] was slightly modified by having numerical scores (1 to 7) to facilitate data analysis. The terms "buy" and "consume" were used to capture consumers' purchasing and consumption intentions, respectively.

2.4. Statistical Analysis and Illustrations

The data were analysed using one-way analysis of variance (ANOVA) to compare the different means. Tukey's test was selected for multiple comparisons with statistical significance at $p < 0.05$. GraphPad Prism (version 10.0.2, San Diego, CA, USA) software was used to conduct statistical analysis and prepare graphs. The nutrient composition of the bread was assessed using Foodworks10 Professional software (Xyris Pty Ltd., Spring Hill, QLD, Australia). Figure illustrations were created using the BioRender.com application.

3. Results

3.1. Bread Formulation

Increasing the spinach concentration in the dough affected the colour, texture and fermentation behaviour of the bread (Figure 3).

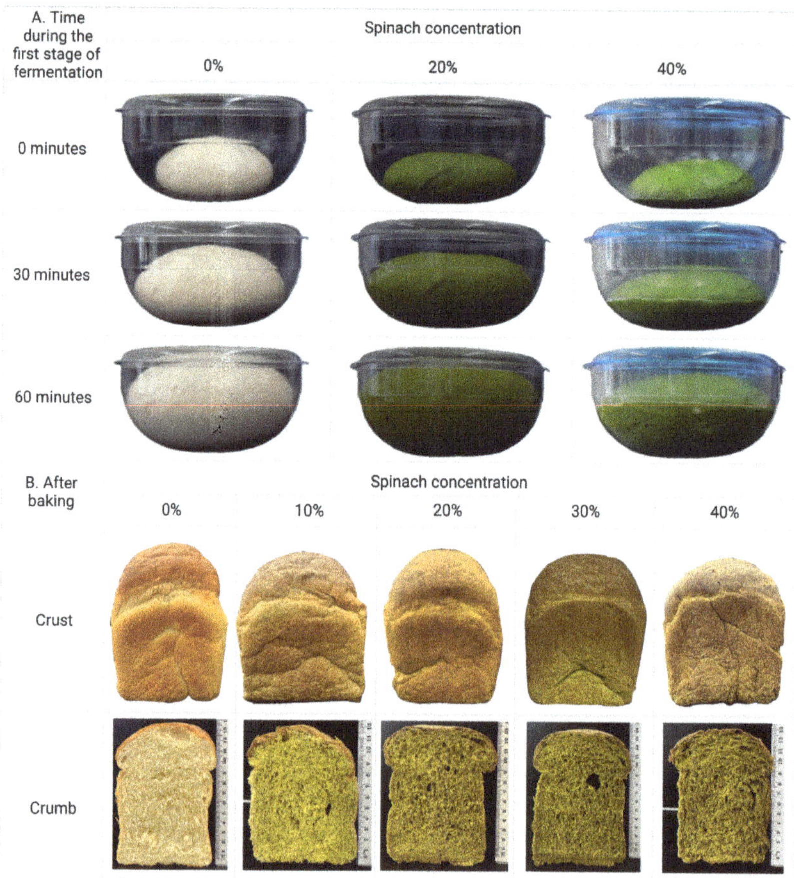

Figure 3. Dough expansion and bread characteristics. (**A**) Dough fermented for 0, 30 and 60 min with 0%, 20% and 40% spinach enrichment. (**B**) Bread crumb and crust characteristics for 0%, 10%, 20%, 30% and 40% spinach-enriched bread. Created with Biorender.com. "https://app.biorender.com/illustrations/66a1a1c7fddd4fb6c3081b2e (accessed on 29 June 2024)".

The dough with 40% spinach had an apparent wetter texture during fermentation than the control, causing some difficulty during the handling and kneading processes. As the spinach concentration increased, there was a noticeable decrease in dough expansion upon fermentation, while an increased green hue in the crumb and crust colour in the baked products was noted (Figure 3).

Table 2 indicates the decrease in yield as the spinach concentration increased. The highest yield was observed in control white bread (84%), followed by 10%, 20% and 30% spinach bread ($p > 0.05$), while the lowest yield was shown in 40% spinach bread ($p < 0.05$). Similarly, the weight was highest in white bread (749 g), followed by 10% and 20% spinach bread ($p > 0.05$), while the 30% and 40% spinach bread had a 4.8% reduction in weight ($p < 0.05$). Despite the reduced dough expansion observed during fermentation, bread height was unaffected by increasing spinach concentration (Figure 4). Spinach enrichment seemed to lower the yields and weights by almost 5% without affecting the height of the baked products.

Table 2. Bread yield percentage and total weight of baked loaf after spinach enrichment.

	Control White Bread	Spinach-Enriched Bread			
		10%	20%	30%	40%
Yield (%)	84 ± 1.5 [a]	82 ± 2.1 [ab]	81 ± 1.5 [ab]	80 ± 1.5 [ab]	80 ± 0.0 [b]
Weight (g)	749 ± 15.7 [a]	726 ± 17.0 [ab]	723 ± 14.8 [ab]	712 ± 12.1 [b]	712 ± 1.5 [b]

Calculated percentage yield. Different letters indicate significant differences at $p < 0.05$; values are presented as mean ± SD, $n = 3$.

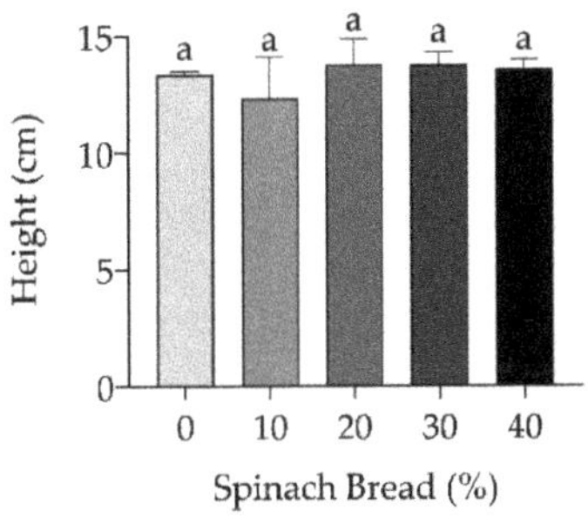

Figure 4. Bread height with and without spinach enrichment. Different letters indicate significant differences at $p < 0.05$. Values are presented as mean ± SD, $n = 3$.

3.2. Nutrient Composition

The nutritional composition of the samples was determined using FoodWorks10 Professional software, which reflects nutrient data sourced from AusFoods, AusBrands, FOODfiles (NZ) and the USDA National Nutrient Database. AusFoods includes nutrient data from AUSNUT 2011-13 and the most current Australian Food Composition Database (AFCD), whereas AusBrands provides data collected from nutrition information panels of commercial products in Australia, linked to AusFoods for a more comprehensive nutrient profile [27,28].

Nutrient analysis of spinach enrichment in bread revealed significant improvements in several key nutritional parameters, including macronutrients, micronutrients and bioactive compounds, compared to white bread (Tables 3 and 4). With increasing spinach concentration, the energy and polyunsaturated fat content decreased by 5% ($p = 0.016$) and 12% ($p < 0.001$), respectively, in the highest spinach enrichment compared to the control, without affecting the total fats, saturated fats and trans fatty acid content throughout. As predicted, the replacement of gluten content in spinach-enriched bread (due to the lesser gluten-fortified bread flour used) increased the protein content by 14% in the highest spinach-enriched bread ($p = 0.0002$).

Table 3. Nutritional composition of spinach and comparative macronutrient, energy and ash content in control white bread and spinach-enriched bread.

Values (/100 g Baked Bread)	Spinach/100 g	Control	10% Spinach	p-Value	20% Spinach	p-Value	30% Spinach	p-Value	40% Spinach	p-Value
Energy (kJ)	82	1053 ± 22	1061± 25	0.7132	1040 ± 21	0.4789	1029 ± 18	0.2118	1003 ± 2 *	0.0163
Protein (g)	2.40	9.04 ± 0.18	9.53 ± 0.22 *	0.0405	9.78 ± 0.19 **	0.0083	10.14 ± 0.17 **	0.0016	10.35 ± 0.02 ***	0.0002
Total fat (g)	0.30	2.26 ± 0.05	2.32 ± 0.06	0.2426	2.32 ± 0.04	0.1757	2.34 ± 0.04	0.0721	2.33 ± 0.01	0.0514
Saturated fat (g)	0.00	0.87 ± 0.02	0.89 ± 0.02	0.3046	0.89 ± 0.02	0.3465	0.90 ± 0.02	0.1926	0.89 ± 0.00	0.2378
Trans Fatty Acids (g)	0.00	0.06 ± 0.01	0.07 ± 0.00	0.1161	0.07 ± 0.00	0.1161	0.07 ± 0.00	0.1161	0.07 ± 0.00	0.1161
Polyunsaturated fat (g)	0.00	0.49 ± 0.01	0.48 ± 0.01	0.2879	0.46 ± 0.01 *	0.0390	0.45 ± 0.01 **	0.0080	0.43 ± 0.00 ***	0.0005
Monounsaturated fat (g)	0.00	0.36 ± 0.01	0.37 ± 0.01	0.2879	0.37 ± 0.01	0.3739	0.37 ± 0.01	0.3739	0.36 ± 0.00	1.0000
Carbohydrate (g)	0.60	47.17 ± 0.97	46.85 ± 1.09	0.7208	45.24 ± 0.89	0.0640	44.07 ± 0.76 *	0.0121	42.22 ± 0.09 ***	0.0009
Sugars (g)	0.60	2.49 ± 0.06	2.59 ± 0.06	0.0946	2.63 ± 0.05 *	0.0335	2.69 ± 0.05 **	0.0085	2.71 ± 0.01 **	0.0023
Starch (g)	0.00	44.68 ± 0.93	44.25 ± 1.03	0.6218	42.61 ± 0.84 *	0.0457	41.38 ± 0.71 **	0.0081	39.51 ± 0.08 ***	0.0007
Dietary fibre (g)	2.50	2.02 ± 0.05	2.33 ± 0.06 **	0.0018	2.58 ± 0.05 ****	0.0001	2.85 ± 0.05 ****	<0.0001	3.10 ± 0.01 ****	<0.0001
Ash (g)	1.60	1.15 ± 0.03	1.37 ± 0.03 ***	0.0006	1.56 ± 0.03 ****	0.0001	1.77 ± 0.03 ****	<0.0001	1.95 ± 0.01 ****	<0.0001

Spinach nutrient data/100 g, adapted from [29]. Control white bread and spinach-enriched bread nutrient profile values are presented as mean ± SD, $n = 3$. Significant differences compared with control bread are denoted by asterisks: * = $p < 0.05$, ** = $p < 0.01$, *** = $p < 0.001$, **** = $p < 0.0001$. Control white bread and spinach-enriched bread data were obtained and analysed using FoodWorks10. The retention factor is applied during the analysis using FoodWorks10 to account for cooking and processing nutrient loss.

Table 4. Nutritional composition of spinach and comparative micronutrient and bioactive compound content in control white bread and spinach-enriched bread.

Values (/100 g Baked Bread)	Spinach/100 g	Control	10% Spinach	p-Value	20% Spinach	p-Value	30% Spinach	p-Value	40% Spinach	p-Value
Thiamin (mg)	0.06	0.28 ± 0.01	0.28 ± 0.01	0.6433	0.28 ± 0.01	1.0000	0.28 ± 0.01	1.0000	0.27 ± 0.00	0.1161
Riboflavin (mg)	0.16	0.00 ± 0.00	0.02 ± 0.00	-	0.04 ± 0.00	-	0.06 ± 0.00	-	0.08 ± 0.00	-
Niacin (mg)	0.40	0.96 ± 0.02	1.02 ± 0.03 *	0.0474	1.04 ± 0.02 **	0.0093	1.08 ± 0.02 **	0.0021	1.10 ± 0.01 ***	0.0004
Niacin eq. (mg)	0.99	2.90 ± 0.06	3.05 ± 0.07 *	0.0435	3.13 ± 0.06 **	0.0094	3.24 ± 0.06 **	0.0020	3.31 ± 0.01 ***	0.0003
Pyridoxine (mg)	0.16	0.05 ± 0.01	0.06 ± 0.01 *	0.0241	0.08 ± 0.00 ***	0.0006	0.10 ± 0.00 ****	0.0001	0.12 ± 0.00 ****	<0.0001
Folic acid (µg)	0.00	38.59 ± 0.80	38.14 ± 0.89	0.5443	36.64 ± 0.73 *	0.0353	35.5 ± 0.61 **	0.0059	33.79 ± 0.07 ***	0.0005
Food folate (µg)	110.0	36.3 ± 0.8	47.4 ± 1.1 ****	0.0001	57.5 ± 1.1 ****	<0.0001	68.4 ± 1.2 ****	<0.0001	78.4 ± 0.2 ****	<0.0001
Total vitamin A eq. (µg)	469.0	11.5 ± 0.2	50.3 ± 1.2 ****	<0.0001	88.8 ± 1.8 ****	<0.0001	128.9 ± 2.2 ****	<0.0001	167.6 ± 0.3 ****	<0.0001
Retinol (µg)	0.00	11.11 ± 0.23	11.46 ± 0.27	0.1643	11.51 ± 0.23	0.0995	11.68 ± 0.20 *	0.0332	11.68 ± 0.03 *	0.0137
β carotene (µg)	1920.0	2.04 ± 0.04	227.5 ± 5.3 ****	<0.0001	452.3 ± 8.9 ****	<0.0001	686.1 ± 11.8 ****	<0.0001	913.0 ± 1.8 ****	<0.0001
Vitamin C (mg)	27.00	0.00 ± 0.00	2.34 ± 0.06 ****	<0.0001	4.67 ± 0.09 ****	<0.0001	7.09 ± 0.12 ****	<0.0001	9.44 ± 0.02 ****	<0.0001
Vitamin E (mg)	1.30	0.09 ± 0.00	0.24 ± 0.01 ****	<0.0001	0.39 ± 0.01 ****	<0.0001	0.55 ± 0.01 ****	<0.0001	0.70 ± 0.00 ****	<0.0001
Tocopherol, α (mg)	1.30	0.09 ± 0.00	0.24 ± 0.01 ****	<0.0001	0.39 ± 0.01 ****	<0.0001	0.55 ± 0.01 ****	<0.0001	0.70 ± 0.00 ****	<0.0001
Iron (mg)	3.20	0.88 ± 0.02	1.28 ± 0.03 ****	<0.0001	1.67 ± 0.03 ****	<0.0001	2.07 ± 0.04 ****	<0.0001	2.45 ± 0.01 ****	<0.0001

Table 4. *Cont.*

Values (/100 g Baked Bread)	Spinach/ 100 g	Control	10% Spinach	p-Value	20% Spinach	p-Value	30% Spinach	p-Value	40% Spinach	p-Value
Zinc (mg)	0.60	0.79 ± 0.02	0.87 ± 0.02 **	0.0080	0.92 ± 0.02 **	0.0013	0.99 ± 0.02 ***	0.0003	1.05 ± 0.01 ****	<0.0001
Calcium (mg)	53.0	16.80 ± 0.34	23.55 ± 0.55 ****	0.0001	29.83 ± 0.58 ****	<0.0001	36.54 ± 0.62 ****	<0.0001	42.80 ± 0.08 ****	<0.0001
Potassium (mg)	570.0	130.3 ± 2.7	200.4 ± 4.7 ****	<0.0001	266.9 ± 5.3 ****	<0.0001	337.4 ± 5.8 ****	<0.0001	403.8 ± 0.8 ****	<0.0001
Magnesium (mg)	68.0	25.15 ± 0.52	33.3 ± 0.77 ****	0.0001	40.80 ± 0.80 ****	<0.0001	48.85 ± 0.84 ****	<0.0001	56.27 ± 0.12 ****	<0.0001
Phosphorus (mg)	46.0	98.1 ± 2.0	104.1 ± 2.4 *	0.0303	107.5 ± 2.1 **	0.0053	112.0 ± 1.9 ***	0.0010	114.9 ± 0.2 ****	0.0001
Sodium (mg)	21.0	267.0 ± 5.6	280.7 ± 6.5	0.0963	284.3 ± 5.6 *	0.0352	290.8 ± 5.0 **	0.0085	293.0 ± 0.6 **	0.0021
Selenium (µg)	0.00	8.21 ± 0.17	8.26 ± 0.20	0.7376	8.09 ± 0.16	0.4310	8.00 ± 0.14	0.1720	7.80 ± 0.02 *	0.0130
Iodine (µg)	0.00	5.69 ± 0.12	5.59 ± 0.13	0.3786	5.34 ± 0.10 *	0.0186	5.14 ± 0.09 **	0.0030	4.86 ± 0.01 ***	0.0003
Linoleic (g)	-	0.45 ± 0.01	0.44 ± 0.01	0.4918	0.43 ± 0.01 *	0.0249	0.42 ± 0.01 **	0.0075	0.40 ± 0.00 ***	0.0010
α-Linolenic (g)	-	0.04 ± 0.00	0.04 ± 0.00	-	0.04 ± 0.01	0.3739	0.04 ± 0.01	0.3739	0.03 ± 0.00	-
Nitrate (mg)	296.0	0.00 ± 0.00	36.6 ± 0.9 ****	<0.0001	73.1 ± 1.4 ****	<0.0001	111.1 ± 1.9 ****	<0.0001	147.9 ± 0.3 ****	<0.0001
Nitrite (mg)	3.80	0.00 ± 0.00	0.46 ± 0.01 ****	<0.0001	0.92 ± 0.02 ****	<0.0001	1.41 ± 0.03 ****	<0.0001	1.87 ± 0.01 ****	<0.0001
Kaempferol (mg)	7.86	0.00 ± 0.00	0.97 ± 0.02 ****	<0.0001	1.94 ± 0.04 ****	<0.0001	2.95 ± 0.05 ****	<0.0001	3.93 ± 0.01 ****	<0.0001
Quercetin (mg)	5.87	0.00 ± 0.00	0.73 ± 0.02 ****	<0.0001	1.45 ± 0.03 ****	<0.0001	2.20 ± 0.04 ****	<0.0001	2.93 ± 0.01 ****	<0.0001
Luteolin (mg)	1.11	0.00 ± 0.00	0.14 ± 0.01 ****	<0.0001	0.28 ± 0.01 ****	<0.0001	0.42 ± 0.01 ****	<0.0001	0.55 ± 0.01 ****	<0.0001
Total polyphenols	248.1	0.00 ± 0.00	30.0 ± 1.0 ****	<0.0001	60.6 ± 1.5 ****	<0.0001	92.7 ± 1.5 ****	<0.0001	123.3 ± 0.6 ****	<0.0001

Spinach nutrient data per 100 g, adapted from [29], except total vitamin E eq. adapted from [30], nitrate and nitrite information adapted from [17], kaempferol, quercetin, luteolin and total polyphenol content was adapted from [11]. Control white bread and spinach-enriched bread nutrient profile values are presented as mean ± SD, $n = 3$. Significant differences compared with control bread are denoted by asterisks: * = $p < 0.05$, ** = $p < 0.01$, *** = $p < 0.001$, **** = $p < 0.0001$. Control white bread and spinach-enriched bread data were obtained and analysed using FoodWorks10, where values for nitrate, nitrite and polyphenols in raw spinach were manually entered into the software. "Niacin eq." accounts for niacin and potential niacin that can be synthesised from the tryptophan content in the food. "Total vitamin A eq." represents the sum of preformed (retinol) and provitamin A carotenoids. Eq.: equivalent. The retention factor is applied during the analysis using FoodWorks10 to account for cooking and processing nutrient losses.

Total carbohydrate content decreased by 7% (p = 0.012) and 10% (p = 0.0009) in the 30% and 40% spinach-enriched bread, respectively. This is evident by a significant 12% (p = 0.0007) decrease in starch content, although a slight increase in sugar in 20%, 30% and 40% spinach bread were noted ($p < 0.05$). Dietary fibre and ash content increased significantly by 53% and 70%, respectively ($p < 0.0001$). The nutrient analysis (Table 3) suggests enhanced nutritional values, with a significant increase in protein, fibre and ash content while decreasing total carbohydrates, especially starch.

Thiamin content remained relatively stable across all samples, indicating that spinach incorporation did not affect this vitamin. Spinach is a good source of riboflavin and niacin, so its addition proportionately increases these nutrients in bread. The pyridoxine and folate levels increased significantly, maximising by a 140% and 116% increase, respectively, as the spinach enrichment reached 40% ($p < 0.05$). Composite antioxidant vitamins A, E and C increased significantly with increasing spinach concentration. Vitamin A and E increased by nearly 15- and 8-fold, respectively ($p < 0.05$), while vitamin C increased from 0 mg to 9.44 mg at the highest spinach concentration ($p < 0.0001$).

Mineral content significantly increased with increasing spinach concentration in bread, where iron increased by ~threefold, while calcium, potassium and magnesium nearly doubled ($p < 0.0001$). At the highest spinach enrichment, zinc and phosphorus elevated to 33% and 17%, respectively ($p < 0.05$). The sodium content was similar between the control and lower spinach enrichment and only increased to 9% in the highest spinach enrichment ($p < 0.05$). Despite the increase in mineral content, selenium decreased significantly by 5% (p = 0.013) in the 40% spinach bread but no difference occurred in the others.

Nitrites and nitrates were significantly increased in all spinach-enriched bread formulations ($p < 0.0001$). The total polyphenols, including kaempferol, quercetin and luteolin levels exhibited pronounced increases from 0 mg/100 g in white bread to a maximum of 123.3 mg/100 g, 3.93 mg/100 g, 2.93 mg/100 g and 0.55 mg/100 g in 40% spinach-enriched bread, respectively ($p < 0.0001$). The enhancement in the nutrient content of the spinach-enriched bread is primarily attributed to the spinach enrichment (Tables 3 and 4), indicating that the addition of spinach significantly boosts the overall nutritional profile.

3.3. Pasting Properties

Pasting properties are measured using viscosity parameters during the heating and cooling of starch to help understand the texture, quality and stability of the final product. In this study, freeze-dried spinach powder was added at 10%, 20%, 30% and 40% and compared with the control during viscosity measurement (Figure 5 and Table 5). The peak, trough, breakdown, final and setback viscosity increased significantly ($p < 0.05$) while the peak time and pasting temperature decreased with increasing spinach concentration from 10% to 40% compared with the control (Table 5).

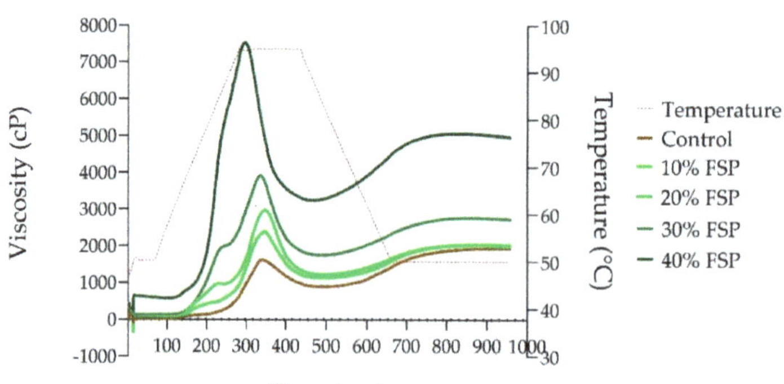

Figure 5. Pasting properties of wheat flour with and without enrichment with 10%, 20%, 30% and 40% freeze-dried spinach powder (FSP). Values are the mean of duplicate samples.

Table 5. Pasting properties of wheat flour starch enriched with freeze-dried spinach powder.

Primary Test	Freeze-Dried Spinach Concentration Added to the Dough				
	Control	10%	20%	30%	40%
Peak viscosity (cP)	1619 ^{c}	2395 ^{c}	2972 ^{bc}	3920 ^{b}	7539 ^{a}
Trough viscosity (cP)	893 ^{c}	1133 ^{bc}	1216 ^{bc}	1742 ^{b}	3237 ^{a}
Breakdown viscosity (cP)	727 ^{d}	1262 ^{cd}	1757 ^{bc}	2179 ^{b}	4302 ^{a}
Final viscosity (cP)	1923 ^{b}	2018 ^{b}	2015 ^{b}	2720 ^{b}	4961 ^{a}
Setback viscosity (cP)	1030 ^{b}	885 ^{b}	799 ^{b}	978 ^{b}	1724 ^{a}
Peak time (min)	5.7 ^{ab}	5.8 ^{a}	5.8 ^{a}	5.6 ^{b}	4.9 ^{c}
Pasting temperature (°C)	85.9 ^{a}	65.9 ^{b}	65.2 ^{b}	65.6 ^{b}	63.3 ^{c}

Different letters indicate significant differences at $p < 0.05$. Values are the mean of duplicate samples.

Peak viscosity, maximum viscosity achieved through the hydration, swelling and disruption of starch granules during heating [31–33] increased gradually with increasing concentrations of spinach powder and reached almost fivefold at 40% ($p < 0.05$). The gradual increase in peak viscosity with an increase in spinach from 10% to 40% suggests the additive and synergistic effects due to swelling of spinach fibres. The stability of the paste became stronger and followed the trend of peak viscosity, gradually increasing with increasing concentrations of spinach powder, maximising to 3.6-fold at 40% ($p < 0.05$). Trough viscosity (or hot paste viscosity) is the minimum viscosity reached after the peak during the cooling phase, representing the stability of the paste to shear [34–36]. With increasing concentrations of spinach powder, breakdown viscosity (peak-trough) gradually increased to almost sixfold ($p < 0.05$). It is noted that the ratio of peak to trough is increased with an increase in spinach powder, suggesting resistance to the breakdown of the paste during the holding and cooling phases in RVA.

Final viscosity reflects the ability of starch to form a viscous gel or paste upon cooling, determined as soluble amylose retrogradation obtained at the end of the cooling phase at 50 °C [33,35,36]. The final viscosity increased gradually with increasing spinach concentration ($p > 0.05$), maximised to 2.6-fold at 40% ($p < 0.05$). Setback viscosity reflects the reassociation of amylose and amylopectin into an ordered structure when starch molecules undergo retrogradation, promoting starch recrystallisation. This is determined by the difference between the final viscosity and trough viscosity [33,35,36]. The setback viscosity did not change with an addition ranging from 10% to 30% spinach powder ($p > 0.05$) but increased significantly in 40% ($p < 0.05$) compared with the control, indicating increased retrogradation and syneresis. It should be noted that in a complex system, i.e., the addition of spinach powder, the final viscosity as well as setback not only affects the function of starch pasting but also is affected by water absorption and retention by additional non-starch components.

Peak time is the time taken to reach peak viscosity, reflecting the rate at which starch granules hydrate, swell and rupture [33,36]. The peak time was similar to the control when 10% and 20% spinach powder was added but decreased significantly at 30% and 40% ($p < 0.05$), indicating that spinach addition accelerates gelatinisation. The pasting temperature determines the point at which viscosity begins to rise (onset of starch gelatinisation), indicating the minimum temperature required to cook a product [33,36]. The pasting temperature decreased significantly ($p < 0.05$) with the addition of spinach powder, indicating that spinach addition interfered with starch swelling, most likely due to providing a physical barrier (e.g., encapsulating or covering the starch) or by limiting the availability of water. It should be noted that spinach powder has a higher water absorption capacity than wheat starch.

3.4. Texture Profile Analysis

Texture profile analysis employs a double-mechanical compression test which helps understand the chewing behaviour of food. Firmness, hardness, springiness, cohesiveness, chewiness, gumminess and resilience are common parameters that make up the texture profile of food products such as bread. Significant variations were observed in the texture profiles of the control and spinach-containing breads (Figure 6).

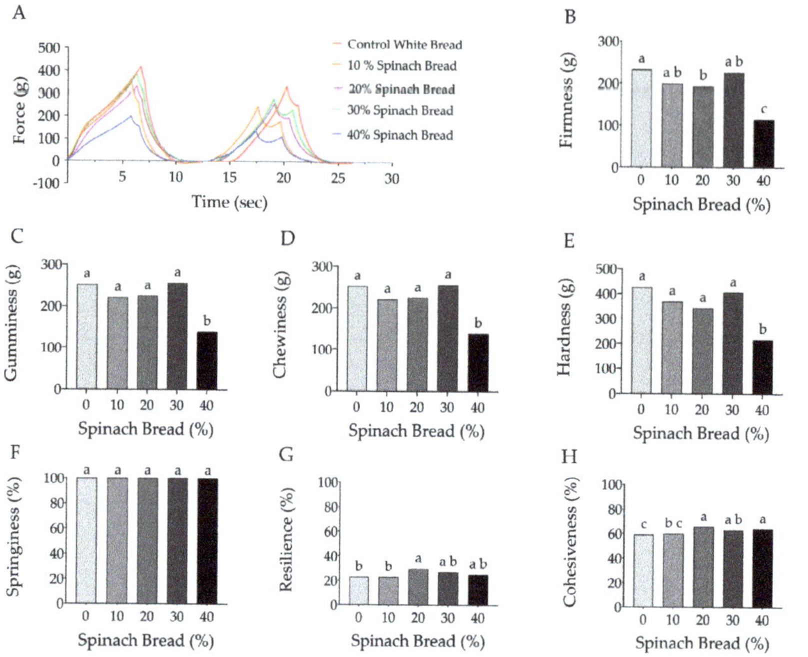

Figure 6. (**A**) Texture profile analysis and texture analysis parameters (**B**) firmness (g), (**C**) gumminess (g), (**D**) chewiness (g), (**E**) hardness (g), (**F**) springiness (%), (**G**) resilience (%), (**H**) cohesiveness (%) of control white bread and spinach-enriched bread at various concentrations from 10% to 40% (*w*/*w*). Different letters indicate significant differences at $p < 0.05$. Values are the mean of duplicate samples.

Firmness is determined by the amount of force required to deform or rupture a food sample [37], indicating a soft texture. Gumminess is used to describe the energy needed to break down a semi-solid food before swallowing, which is calculated by multiplying the hardness and cohesiveness of the food [38]. Chewiness is determined by the product of hardness, cohesiveness and springiness, which together represent the level of effort required to chew food to a consistency that is ideal for swallowing. [39,40]. Crumb hardness is measured by the maximum force during the first compression [41], reflecting the compression force between the molars (using incisors to bite through) and the compression force between the tongue and palate [39,42].

The control bread had the highest firmness, gumminess, chewiness and hardness, which decreased gradually with the addition of spinach, and the lowest at 40% ($p < 0.05$), reflecting the positive influence of spinach on bread softness and chewability.

Springiness and resilience are reverse measures when breadcrumbs are compressed by a tool. Springiness was similar in all samples with and without spinach ($p < 0.05$), indicating that spinach addition did not affect the deformation of bread crumb shapes when compressed. This means that the addition of spinach to bread did not affect its viscoelastic properties. Resilience determines elasticity, measured by the ability of a food to quickly return to its original height after being compressed [33]. Maximum resilience (30%) was shown by 20% spinach bread, which was significantly lower in the control bread and 10% spinach bread, but not different to the 30% and 40% spinach bread ($p > 0.05$). This indicates that moderate to high spinach enrichment improves elasticity.

Cohesiveness is a measure of how well the product withstands a second deformation relative to the first deformation [39], which is somewhat affected by resilience. The 20% spinach bread showed the highest cohesiveness, similarly, shown by 30% and 40% spinach bread, but significantly lowered in the control white bread and 10% spinach bread ($p < 0.05$). This indicates a stronger ability to hold the internal structure together at 20% spinach concentration.

3.5. Colour Analysis

Colour often influences consumers' acceptance and liking of certain foods. In spinach, the presence of naturally coloured compounds (e.g., yellow, red and orange carotenoids) is often masked by abundant green chlorophyll [43]. In this study, colour parameters were measured using the values of lightness (L*), redness (a*), yellowness (b*) and overall colour difference (ΔE*) to assess their correlations with consumer perception towards the bread varieties. Colour analysis reveals significant changes in colour parameters with the addition of spinach (10%, 20%, 30% and 40%) compared to the control white bread (Figure 7).

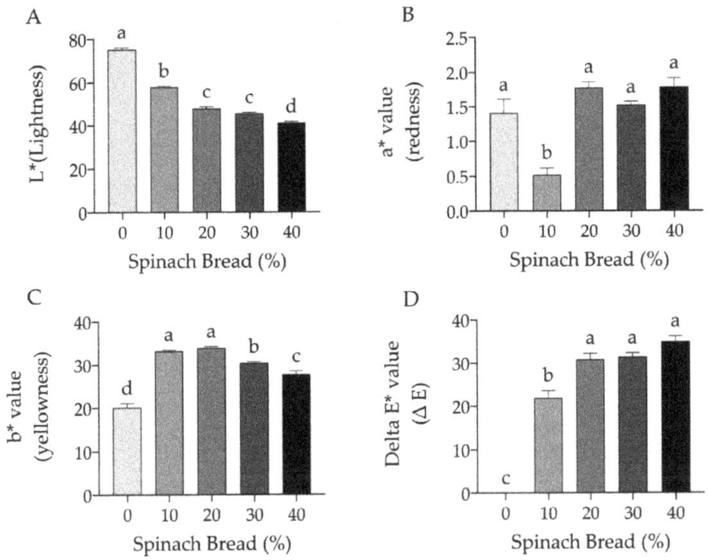

Figure 7. The colour analysis of bread with and without the addition of spinach at 10%, 20%, 30% and 40% (*w*/*w*). (**A**) L* value: lightness/brightness (0: black to 100:white), (**B**) a* value: redness (+a*) or greenness (−a*), (**C**) b* value: yellowness (+b*) or blueness (−b*) and (**D**) ΔE* value: overall colour difference. Different letters indicate significant differences at $p < 0.05$. Values are mean ± SEM, $n = 3$.

Increasing spinach content led to a significant decrease in the lightness (L*) of the bread ($p < 0.05$) (Figure 7A). White bread showed the brightest colour, which gradually decreased with increasing spinach concentration. The highest spinach concentration (40%) resulted in the darkest bread (also reflected in Figure 3). Alternatively, redness (a*) showed no difference with or without the addition of spinach except for 10% spinach bread ($p < 0.05$) (Figure 7B). The decrease in redness at 10% spinach enrichment may be due to the dilution effect of carotenoids, resulting in a more prominent green colour from chlorophyll. The yellowness (b*) was highest at the 10% and 20% spinach bread, gradually decreasing with increasing concentration of spinach at 30% and 40% and lowest in the control white bread ($p < 0.05$) (Figure 7C). The overall colour difference (ΔE*) increased significantly with increasing concentrations of spinach compared to white bread ($p < 0.05$) (Figure 7D).

3.6. Sensory Analysis

The sensory analysis of the breads was analysed in two ways: a 9-point hedonic scale test to measure their levels of preference for certain attributes, and a 7-point consumers' future purchasing or consuming intention test. The hedonic scale test reveals the consumer preferences for bread enriched with and without spinach (Figure 8). Despite the analysis conducted in a red-lighting room presented as coded samples to conceal the colour, the highest liking to appearance, odour and taste was demonstrated by the white bread with no differences in texture acceptance between groups ($p > 0.05$). The bread with 10% spinach showed similar acceptance in appearance, odour and taste, while the 20% spinach addition had similar appearance and taste acceptance to the control bread ($p > 0.05$). The overall liking only decreased with the addition of 40% spinach ($p < 0.05$).

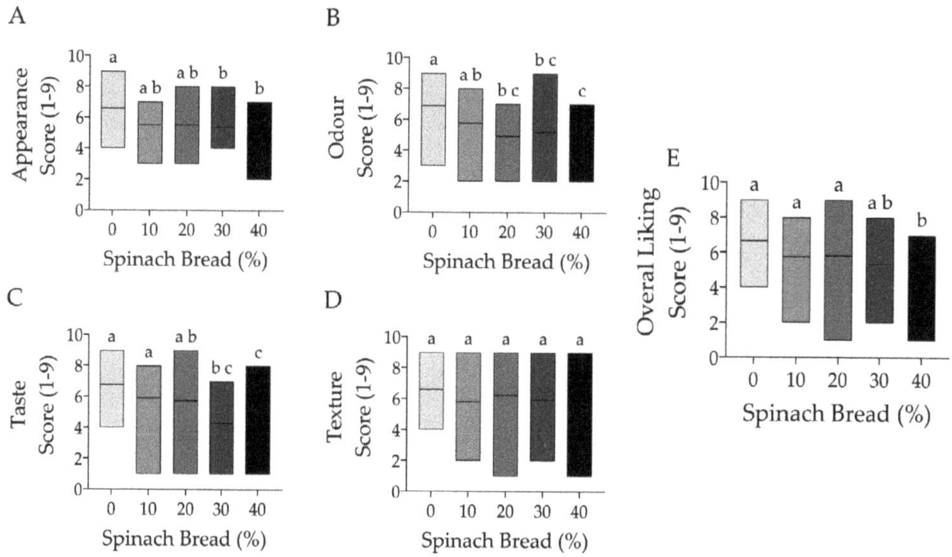

Figure 8. Sensory analysis of control white bread compared with spinach-enriched bread at varying concentrations of 10–40% (*w*/*w*), assessing (**A**) appearance, (**B**) odour, (**C**) taste, (**D**) texture and (**E**) overall liking between scores 1–9. Different letters indicate significant differences at $p < 0.05$. The middle line in the bar graph represents the mean ($n = 21$).

Despite a positive hedonic rating for the spinach-enriched bread, particularly at 10% and 20%, the panellists provided several comments and recommendations to improve the spinach-enriched bread, revealing various consumer perceptions for future product improvement (Table 6). White bread was the most preferred option based on positive feedback, while the spinach-enriched bread received criticism for its odour and bitterness. The negative comments intensified with higher spinach concentrations, with 40% being the most disliked, having the most recommendations, due to the highest degree of bitterness, unpleasant aftertaste and odour. Surprisingly, participants commented on the dry texture of the control white bread and 10% spinach bread but not on the 20%, 30% and 40% spinach-enriched bread. The lack of saltiness is commented on throughout; however, sourness and bitterness are more pronounced in the spinach-enriched bread, coupled with a strongly disliked odour.

Table 6. Panellist comments and recommendations during sensory analysis.

Spinach Bread	Comments and Recommendations
0% (286)	"I love this product" "a bit dry" "a bit bitter and bland, so more salt"
10% (492)	"a bit crumbly" "good balance of taste of bread and spinach" "the smell is still quite strong" "the product is a bit dry, lacks moisture" "feels stale" "bitter and bland, it also tastes eggy" "overall flavour needs improving, maybe more salt" "reduce dryness"
20% (219)	"similar to 492 but slightly more sour" "less spinach taste" "the smell of the bread needs to be improved" "very distinct flavour" "add salt" "bitter and bland"
30% (537)	"still a slight taste of spinach that overpowers the bread" "strong aftertaste" "the smell is quite strong, decrease it" "it is bitter" "very bitter taste"
40% (153)	"a little too fluffy" "strong leafy taste, mouldy odour" "this one has an unpleasant aftertaste" "masking the smell of the spinach" "less spinach taste" "the herbal smell of the bread is quite strong, it might need to be improved, (make it less)" "it is bitter so check if spoilt and if not add more salt to enhance taste" "it is bitter" "very strong taste of spirulina, taste like grass. The texture of the bread is lovely, but need to mask the flavour" "too earthy" "make it less bitter"

Note: Samples were presented as randomised and blinded 3-digit codes.

The consumer's future purchasing or consumption intention test uses a scale of 1 to 7, scoring from the lowest (1) to the highest (7) frequency of willingness to purchase or consume the product. A high mean score (4.62) is received by the control white bread that is not significantly different from spinach-enriched bread at 10% and 20%, scoring 3.86 and 3.90, respectively, with a slight decrease at 30% ($p > 0.05$) (Figure 9). This means that moderate spinach enrichment (20–30%) is well accepted and participants would still purchase or consume the bread frequently. A significant reduction in purchasing or consuming score is most obvious in 40% spinach-enriched bread ($p < 0.05$) and is consistent with the most disliked type observed from the hedonic rating test. Disliking certain characteristics of the bread influences the frequency of consumers' intention to purchase or consume in the future.

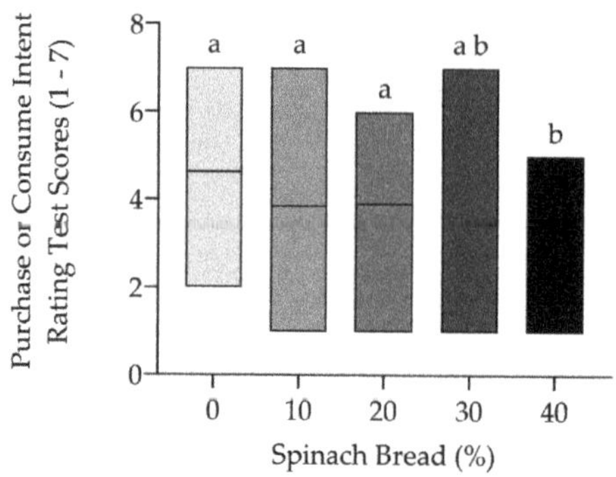

Figure 9. Purchase or consume rating test of control white bread compared to spinach-enriched bread at varying concentrations from 10% to 40% (*w*/*w*) between scores 1–7. Different letters indicate significant differences at $p < 0.05$. The middle line in the bar graph represents the mean ($n = 21$).

4. Discussion

Waste valorisation has gained significant attention in the food industry due to its dual benefits of reusing sustainable ingredients and enhancing the nutritional profile of food products. Incorporating spinach, a nutrient-rich green leafy vegetable, into staple bread is one example of waste valorisation. Several plant ingredients, such as banana, mung bean and moringa leaves, have been used earlier to increase the nutrient composition of bread [44–46]. Our study indicates that spinach-enriched bread exhibits improved nutritional qualities, with acceptable physical and sensory attributes, suggesting potential for innovative and sustainable food products.

Achieving the best formulation for the bread dough was challenging due to the water present in the raw spinach that interfered with the initial product development process. Although several others have formulated bread using freeze-dried vegetables [47,48], in our study, fresh spinach was used to minimise the losses of processing-sensitive nutrients, particularly polyphenols. In addition, the use of frozen spinach developed an unpleasant green hue on the crust and crumb (unpublished results) and was, therefore, not pursued further.

Dough with the highest amount of added spinach showed reduced expansion (5%) upon fermentation due to high fibre content. It is known that fibre-rich doughs have high water absorbability and reduced fermentation tolerance, resulting in shorter doughs [49]. Fibre disrupts the gluten network by interfering and competing with starch and gluten for hydration, resulting in compromised strength and cohesiveness of the dough shown by greater stickiness during handling and kneading processes [50]. Hence, with increasing fibre content in spinach, more prominently in 40% enrichment, greater stickiness reduced the total weight of dough due to losses on the surface, which contributes to lesser bread yield when baked. Moreover, the fibre in spinach enhanced the pasting properties of starch, evident from the increased viscosity parameters. The fibre in the starch matrix resulted in increased viscosity, porosity and stability, reflected by a softer texture. This may be due to the high water absorption and holding ability of fibres compared to starch. Further, the physical degradation (shear scission) of fibres is comparatively lower than that of starch polymer, leading to the increase in trough and final viscosity proportionate to the addition of spinach powder. Additionally, the high water absorption, water-holding capacity and the ability to form a networked gel structure of these fibres [33,51] further contributed to the softness of baked bread without being dry. Similar pasting behaviour was also observed with other high-fibre ingredients, such as oats, peas, lemons, apples [33] and kiwi [51]. Along with fibres, the polyphenols in vegetables interact with the starch matrix to increase viscosity in the starch system [52]. The increased gelatinisation characteristics could be due to the expansion of amylopectin and the molecular dissolution of amylose in the starch-polyphenol system at high temperatures [21]. This means that both the fibre and polyphenols in spinach can potentially influence the gelatinisation process, affecting the pasting properties of bread, and altering the sensory properties of the final products.

The resulting baked bread had similar physical characteristics with or without spinach enrichment owing to the extra added gluten, unlike previous reports, which did not have extra gluten [44–46]. When wheat flour is mixed with water, gluten proteins help create a cohesive dough that traps gases during proofing, essential for producing desirable structural characteristics of soft-leaven bread when heated. Without gluten, doughs lack cohesion and elasticity, affecting the final rheological properties of the bread [53]. In this study, we propose that the interaction between gluten proteins and spinach fibre improved the final texture through noncovalent bonds [54–57], resulting in a denser crumb porosity and better water-holding capacity and firmness with reduced chewiness.

Naturally coloured bioactive compounds such as green chlorophyll and red–orange–yellow carotenoids are prominent in spinach leaves [43], and so the amount of spinach added to the bread affected the colour of the final crumb and crust of the bread. The decrease in lightness values in the spinach bread can be attributed to the degradation or conversion of green chlorophylls (chlorophyll a and b) into olive-brown products (such as pheophytins, pheophorbides, pyropheophytins and pyropheophorbides) during baking [43, 58]. The redness at a lower spinach enrichment level is masked by a high chlorophyll content, showing a significant reduction in 10% spinach bread, while higher stability of carotenoids compared to chlorophylls was observed during heat treatment, as reported earlier [43]. Similar results were not seen when powdered vegetables were used [18,57,59], demonstrating losses in bioactive compounds when fresh ingredients are not used. In contrast, the decrease in yellowness is less affected by green chlorophyll, probably due to their higher abundance than the thermal-sensitive red pigments [43] or produced from the Maillard reactive products following the interactions between amino acids and reducing sugars, similarly found earlier in spinach-enriched flatbread [60]. The green hue was most preferred at 10 or 20% spinach concentrations, while further increases were disliked in appearance and overall liking, consistent with previous studies when spinach and moringa powders were introduced into bread [18,57,60,61].

In general, spinach enrichment affected the taste and odour of the final products. While spinach at lower levels (10% or 20%) did not significantly affect the sensory attributes and purchasing or consumption rating of the bread, higher concentrations (30% or 40%) resulted in lesser acceptance, consistent with several other reports [18,44,57,60]. Bitterness is apparent with increasing spinach concentration, confirming the astringency of spinach polyphenols [44]. Polyphenols from spinach have been shown to interact with the taste receptors T2R39 and T2R60 [62], and the strong grassy and sulphurous odour of spinach is partly due to dimethyl sulphide, methanethiol and dimethyl trisulphide [63–65]. Nevertheless, spinach enrichment at 20% demonstrated the highest acceptance of most sensory characteristics, with few differences from white bread.

The incorporation of spinach in the bread enhanced its nutritional value. While the reduced starch and carbohydrate levels resulted from the reduced flour amount in the formulation, the slight increase in protein and sugar in spinach-enriched bread reflected following the protein and sugar present in spinach at 2.4 g/100 g FW and 0.6 g/100 g FW [29]. Fresh spinach is rich in the antioxidant vitamins A (469 μg/100 g FW) [30], C (28.1 mg/100 g FW) and E (1.3 mg/100 g FW) [29], as well as nitrates, nitrates and polyphenols. Hence, these vitamins become substantially increased upon enrichment, supporting a potential health benefit [66]. Previous incorporation of vegetables into bread [48,67–69] consistently showed increased protein, fibre, phenolic compounds and several micronutrients. While there may be some antagonistic effects due to the presence of compounds like phytic acid and oxalates [70], the overall nutritional benefits of spinach-enriched bread outweigh these effects. The synergistic effects, particularly the enhanced micronutrient, fibre and polyphenol content make spinach-enriched bread a nutritionally superior option compared to regular white bread.

While WHO recommends a daily intake of 400 g of fruits and vegetables to improve health [2], spinach enrichment at 20% contributes to a ~4% increase in vegetable consumption for every slice of bread, together with a several-fold increase in the intake of micronutrients and bioactive compounds, particularly folate, calcium, iron, nitrate, nitrate and polyphenols. The intake of folate, iron and vitamin A is often deficient in some sections of the population globally, causing anaemia and other health issues [71], suggesting the new bread formulation may benefit individuals at risk of deficiency in meeting their nutritional needs. For instance, consuming an average of 76 g/day from two slices of spinach-enriched bread [72] would increase the daily values to 10–23% and 15% for iron and folate, respectively, instead of only 4–8% and 7% from white bread (where the recommended daily intake values for Australian adults are 8–18 mg and 400 µg for iron and folate, respectively [73]).

Nitrates and nitrites regulate blood pressure and lower hypertension, aiding cardiovascular health and non-communicable disease prevention [74–77], while polyphenols, a group of bioactive compounds in fruits and vegetables, offer potential health benefits. It is worth noting that certain flavonoids can inhibit carbohydrate-digesting enzymes, α-amylase and α-glucosidase, which may contribute to better glycaemic control and the management of diabetes [12,78,79]. For example, quercetagetin, initially isolated from spinach leaves [10], exhibited substantial inhibitory action against sucrase, maltase and isomaltase [12]. Ultimately, spinach-enriched bread addresses nutrient gaps in the diet and supports better health outcomes.

5. Conclusions

The incorporation of fresh spinach into bread significantly enhanced the nutrient composition, physical characteristics and sensory quality of bread, and at 20% *w*/*w*, it was acceptable to consumers. Spinach enrichment increased protein, fibre, iron, folate, niacin, pyridoxine, riboflavin, magnesium, potassium, zinc, calcium, vitamins A, C, and E, nitrate, nitrite and polyphenols. The improved pasting properties of spinach-enriched bread, characterised by increased viscosities and decreased pasting time and temperature, contributed to a more porous bread structure and enhanced textural properties. The interactions between spinach fibres and gluten proteins, along with gluten added to compensate for flour substitution, played a key role in these improvements. Sensory evaluation revealed that higher spinach concentrations led to decreased visual appeal, less preferred taste, odour, and overall liking, with a decrease in purchase/consumption intention. However, bread with 20% spinach was mostly comparable to white bread in terms of sensory attributes. In addition, this approach offers a sustainable solution for the valorisation of food waste, contributing to both consumer health and environmental sustainability.

Supplementary Materials: The following supporting information can be downloaded at https://www.mdpi.com/article/10.3390/foods13152401/s1, Video S1: Timelapse of dough fermentation for control and dough enriched with 20% and 40% spinach. Research Data S1: Raw data supporting reported results.

Author Contributions: Conceptualisation, G.W. and E.B.; data curation, R.V.P. and E.B.; formal analysis, R.V.P. and E.B.; funding acquisition, G.W. and E.B.; investigation, R.V.P. and E.B.; methodology, R.V.P., S.D., G.W. and E.B.; project administration, E.B.; resources, G.W. and E.B.; supervision, G.W. and E.B.; validation, R.V.P., G.W. and E.B.; visualization, R.V.P. and E.B.; writing—original draft, R.V.P. and E.B.; writing—review and editing, R.V.P., S.D., G.W. and E.B. All authors have read and agreed to the published version of the manuscript.

Funding: This research received funding from Monash University.

Institutional Review Board Statement: The ethics was approved by the Monash University Human Research Ethics Committee (MUHREC #26497).

Informed Consent Statement: Informed consent was obtained from all subjects involved in the study.

Data Availability Statement: The original contributions presented in the study are included in the article/Supplementary Material; further inquiries can be directed to the corresponding author.

Acknowledgments: The authors would like to thank Louise Bennett for her guidance, Dilini Perera for her technical assistance and the participants who volunteered for the sensory analysis.

Conflicts of Interest: The authors declare no conflict of interest.

References

1. Sarker, A.; Ahmmed, R.; Ahsan, S.M.; Rana, J.; Ghosh, M.K.; Nandi, R. A Comprehensive Review of Food Waste Valorization for the Sustainable Management of Global Food Waste. *Sustain. Food Technol.* **2024**, *2*, 48–69. [CrossRef]
2. World Health Organisation. Increasing Fruit and Vegetable Consumption to Reduce the Risk of Noncommunicable Diseases. Available online: https://www.who.int/tools/elena/interventions/fruit-vegetables-ncds#:~:text=WHO%20Recommendations, the%20risk%20of%20certain%20NCDs (accessed on 18 January 2024).
3. Petersen, K.S.; Kris-Etherton, P.M. Diet Quality Assessment and the Relationship between Diet Quality and Cardiovascular Disease Risk. *Nutrients* **2021**, *13*, 4305. [CrossRef] [PubMed]
4. Shang, X.; Liu, J.; Zhu, Z.; Zhang, X.; Huang, Y.; Liu, S.; Wang, W.; Zhang, X.; Tang, S.; Hu, Y.; et al. Healthy Dietary Patterns and the Risk of Individual Chronic Diseases in Community-Dwelling Adults. *Nat. Commun.* **2023**, *14*, 6704. [CrossRef] [PubMed]
5. Sharifi-Rad, J.; Rodrigues, C.F.; Sharopov, F.; Docea, A.O.; Can Karaca, A.; Sharifi-Rad, M.; Kahveci Karıncaoglu, D.; Gülseren, G.; Şenol, E.; Demircan, E.; et al. Diet, Lifestyle and Cardiovascular Diseases: Linking Pathophysiology to Cardioprotective Effects of Natural Bioactive Compounds. *Int. J. Environ. Res. Public Health* **2020**, *17*, 2326. [CrossRef] [PubMed]
6. Cicero, A.F.G.; Fogacci, F.; Stoian, A.P.; Vrablik, M.; Al Rasadi, K.; Banach, M.; Toth, P.P.; Rizzo, M. Nutraceuticals in the Management of Dyslipidemia: Which, When, and for Whom? Could Nutraceuticals Help Low-Risk Individuals with Non-Optimal Lipid Levels? *Curr. Atheroscler. Rep.* **2021**, *23*, 57. [CrossRef]
7. Iqbal, I.; Wilairatana, P.; Saqib, F.; Nasir, B.; Wahid, M.; Latif, M.F.; Iqbal, A.; Naz, R.; Mubarak, M.S. Plant Polyphenols and Their Potential Benefits on Cardiovascular Health: A Review. *Molecules* **2023**, *28*, 6403. [CrossRef] [PubMed]
8. Roughani, A.; Miri, S.M. Spinach: An Important Green Leafy Vegetable and Medicinal Herb. In Proceedings of the 2nd International Conference on Medicinal Plants, Organic Farming, Natural and Pharmaceutical Ingredients, Mashad, Iran, 13 March 2019.
9. Sri Lasya, C.; Author Chokkara Sri Lasya, C. Spinach and Its Health Benefits: A Review. *Pharma Innov. J.* **2022**, *11*, 1232–1239.
10. Zane, A.; Wender, S.H. Flavonols in Spinach Leaves. *J. Org. Chem.* **1961**, *26*, 4718–4719. [CrossRef]
11. Phenol-Explorer Food Composition-Spinach, Raw. Available online: http://phenol-explorer.eu/contents/food/272#folin-assay (accessed on 28 January 2024).
12. Barber, E.; Houghton, M.J.; Williamson, G. Flavonoids as Human Intestinal α-Glucosidase Inhibitors. *Foods* **2021**, *10*, 1939. [CrossRef]
13. Liu, R.H. Health-Promoting Components of Fruits and Vegetables in the Diet. *Adv. Nutr.* **2013**, *4*, 384S–392S. [CrossRef]
14. Frankowska, A.; Jeswani, H.K.; Azapagic, A. Environmental Impacts of Vegetables Consumption in the UK. *Sci. Total Environ.* **2019**, *682*, 80–105. [CrossRef] [PubMed]
15. Statista Bread—Worldwide. Available online: https://www.statista.com/outlook/cmo/food/bread-cereal-products/bread/worldwide (accessed on 28 January 2024).
16. Willett, W.; Manson, J.; Liu, S. Glycemic Index, Glycemic Load, and Risk of Type 2 Diabetes. *Am. J. Clin. Nutr.* **2002**, *76*, 274S–280S. [CrossRef] [PubMed]
17. Food Standards Australia New Zealand Survey of Nitrates and Nitrites in Food and Beverages in Australia. Available online: https://www.foodstandards.gov.au/science-data/surveillance/surveyofnitrates (accessed on 22 June 2024).
18. Junejo, S.A.; Rashid, A.; Yang, L.; Xu, Y.; Kraithong, S.; Zhou, Y. Effects of Spinach Powder on the Physicochemical and Antioxidant Properties of Durum Wheat Bread. *LWT* **2021**, *150*, 112058. [CrossRef]
19. American Association of Cereal Chemists. *Approved Methods of the AACC–Method 74-09*; AACC: St Paul, MN, USA, 1988.
20. Baker, A.E.; Walker, C.E.; Kemp, K. An Optimum Compression Depth for Measuring Bread Crumb Firmness. In *Cereal Chemistry*; American Association of Cereal Chemists: St. Paul, MN, USA, 1988; Volume 65, pp. 302–307.
21. Chou, S.; Meng, X.; Cui, H.; Zhang, S.; Wang, H.; Li, B. Rheological and Pasting Properties of Maize, Wheat and Rice Starch as Affected by Apple Polyphenols. *Int. J. Food Prop.* **2019**, *22*, 1786–1798. [CrossRef]
22. Han, X.; Zhang, M.; Zhang, R.; Huang, L.; Jia, X.; Huang, F.; Liu, L. Physicochemical Interactions between Rice Starch and Different Polyphenols and Structural Characterization of Their Complexes. *LWT* **2020**, *125*, 109227. [CrossRef]
23. *ISO 6658:2017*; Sensory Analysis—Methodology—General Guidance 2017. ISO: Geneva, Switzerland, 2017.
24. Mammasse, N.; Schlich, P. Adequate Number of Consumers in a Liking Test. Insights from Resampling in Seven Studies. *Food Qual. Prefer.* **2014**, *31*, 124–128. [CrossRef]
25. Nyitrai, Á.; Urbin, Á.; Nagy, B.V.; Sipos, L. Novel Approach in Sensory Color Masking: Effects of Colored Environments on Chocolates with Different Cocoa Content. *Food Qual. Prefer.* **2022**, *95*, 104363. [CrossRef]
26. Singh-Ackbarali, D.; Maharaj, R. Sensory Evaluation as a Tool in Determining Acceptability of Innovative Products Developed by Undergraduate Students in Food Science and Technology at The University of Trinidad and Tobago. *J. Curric. Teach.* **2014**, *3*, 10–27. [CrossRef]
27. Food Standards Australia and New Zealand (FSANZ) Food Composition Databases. Available online: https://www.foodstandards.gov.au/science-data/food-composition-databases (accessed on 7 June 2024).

28. Xyris FoodWorks Introduction to Dietary Analysis with FoodWorks 10. Available online: https://support.xyris.com.au/hc/en-us/article_attachments/360002775975 (accessed on 7 June 2024).
29. Food Standards Australia and New Zealand Australian Food Composition Database—Release 2.0. Available online: https://afcd.foodstandards.gov.au/fooddetails.aspx?PFKID=F008761 (accessed on 22 July 2024).
30. U.S. Department of Agriculture Spinach, Raw. Available online: https://fdc.nal.usda.gov/fdc-app.html#/food-details/168462/nutrients (accessed on 19 May 2024).
31. Kumar, R.; Khatkar, B.S. Thermal, Pasting and Morphological Properties of Starch Granules of Wheat (*Triticum aestivum* L.) Varieties. *J. Food Sci. Technol.* **2017**, *54*, 2403–2410. [CrossRef]
32. Mohd Shukri, A.; Cheng, L.-H. The Properties of Different Starches under the Influence of Glucono-Delta-Lactone at Different Concentrations. *Foods* **2023**, *12*, 1770. [CrossRef]
33. Yildiz, Ö.; Yurt, B.; Baştürk, A.; Toker, Ö.S.; Yilmaz, M.T.; Karaman, S.; Dağlıoğlu, O. Pasting Properties, Texture Profile and Stress–Relaxation Behavior of Wheat Starch/Dietary Fiber Systems. *Food Res. Int.* **2013**, *53*, 278–290. [CrossRef]
34. Ragaee, S.; Abdel-Aal, E.-S.M. Pasting Properties of Starch and Protein in Selected Cereals and Quality of Their Food Products. *Food Chem.* **2006**, *95*, 9–18. [CrossRef]
35. Kaushal, P.; Kumar, V.; Sharma, H.K. Comparative Study of Physicochemical, Functional, Antinutritional and Pasting Properties of Taro (*Colocasia esculenta*), Rice (*Oryza sativa*) Flour, Pigeonpea (*Cajanus cajan*) Flour and Their Blends. *LWT-Food Sci. Technol.* **2012**, *48*, 59–68. [CrossRef]
36. Pacheco, A.F.C.; Pacheco, F.C.; dos Santos, F.R.; Cunha, J.S.; Paiva, P.H.C. Starch Paste Properties. In *Methods and Protocols in Food Science*; Banger, S.P., Ed.; Springer Science+Business Media: Berlin/Heidelberg, Germany, 2024; pp. 123–143.
37. Fagundes, G.A.; Rocha, M.; Salas-Mellado, M.M. Improvement of Protein Content and Effect on Technological Properties of Wheat Bread with the Addition by Cobia (*Rachycentron canadum*). *Food Res.* **2018**, *2*, 221–227. [CrossRef]
38. Sari, K.I.; Rafisa, A. Chewing and Swallowing Patterns for Different Food Textures in Healthy Subjects. *Int. J. Dent.* **2023**, *2023*, 6709350. [CrossRef]
39. Guiné, R.P.F.; Roque, A.R.F.; Seiça, F.F.A.; Batista, C.E.O. Effect of Chemical Pretreatments on the Physical Properties of Kiwi. *ETP Int. J. Food Eng.* **2016**, *2*, 90–95. [CrossRef]
40. Paredes, J.; Cortizo-Lacalle, D.; Imaz, A.M.; Aldazabal, J.; Vila, M. Application of Texture Analysis Methods for the Characterization of Cultured Meat. *Sci. Rep.* **2022**, *12*, 3898. [CrossRef]
41. Trịnh, K.T.; Glasgow, S. On the Texture Profile Analysis Test. In Proceedings of the Chemeca 2012, Wellington, New Zealand, 23–26 September 2012.
42. Szczesniak, A.S. Texture Is a Sensory Property. *Food Qual. Prefer.* **2002**, *13*, 215–225. [CrossRef]
43. Kidmose, U.; Edelenbos, M.; Christensen, L.P.; Hegelund, E. Chromatographic Determination of Changes in Pigments in Spinach (*Spinacia oleracea* L.) During Processing. *J. Chromatogr. Sci.* **2005**, *43*, 466–472. [CrossRef]
44. Waseem, M.; Akhtar, S.; Manzoor, M.F.; Mirani, A.A.; Ali, Z.; Ismail, T.; Ahmad, N.; Karrar, E. Nutritional Characterization and Food Value Addition Properties of Dehydrated Spinach Powder. *Food Sci. Nutr.* **2021**, *9*, 1213–1221. [CrossRef] [PubMed]
45. Sengev, A.I.; Abu, J.O.; Gernah, D.I. Effect of *Moringa oleifera* Leaf Powder Supplementation on Some Quality Characteristics of Wheat Bread. *Food Nutr. Sci.* **2013**, *4*, 270–275. [CrossRef]
46. Okakpu, K.G.; Offia-Olua, B.I.; Okakpu, C.J.; Okpara, C.M. Quality Characteristics of Bread Made from Flour Blends of Wheat, Cooking Banana and Mungbean. *J. Adv. Food Sci. Technol.* **2023**, *10*, 9–15. [CrossRef]
47. Begum, R.; Chowdhury, M.A.F.; Hasan, M.R.; Rahman, M.F.; Rahman, M.H.; Alim, M.A. Efficacy of Freeze-Dried Carrot Pomace Powder in Improving the Quality of Wheat Bread. *Food Res.* **2023**, *7*, 11–22. [CrossRef]
48. Ranawana, V.; Campbell, F.; Bestwick, C.; Nicol, P.; Milne, L.; Duthie, G.; Raikos, V. Breads Fortified with Freeze-Dried Vegetables: Quality and Nutritional Attributes. Part II: Breads Not Containing Oil as an Ingredient. *Foods* **2016**, *5*, 62. [CrossRef]
49. Wang, J.; Rosell, C.M.; Benedito de Barber, C. Effect of the Addition of Different Fibres on Wheat Dough Performance and Bread Quality. *Food Chem.* **2002**, *79*, 221–226. [CrossRef]
50. Ficco, D.B.M.; Muccilli, S.; Padalino, L.; Giannone, V.; Lecce, L.; Giovanniello, V.; Del Nobile, M.A.; De Vita, P.; Spina, A. Durum Wheat Breads "high in Fibre" and with Reduced in Vitro Glycaemic Response Obtained by Partial Semolina Replacement with Minor Cereals and Pulses. *J. Food Sci. Technol.* **2018**, *55*, 4458–4467. [CrossRef]
51. Wang, Y.; Pan, Y.; Zhou, C.; Li, W.; Wang, K. Effects of Kiwifruit Dietary Fibers on Pasting Properties and In Vitro Starch Digestibility of Wheat Starch. *Nutrients* **2024**, *16*, 749. [CrossRef]
52. Zhu, F. Interactions between Starch and Phenolic Compound. *Trends Food Sci. Technol.* **2015**, *43*, 129–143. [CrossRef]
53. Cappelli, A.; Oliva, N.; Cini, E. A Systematic Review of Gluten-Free Dough and Bread: Dough Rheology, Bread Characteristics, and Improvement Strategies. *Appl. Sci.* **2020**, *10*, 6559. [CrossRef]
54. Zhou, Y.; Dhital, S.; Zhao, C.; Ye, F.; Chen, J.; Zhao, G. Dietary Fiber-Gluten Protein Interaction in Wheat Flour Dough: Analysis, Consequences and Proposed Mechanisms. *Food Hydrocoll.* **2021**, *111*, 106203. [CrossRef]
55. Gómez, M.; Ronda, F.; Blanco, C.A.; Caballero, P.A.; Apesteguía, A. Effect of Dietary Fibre on Dough Rheology and Bread Quality. *Eur. Food Res. Technol.* **2003**, *216*, 51–56. [CrossRef]
56. Amoah, I.; Cairncross, C.; Rush, E. Swallowing and Liking of Vegetable-Enriched Bread Compared With Commercial Breads as Evaluated by Older Adults. *Front. Nutr.* **2021**, *7*, 599737. [CrossRef] [PubMed]

57. Zhou, Q.; Liang, W.; Wan, J.; Wang, M. Spinach (*Spinacia oleracea*) Microgreen Prevents the Formation of Advanced Glycation End Products in Model Systems and Breads. *Curr. Res. Food Sci.* **2023**, *6*, 100490. [CrossRef] [PubMed]
58. Nisha, P.; Singhal, R.S.; Pandit, A.B. A Study on the Degradation Kinetics of Visual Green Colour in Spinach (Spinacea Oleracea L.) and the Effect of Salt Therein. *J. Food Eng.* **2004**, *64*, 135–142. [CrossRef]
59. Mohtarami, F.; Esmaiili, M.; Nouraddini, M.; Ostad, I. Fortification of Simit Bread with Spinach Powder: Evaluation of Physico-Chemical, Textural, and Sensorial Properties. *Food Sci. Technol.* **2022**, *18*, 225–236. [CrossRef]
60. Khan, M.A.; Mahesh, C.; Semwal, A.D.; Sharma, G.K. Effect of Spinach Powder on Physico-Chemical, Rheological, Nutritional and Sensory Characteristics of Chapati Premixes. *J. Food Sci. Technol.* **2015**, *52*, 2359–2365. [CrossRef] [PubMed]
61. Kiranawati, T.M.; Hidayati, L.; Saputri, A.M.J.; Sari, R.A.; Susanto, H. The Analysis of Bread Quality from Moringa Oleifera (Kelor) Leaf Flour. In Proceedings of the International Conference on Life Sciences and Technology, Malang, Indonesia, 29 September 2020; AIP Publishing: Melville, NY, USA, 2021; p. 030023.
62. Osakabe, N.; Shimizu, T.; Fujii, Y.; Fushimi, T.; Calabrese, V. Sensory Nutrition and Bitterness and Astringency of Polyphenols. *Biomolecules* **2024**, *14*, 234. [CrossRef] [PubMed]
63. Kebede, B.T.; Grauwet, T.; Tabilo-Munizaga, G.; Palmers, S.; Vervoort, L.; Hendrickx, M.; Van Loey, A. Headspace Components That Discriminate between Thermal and High Pressure High Temperature Treated Green Vegetables: Identification and Linkage to Possible Process-Induced Chemical Changes. *Food Chem.* **2013**, *141*, 1603–1613. [CrossRef] [PubMed]
64. Masanetz, C.; Guth, H.; Grosch, W. Fishy and Hay-like off-Flavours of Dry Spinach. *Z. Leb. Und-Forsch. A* **1998**, *206*, 108–113. [CrossRef]
65. Díaz-Mula, H.M.; Marín, A.; Jordán, M.J.; Gil, M.I. Off-Odor Compounds Responsible for Quality Loss of Minimally Processed Baby Spinach Stored under MA of Low O_2 and High CO_2 Using GC–MS and Olfactometry Techniques. *Postharvest Biol. Technol.* **2017**, *129*, 129–135. [CrossRef]
66. Mehta, V.; Desai, N.; Perwez, A.; Nemade, D.; Dawoodi, S.; Zaman, S. Bin ACE Alzheimer's: The Role of Vitamin A, C and E (ACE) in Oxidative Stress Induced Alzheimer's Disease. *J. Med. Res. Innov.* **2017**, *2*, e000086. [CrossRef]
67. Purić, M.; Rabrenović, B.; Rac, V.; Pezo, L.; Tomašević, I.; Demin, M. Application of Defatted Apple Seed Cakes as a By-Product for the Enrichment of Wheat Bread. *LWT* **2020**, *130*, 109391. [CrossRef]
68. Czarnowska-Kujawska, M.; Starowicz, M.; Barišić, V.; Kujawski, W. Health-Promoting Nutrients and Potential Bioaccessibility of Breads Enriched with Fresh Kale and Spinach. *Foods* **2022**, *11*, 3414. [CrossRef]
69. Galla, N.R.; Pamidighantam, P.R.; Karakala, B.; Gurusiddaiah, M.R.; Akula, S. Nutritional, Textural and Sensory Quality of Biscuits Supplemented with Spinach (*Spinacia oleracea* L.). *Int. J. Gastron. Food Sci.* **2017**, *7*, 20–26. [CrossRef]
70. Lee, H.W.; Bi, X.; Henry, C.J. Assessment of Oxalates and Phytic Acid Contents in Asian Green Leafy Vegetables: Dietary Recommendations. *Food Humanit.* **2023**, *1*, 1223–1228. [CrossRef]
71. World Health Organisation. *Accelerating Anaemia Reduction: A Comprehensive Framework for Action*; WHO: Geneva, Switzerland, 2023.
72. Grafenauer, S.; Curtain, F. An Audit of Australian Bread with a Focus on Loaf Breads and Whole Grain. *Nutrients* **2018**, *10*, 1106. [CrossRef]
73. National Health and Medical Research Council. *Nutrient Reference Values for Australia and New Zealand Executive Summary*; National Health and Medical Research Council: Canberra, Australia, 2006.
74. Kapil, V.; Khambata, R.S.; Robertson, A.; Caulfield, M.J.; Ahluwalia, A. Dietary Nitrate Provides Sustained Blood Pressure Lowering in Hypertensive Patients. *Hypertension* **2015**, *65*, 320–327. [CrossRef]
75. Bondonno, C.P.; Blekkenhorst, L.C.; Liu, A.H.; Bondonno, N.P.; Ward, N.C.; Croft, K.D.; Hodgson, J.M. Vegetable-Derived Bioactive Nitrate and Cardiovascular Health. *Mol. Aspects Med.* **2018**, *61*, 83–91. [CrossRef]
76. Bondonno, C.P.; Yang, X.; Croft, K.D.; Considine, M.J.; Ward, N.C.; Rich, L.; Puddey, I.B.; Swinny, E.; Mubarak, A.; Hodgson, J.M. Flavonoid-Rich Apples and Nitrate-Rich Spinach Augment Nitric Oxide Status and Improve Endothelial Function in Healthy Men and Women: A Randomized Controlled Trial. *Free Radic. Biol. Med.* **2012**, *52*, 95–102. [CrossRef]
77. Jonvik, K.L.; Nyakayiru, J.; Pinckaers, P.J.; Senden, J.M.; van Loon, L.J.; Verdijk, L.B. Nitrate-Rich Vegetables Increase Plasma Nitrate and Nitrite Concentrations and Lower Blood Pressure in Healthy Adults. *J. Nutr.* **2016**, *146*, 986–993. [CrossRef]
78. Zhu, J.; Chen, C.; Zhang, B.; Huang, Q. The Inhibitory Effects of Flavonoids on α-Amylase and α-Glucosidase. *Crit. Rev. Food Sci. Nutr.* **2020**, *60*, 695–708. [CrossRef]
79. Tadera, K.; Minami, Y.; Takamatsu, K.; Matsuoka, T. Inhibition of α-Glucosidase and α-Amylase by Flavonoids. *J. Nutr. Sci. Vitaminol.* **2006**, *52*, 149–153. [CrossRef]

foods

MDPI

Article

The Impact of Selected Ingredients on the Predicted Glycemic Index and Technological Properties of Bread

Ilaria Pasqualoni, Roberta Tolve, Barbara Simonato * and Federico Bianchi

Department of Biotechnology, University of Verona, Strada Le Grazie 15, 37134 Verona, Italy; ilaria.pasqualoni@univr.it (I.P.); roberta.tolve@univr.it (R.T.); federico.bianchi_02@univr.it (F.B.)

* Correspondence: barbara.simonato@univr.it

Abstract: Bread, a staple food consumed worldwide, plays a pivotal role in nutrition. Nevertheless, it is to be underlined that white bread is classified as a high glycemic index food, and its frequent consumption can lead to rapid increases in blood glucose, potentially causing metabolic stress and contributing to insulin resistance and type 2 diabetes. So, there is a growing interest in bread formulations with ingredients that can lower its GI. With this view, bread was formulated, substituting wheat with chickpea flour, red chicory powder, and three distinct types of resistant starch. The results showed the different resistant starches' impacts on the glycemic index reduction. Specifically, chemically modified tapioca RS IV produced a bread formulation with a low predicted glycemic index (pGI < 55). Retrograded starch from tapioca (RS III) allows the bread to reach a pGI value of 55, the upper value for classifying a food as low pGI. The retrograded starch from corn (RS III) allows a decrease in the bread's glycemic index, but the product is still classified as 'high pGI' (>70). Moreover, the addition of by-products rich in polyphenols contributes to a lowering of the pGI. Concerning the technological parameters, the outcome revealed an increase in the moisture content across all the newly formulated samples compared to the control. At the same time, the volume and specific volume showed a decrease. The newly formulated samples exhibited a higher baking loss, particularly when incorporating resistant starch, which increased the hardness and chewiness with decreased cohesiveness. In conclusion, incorporating chickpea flour, red chicory powder, and tapioca-resistant starch (RS III and IV) offers a promising strategy for producing high-fiber bread with a low glycemic index, catering to health-conscious consumers.

Keywords: low glycemic index; chickpea flour; dietary fiber; resistant starches; red chicory by-product powder

1. Introduction

Bread plays a pivotal role in the longstanding dietary traditions of the Mediterranean region, serving as a cornerstone in culinary practices. Traditionally, bread is made with gluten-rich wheat flour, contributing to its distinctive texture and palatability. Gluten's structure is maintained by non-covalent interactions (hydrogen bonds, ionic bonds, hydrophobic interactions) and a small proportion of disulfide-type covalent bonds (about 2%). These interactions are crucial for protein aggregation, influencing the unique properties of commercial products. Wheat gluten's characteristics derive from gliadins (providing viscosity) and glutenins (providing elasticity). Gliadins, monomeric proteins with molecular weights between 3 and 8×10^4 Da, contain intra-chain disulfide bridges, while glutenins, polymeric proteins with molecular weights between 10^5 and 10^7 Da, have both intra- and inter-chain disulfide bonds, contributing to gluten's rigidity and elasticity [1]. Wheat bread is typically classified as a high glycemic index food. The glycemic index (GI) measures how quickly a food can raise blood glucose levels and reflects the quality of carbohydrates in the diet. Specifically, the rate of enzymatic digestion of starch into glucose may vary from rapid to slow. Starch resistant to enzymatic hydrolysis is also known [2]. Frequent consumption of high-GI foods can lead to rapid increases in blood glucose, potentially causing metabolic stress and contributing to insulin resistance and type 2 diabetes [3]. Nowadays, increasing consumer awareness about health and wellness is transforming the food industry. As a result, there is a growing interest in bread formulations with ingredients that can lower its GI, such as different types of legume flours, dietary fibers (DF), proteins, and phenolic compounds that can enhance the bread's nutritional profile [4,5]. DF contributes to various physiological effects, including decreased intestinal transit time, increased stool bulk, reduced postprandial blood glucose and insulin levels, gastric acid buffering, and the prevention of constipation [6,7].

Incorporating DF into food products offers various health benefits and functional attributes [8]. An ample spectrum of fiber can be selected to enrich foods; among them, resistant starches (RSs) have been evaluated and considered in the present study.

RSs appeared to be a sensible alternative for developing functional bread and other starch-based foods due to their ability to reduce the caloric value of food products and to decrease the predicted glycemic index. RSs, being non-digestible, are fermented in the colon by the microbiota to produce short-chain fatty acids like acetic, propionic, and butyric acids, which favorably select intestinal microflora; reduce cholesterol, triglycerides, and urea levels; and prevent gut cancer [9]. RSs are classified into five types: physically inaccessible starch in undamaged plant cells (RS I), raw starch from certain plants like potatoes (RS II), retrograded starch forming water-insoluble structures (RS III), chemically or physically modified starch (RS IV), and amylose-lipid complexes (RS V) [10].

In addition to DF, phenolic compounds (PCs) can modulate starch digestibility by directly inhibiting α-glycosidase digestive enzymes and/or forming complexes with starch [11]. In this view, enhancing the nutritional profile of functional bread using red chicory by-products rich in PCs and DF aligns with the EU's Green Deal initiatives [12,13].

Finally, chickpea flour (CPF), similar to other legumes, is a rich source of high-quality proteins and dietary fiber, which can help lower the glycemic response. Several factors contribute to this, including the rigidity of cotyledonary cell walls, legume starches' low enzyme susceptibility, and polyphenols and other α-amylase inhibitors [14]. Thus, replacing wheat flour with chickpea flour can reduce the GI of wheat bread [15]. It is essential to highlight that including various ingredients, at the expense of wheat flour and its gluten content, also impacts the bread's technological characteristics. While many researchers have assessed the nutritional benefits of bread supplemented with resistant starch, agrifood by-products rich in phenolic compounds, and legume flour, none have explicitly focused on how these ingredients' single and simultaneous presence impact the resulting bread's technological and glycemic index characteristics.

This study aimed to understand the effect of red chicory powder, chickpea flour, and resistant starches on bread glycemic index reduction and its technological properties.

2. Materials and Methods

2.1. Ingredients and Breadmaking

Common white wheat flour was kindly supplied by Macinazione Lendinara SpA (Arcole, Italy). The wheat flour label detailed the following composition: 343 kcal/100 g, fat 1.2 g/100 g, total carbohydrates 70 g/100 g, protein 12 g/100 g, and dietary fiber 2.3 g/100 g. Red Chicory leaves (*Cichorium intybus* L.) were provided by GEOFUR, a local consortium of Verona (Legnago, Italy). The leaves were dried in an oven at 50 °C for 16 h, ground with Grindomix GM 200 (Retsch, Haan, Germany), and went through a sieve with meshes of 0.2 mm to achieve a fine powder that has the following nutritional characteristics: 240 kcal/100 g, total carbohydrates 23.5 g/100 g, fat 3.3 g/100 g, fiber 27.4 g/100 g, and protein 15.8 g/100 g. The chickpea flour was purchased from Molino Rossetto (Padova, Italy), with the following composition: 355 kcal/100 g, fat 6.6 g/100 g, total carbohydrates 4.1 g/100 g, fiber 9 g/100 g, protein 21 g/100 g, and salt 0.03 g/100 g. Gluten, RSs, and guar gum were provided by ABS Food srl (Padova, Italy), and other ingredients such as sugar, salt, and dried yeast were purchased from a local market. The breads were prepared with a commercial breadmaking machine (Panexpress 750 Metal, model 0132/00—Ariete, Trieste, Italy). Doughs were mixed with a blender (Kenwood, Chef XL KVL4110S, Treviso, Italy), fermented for 3 h at room temperature, and then cooked at 170 °C for 2 h. The overview of the experimental bread formulations and analyses are reported in Figure 1, while the sample formulations expressed as percentages are reported in Table 1.

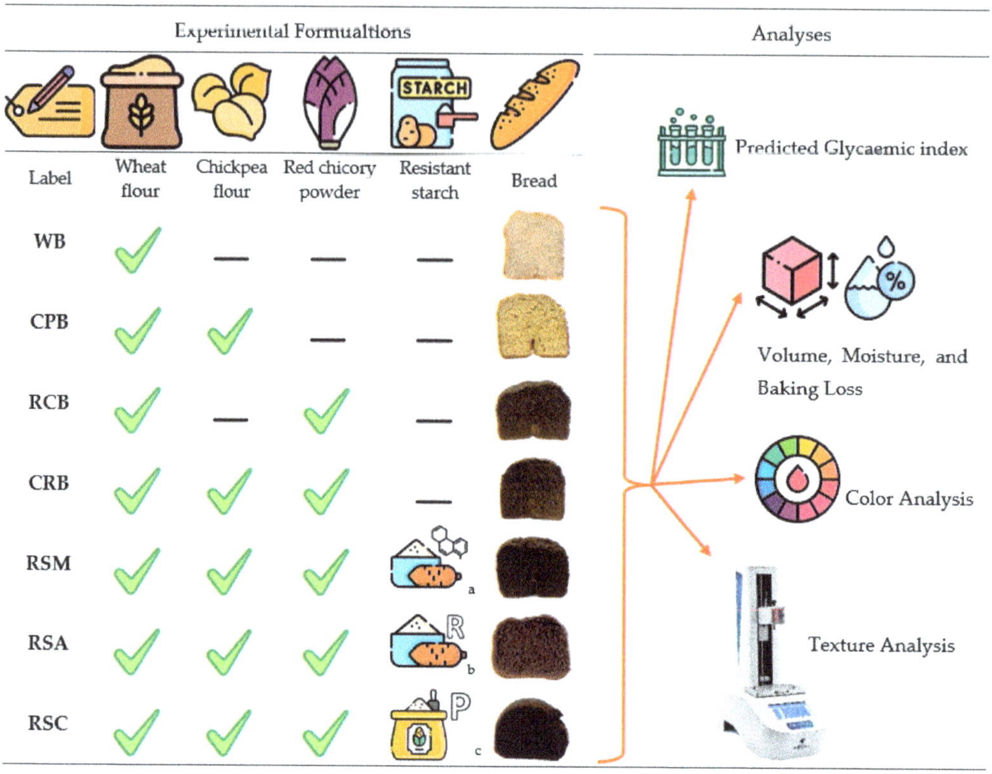

Figure 1. Summary of the experimental bread formulations and analyses performed in this study. a chemically modified tapioca resistant starch. b retrograded tapioca resistant starch. c pregelatinized corn resistant starch.

Table 1. Control and experimental bread formulated with chickpea flour, red chicory powder, and resistant starch.

Sample	Wheat Flour (%)	Chickpea Flour (%)	Red Chicory Powder (%)	Dried Yeast (%)	Sugar (%)	Salt (%)	Guar Gum (%)	Native Gluten (%)	Resistant Starch (%)
WB	86	-	-	2	1	1	5	5	-
CPB	56	30	-	2	1	1	5	5	-
RCB	76	-	10	2	1	1	5	5	-
CRB	46	30	10	2	1	1	5	5	-
RSM	31	30	10	2	1	1	5	5	15
RSA	31	30	10	2	1	1	5	5	15
RSC	31	30	10	2	1	1	5	5	15

WB: white bread; CPB: bread with 30% chickpea flour; RCB: bread with 10% red chicory powder; CRB: bread with 30% chickpea flour and 10% red chicory powder; RSM: bread with 30% chickpea flour, 10% red chicory powder, and 15% chemically modified tapioca resistant starch; RSA: bread with 30% chickpea flour, 10% red chicory powder, and 15% retrograded tapioca resistant starch; RSC: bread with 30% chickpea flour, 10% red chicory powder, and 15% pregelatinized corn resistant starch.

2.2. Bread Hydrolysis Index and Predicted Glycemic Index

The hydrolysis index was determined following the methodology of Bianchi et al. [13]. One hundred milligrams of bread was incubated in glass vials at 37 °C. During the incubation, 4 mL of maleic buffer (pH 6) containing 40 mg of pancreatic amylase (3000 U/mg) and 4 mL of amyloglucosidase solution (300 U/mL) from Megazyme (Megazyme, Wicklow, Ireland) were added. The reaction was halted at 0, 30, 60, 120, and 180 min with 4 mL of pure ethanol introduction; the solution was then centrifuged at 1500× *g* for 10 min. The assessment of D-glucose was determined spectrophotometrically at 510 nm using a glucose oxidase kit (GOPOD, Megazyme, Wicklow, Ireland). The hydrolysis index was defined as the percentage ratio of the area under the curve representing pasta sample hydrolysis (0–180 min) to the area under the curve of a reference white bread. The pGI was calculated according to Granfeldt et al. [16] using a specific Formula (1).

$$pGI = 8.198 + 0.862 \times HI \quad (1)$$

2.3. Bread Quality Characteristics

2.3.1. Moisture Content, Volume, Specific Volume, and Baking Loss

The moisture content was measured using the AACC method 44-15A [17]. The bread volume was determined by seed displacement [17], and the specific volume (cm^3/g) was determined by dividing the volume by the weight of the samples. The baking loss (BL) was determined as the difference in mass between the dough and the baked loaves [7].

2.3.2. Texture Profile Analysis of Bread

The texture profile analysis was performed using a TA-XT plus Texture Analyser (TX-700, Lamy Rheology, Champagne au Mont d'Or, France) with a 5 kg load cell to determine the hardness, cohesiveness, and chewiness. As described by Tolve et al. [7], two slices of bread (20 mm × 20 mm) were overlapped and subjected to a double compression using a metal cylindrical probe (25 mm diameter), operating at a speed of 1 mm/s with a deformation percentage of 50%. Hardness was calculated as the maximum force of the 1st cycle of compression, cohesiveness was the ratio of the work of the first compression and the second, and chewiness was a product of hardness, cohesiveness, and springiness, where springiness was the ratio between the distance of the 2nd and 1st compression.

2.3.3. Color Analysis

The color was measured by a reflectance colorimeter (illuminant D65) (Minolta Chroma meter CR-300, Osaka, Japan) based on the color system CIE—L* a* b*. Analyses were performed at five different points within the crumb and the crust area.

2.4. Statistical Analysis

The analyses were carried out in triplicate. All data are expressed as the mean values ± standard deviation. XLSTAT (Version 2406 Build 16.0.17726.20078, Addinsoft, Paris, France) was used for analyzing the data using the analysis of variance (ANOVA) with the post hoc Tukey's test at $p < 0.05$.

3. Results and Discussion

3.1. Predicted Glycemic Index of Breads

Foods are categorized based on their GI as low (<55), medium (56–69), or high (>70) [18]. White bread is commonly used as a reference food, having a GI value of 100. To develop low GI bread, it is crucial to understand to what extent the inclusion of different ingredients affects the bread's pGI. The sole incorporation of chickpea flour leads to a pGI reduction of 35.7% compared to the control (Figure 2) and, according to the proposed classification, allows for obtaining medium pGI loaves. The failure to achieve a low pGI bread using chickpea flour alone can be attributed to the interplay of several factors. The low incorporation level of chickpea flour and the small particle size of the flour (leading to an increased starch digestion rate) compared to those of whole chickpeas may have resulted in the lack of a significant difference in the glycemic response, as previously reported by Johnson et al. [19]. However, it is generally recognized that the chickpea flour's ability to decrease the pGI is because its protein content limits the availability of starch for α-amylase by "encapsulating" it in its structures [20,21]. In addition, legume flours seem to slow carbohydrate absorption due to their content of resistant starch and amylose [22,23]. Previous studies reported that legumes like beans, chickpeas, and lentils moderate the postprandial glucose response, lowering the diet's GI. Similar results were obtained by Goñi and Valentín-Gamazo and Zafar et al. [14,15] in pasta fortified with chickpea flour. The use of red chicory powder alone reduced the pGI by 31.6%. Polyphenols from red chicory powder could reduce starch digestibility in vitro by inhibiting starch hydrolyzing enzymes. Polyphenols can also interact with starch through non covalent interactions during cooking, forming starch complexes with partial enzymatic accessibility [11]. Accordingly, Hanhineva et al. [24] demonstrated that different classes of polyphenols can inhibit the in vitro activities of α-amylase and α-glucosidase. Furthermore, polyphenols can limit the rate and degree of starch gelatinization due to hydrogen bonds between phenol and amylose molecules [25]. The same observation was reported by Rocchetti et al. [26] in their studies on fortified bread with grape pomace.

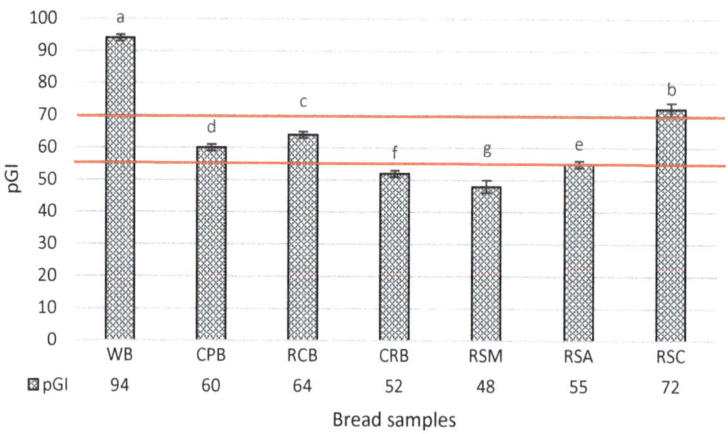

Figure 2. Predicted glycemic index of bread samples. The values are reported below the sample labels. Values with different superscripts are significantly different at $p < 0.05$. pGI: predicted glycemic index; WB: white bread; CPB: bread with 30% chickpea flour; RCB: bread with 10% red chicory powder; CRB: bread with 30% chickpea flour and 10% red chicory powder; RSM: bread with 30% chickpea flour, 10% red chicory powder, and 15% chemically modified tapioca resistant starch; RSA: bread with 30% chickpea flour, 10% red chicory powder, and 15% retrograded tapioca resistant starch; RSC: bread with 30% chickpea flour, 10% red chicory powder, and 15% pregelatinized corn resistant starch. The horizontal lines indicate the values of low GI (up to 55), medium GI (ranging from 56 to 69), and high GI (above 70).

Nevertheless, the inclusion of red chicory powder allows for obtaining a medium pGI bread. The combination of chickpea flour and red chicory powder further reduces the pGI by 44.1% compared to the control, proving the possibility of exploiting the combined effect of two ingredients to reduce the pGI of bread.

It is widely acknowledged that adding resistant starch can lower a bread's glycemic index. However, it is crucial to consider that different types of resistant starch and their various origins can affect pGI differently. The main distinctions arise from the resistant starch origins and production methods, as reviewed by Arp et al. [27]. Raigond et al. [9] point out that RS III could be partially and slowly digested in the small intestine. In contrast, RS IV resists enzymatic hydrolysis both in vitro and in the small intestine, as the typical arrangement of starch chains is altered by several substituents (cross-linking) that bind to amylose and amylopectin. Their presence and the resulting spatial changes in the chain can hinder the action of amylolytic enzymes, preventing their regular activity [28]. Highly cross-linked starches lose their swelling abilities during cooking, making their granules less sensitive to enzymes and thus resistant to amylase hydrolysis [29]. Adding RS III from corn in our experimental bread produced a high pGI sample. Instead, RS III from tapioca performed better than the one from corn, and the obtained RSA samples reached a pGI of 55, the upper limit for considering a food with a low glycemic index. The differences observed between the samples made with retrograded resistant starch from corn and tapioca could be attributed to the higher initial content of resistant starch in tapioca flour compared to corn flour, as reported by Martín Bernabé et al. [30]. This observation could also indicate partial starch gelatinization during baking [9].

The best results regarding glycemic index lowering were achieved with chemically modified resistant starch (RS IV), obtained through cross-linking with chemical reagents. Bread prepared with RS IV had a significantly decreased pGI, and it helped classify the obtained bread as a low predicted glycemic index food. The rate and extent of digestion are influenced by a large number of factors, all interconnected, which complicates the understanding of the resistant nature of starch granules [9].

The lowering of pGI in the RSM sample with added RS IV is likely due to the chemical modifications and cross-linkages that alter the structure and composition of the starch granules, increasing their resistance to amylolytic enzymes. These modifications, occurring during the industrial process to produce RS IV, create spatial changes in the starch chain that interfere with enzyme access, impeding their regular activity [28].

3.2. Bread Technological Properties

The technological properties of control and experimental bread in terms of the moisture content, volume, specific volume, and baking loss (BL) are reported in Table 2. As expected, using different ingredients significantly impacts the technological characteristics of bread.

Table 2. Technological properties of control and experimental bread with chickpea flour, red chicory powder, and resistant starch.

Sample	Moisture Content (%)	Volume (cm^3)	Specific Volume (cm^3/g)	Baking Loss (%)
WB	29.25 ± 1.43 [d]	1977 ± 8 [a]	3.53 ± 0.03 [a]	13.74 ± 0.00 [b]
CPB	55.70 ± 1.41 [a]	1863 ± 2 [c]	3.10 ± 0.09 [c]	12.24 ± 0.14 [c]
RCB	47.27 ± 1.16 [c]	1945 ± 2 [b]	3.15 ± 0.03 [b]	11.89 ± 0.10 [c]
CRB	54.79 ± 0.27 [a]	1635 ± 2 [d]	2.58 ± 0.02 [d]	10.63 ± 0.07 [d]
RSM	47.30 ± 1.02 [c]	1626 ± 4 [e]	2.58 ± 0.08 [d]	14.58 ± 0.38 [a]
RSA	51.15 ± 0.11 [b]	973 ± 2 [g]	1.55 ± 0.05 [f]	14.89 ± 0.11 [a]
RSC	47.78 ± 0.25 [c]	1053 ± 1 [f]	1.66 ± 0.04 [e]	14.47 ± 0.36 [a]

Values with different superscripts within the same column significantly differ at $p < 0.05$. WB: white bread; CPB: bread with 30% chickpea flour; RCB: bread with 10% red chicory powder; CRB: bread with 30% chickpea flour and 10% of red chicory powder; RSM: bread with 30% chickpea flour, 10% red chicory powder, and 15% chemically modified tapioca resistant starch; RSA: bread with 30% chickpea flour, 10% red chicory powder, and 15% retrograded tapioca resistant starch; RSC: bread with 30% chickpea flour, 10% red chicory powder, and 15% pregelatinized corn resistant starch.

Overall, experimental bread showed a significant increase in moisture content compared to the control sample due to the fiber addition from chickpea flour, red chicory powder, and RSs that retain water molecules [31].

Additionally, as expected, the substitution of wheat flour with a high dietary fiber ingredient induces a significant decrease in volume and specific volume compared to the control sample. The loss in volume can be attributed to a reduction in the gluten concentration, causing lower gas retention and a concurrent rise in fiber content. Fiber particles can potentially hinder optimal gluten development by intersecting and disrupting gluten strands, thereby inhibiting the formation of a cohesive viscoelastic network and weakening the dough [32,33]. The interaction between dietary fibers and gluten proteins involves complex mechanisms, including changes in disulfide bonds and the secondary structures of the proteins. Indeed, dietary fibers can impact gluten by affecting the α-helix and β-sheet and influencing high-order intermolecular structures through steric hindrance, hydrogen bonding, hydrophobic interactions, and occasionally electrostatic interactions [34]. Moreover, polyphenols from red chicory powder can weaken the gluten strands through their redox reaction on protein disulfide cross-links, thereby reducing the size of the network. As a further effect, polyphenols might reduce the elasticity and impact the final volume of the dough and bread [35].

Furthermore, Saaed et al. [36] reported that roasted chickpea flour lowered the gluten development of biscuits due to water absorption competition between the chickpea components (namely, starches, proteins, and fibers) and wheat protein, thus causing insufficient hydration of the gluten matrix.

Moreover, the decrease in volume in the new bread formulation could also be ascribed to yeast's inability to utilize RS during fermentation, which slows down the process [37]. The BL values were higher in the newly formulated bread with RSs than in the control and the other samples without resistant starch. The decrease observed in BL may stem from the abundance of RSs that tend to bind water molecules via hydrogen bonding mechanisms [38].

3.3. Texture Profile Analysis

Texture attributes regarding the hardness, cohesiveness, and chewiness are reported in Table 3.

Table 3. Texture analysis of control and experimental bread with chickpea flour, red chicory powder, and resistant starch.

Sample	Hardness (N)	Cohesiveness (N)	Chewiness (N)
WB	5.70 ± 0.97 d	0.85 ± 0.03 a	4.56 ± 0.92 c
CPB	16.17 ± 3.79 c	0.70 ± 0.04 b	11.20 ± 2.71 b
RCB	20.29 ± 3.32 bc	0.63 ± 0.04 b	12.63 ± 2.64 b
CRB	22.05 ± 2.12 b	0.50 ± 0.08 c	10.90 ± 1.9 b
RSM	23.81 ± 1.92 b	0.46 ± 0.06 c	10.90 ± 0.89 b
RSA	21.40 ± 2.23 b	0.46 ± 0.3 c	9.96 ± 1.24 b
RSC	49.62 ± 1.72 a	0.43 ± 0.06 c	19.65 ± 2.42 a

Values with different superscripts within the same column significantly differ at $p < 0.05$. WB: white bread; CPB: bread with 30% chickpea flour; RCB: bread with 10% red chicory powder; CRB: bread with 30% chickpea flour and 10% red chicory powder; RSM: bread with 30% chickpea flour, 10% red chicory powder, and 15% chemically modified tapioca resistant starch; RSA: bread with 30% chickpea flour, 10% red chicory powder, and 15% retrograded tapioca resistant starch; RSC: bread with 30% chickpea flour, 10% red chicory powder, and 15% pregelatinized corn resistant starch.

The hardness of the fortified bread increased significantly compared to the control, likely due to reduced gas retention by the gluten network, resulting in a densely packed bread structure [39]. On the contrary, the cohesiveness, which measures the crumb's internal resistance and its ability to deform before fracturing, decreased in the experimental bread, with the CPB sample exhibiting the highest cohesiveness among the experimental types. A lower cohesiveness indicates a higher susceptibility to rupture, suggesting that fortified bread disintegrates more during chewing and has reduced elasticity and a denser microstructure [40]. A high cohesiveness indicates a better ability to form a bolus rather than crumbling during mastication [34]. The chewiness, reflecting the energy required to masticate food to a ready-to-swallow state, significantly increased with flour substitution. Other studies have also reported that adding dietary fiber significantly increases chewiness in fortified bread [41,42].

3.4. Color Analysis

Table 4 reports the results obtained from color analysis using the CIE-LAB colorimetric system for various samples, with parameters of lightness (L*), red/green hue (a*), and yellow/blue hue (b*), as well as the crust and crumb sides of the bread.

Table 4. Color analysis of control bread and experimental bread with chickpea flour, red chicory powder, and resistant starch and crust side and crumb side of bread samples.

Sample	Crust				Crumb			
	L*	a*	b*		L*	a*	b*	
WB	51.92 ± 2.99 [a]	11.94 ± 0.99 [a]	29.45 ± 2.80 [a]		70.85 ± 2.79 [a]	0.81 ± 0.23 [e]	17.13 ± 1.12 [b]	
CPB	39.22 ± 2.30 [b]	13.67 ± 2.17 [a]	27.61 ± 5.45 [a]		62.00 ± 1.52 [b]	0.84 ± 0.69 [d]	26.13 ± 1.70 [a]	
RCB	31.97 ± 5.10 [c]	12.08 ± 2.58 [a]	15.66 ± 4.47 [bc]		28.63 ± 0.48 [d]	8.52 ± 0.30 [b]	10.58 ± 0.69 [c]	
CRB	29.89 ± 2.90 [cd]	12.34 ± 2.23 [a]	16.76 ± 3.03 [b]		35.88 ± 1.19 [c]	6.23 ± 0.17 [c]	10.4 ± 0.59 [c]	
RSM	26.97 ± 2.28 [d]	12.67 ± 1.75 [a]	11.76 ± 1.87 [c]		26.29 ± 2.00 [d]	8.37 ± 0.20 [b]	9.25 ± 1.60 [cd]	
RSA	32.227 ± 2.93 [c]	12.52 ± 2.00 [a]	15.65 ± 2.56 [bc]		22.72 ± 1.25 [e]	8.47 ± 0.52 [b]	8.49 ± 1.04 [de]	
RSC	34.16 ± 4.55 [c]	12.74 ± 1.67 [a]	15.76 ± 2.61 [bc]		26.30 ± 1.47 [d]	9.56 ± 0.33 [a]	7.45 ± 0.92 [e]	

Values with different superscripts within the same column significantly differ at $p < 0.05$. L*: lightness; a*: redness; b*: yellowness; WB: white bread; CPB: bread with 30% chickpea flour; RCB: bread with 10% red chicory powder; CRB: bread with 30% chickpea flour and 10% red chicory powder; RSM: bread with 30% chickpea flour, 10% red chicory powder, and 15% chemically modified tapioca resistant starch; RSA: bread with 30% chickpea flour, 10% red chicory powder, and 15% retrograded tapioca resistant starch; RSC: bread with 30% chickpea flour, 10% red chicory powder, and 15% pregelatinized corn resistant starch.

The lightness parameter of the crust and crumb significantly decreases in the experimental samples, especially with the substitution with red chicory powder, which changes the color of the samples; the decrease is also visible in the CPB sample, albeit less markedly.

Instead, the a* red index was higher in fortified samples with chicory (RCB, CRB, RSM, RSA, RSC) than in the control due to the high pigmentation of red chicory powder. This redder color is evident at the level of the crumb. On the other hand, the coloration due to the Maillard reaction at the crust level mitigates this effect.

Consequently, for the same reason, the b* yellow index decreases in the crumb and crust in all fortified samples with chicory powder. In the CPB sample, the b* of the crumb increased compared to the control sample due to the typical yellow pigmentation of chickpea flour.

4. Conclusions

The results of this study demonstrate the importance of ingredient selection in producing low-glycemic index bread. Specifically, chickpea flour, red chicory by-product powder, and resistant starch contribute to reducing the glycemic index in bread. The sole inclusion of chickpea flour or red chicory powder reduces the pGI by 35.7% and 31.6%, respectively, while their combination results in a 44.1% reduction. Despite these reductions, the resulting breads still have a medium GI. The use of resistant starch is known to reduce the pGI of foods when incorporated as an ingredient. However, the results highlight how the starch's type and origin significantly affect this reduction. Bread obtained by adding retrograded RS III from corn produced a high pGI sample. In contrast, the incorporation of retrograded RS III from tapioca had a better performance, giving the bread a pGI of 55, the upper limit value for considering a food with a low glycemic index. Chemically modified RS IV from tapioca significantly decreased the bread's pGI, classifying the product as a low glycemic index. The lowering of the pGI using RS IV addition was probably due to chemical modifications and cross-linkages that change the structure and composition of the starch granules, increasing their resistance to amylolytic enzymes. Although all the experimental breads are different regarding their technological and texture properties, the one obtained using the retrograded corn resistant starch has a greater impact on both the technological and texture parameters. In conclusion, chickpea flour, red chicory by-product powder, and resistant starch are intriguing ingredients for producing functional bread fortified with dietary fiber. These ingredient selections could contribute to positive effects on human health for the lower predicted glycemic index and, in the end, enhance the economic value of the red chicory food chain with the reuse of a by-product.

Author Contributions: Conceptualization, I.P. and B.S.; methodology, I.P. and B.S.; software, R.T. and F.B.; validation, R.T. and F.B.; formal analysis, I.P.; writing—original draft preparation, I.P.; writing—review and editing, R.T. and B.S. All authors have read and agreed to the published version of the manuscript.

Funding: This research received no external funding.

Institutional Review Board Statement: Not applicable.

Informed Consent Statement: Not applicable.

Data Availability Statement: The original contributions presented in the study are included in the article, further inquiries can be directed to the corresponding author.

Acknowledgments: This study was carried out within the Interconnected Nord-Est Innovation Ecosystem (iNEST) and received funding from the European Union Next-GenerationEU (PIANO NAZIONALE DI RIPRESA E RESILIENZA (PNRR)—MISSIONE 4 COMPONENTE 2, INVESTIMENTO 1.5—DD 1058 23/06/2022, ECS00000043). This manuscript reflects only the authors' views and opinions; neither the European Union nor the European Commission can be considered responsible for them.

Conflicts of Interest: The authors declare no conflicts of interest.

References

1. Tapia-Hernández, J.A.; Del-Toro-Sánchez, C.L.; Cinco-Moroyoqui, F.J.; Juárez-Onofre, J.E.; Ruiz-Cruz, S.; Carvajal-Millan, E.; López-Ahumada, G.A.; Castro-Enriquez, D.D.; Barreras-Urbina, C.G.; Rodríguez-Felix, F. Prolamins from Cereal By-Products: Classification, Extraction, Characterization and Its Applications in Micro- and Nanofabrication. *Trends Food Sci. Technol.* **2019**, *90*, 111–132. [CrossRef]
2. Kurek, M.A.; Wyrwisz, J.; Karp, S.; Wierzbicka, A. Effect of Fiber Sources on Fatty Acids Profile, Glycemic Index, and Phenolic Compound Content of In Vitro Digested Fortified Wheat Bread. *J. Food Sci. Technol.* **2018**, *55*, 1632–1640. [CrossRef] [PubMed]
3. Papakonstantinou, E.; Oikonomou, C.; Nychas, G.; Dimitriadis, G.D. Effects of diet, lifestyle, chrononutrition and alternative dietary interventions on postprandial glycemia and insulin resistance. *Nutrients* **2022**, *14*, 823. [PubMed]
4. Spina, A.; Brighina, S.; Muccilli, S.; Mazzaglia, A.; Fabroni, S.; Fallico, B.; Rapisarda, P.; Arena, E. Wholegrain Durum Wheat Bread Fortified with Citrus Fibers: Evaluation of Quality Parameters during Long Storage. *Front. Nutr.* **2019**, *6*, 13. [CrossRef] [PubMed]
5. Gómez, M.; Martinez, M.M. Fruit and Vegetable By-Products as Novel Ingredients to Improve the Nutritional Quality of Baked Goods. *Crit. Rev. Food Sci. Nutr.* **2018**, *58*, 2119–2135. [CrossRef] [PubMed]
6. Sivam, A.S.; Sun-Waterhouse, D.; Quek, S.Y.; Perera, C.O. Properties of Bread Dough with Added Fiber Polysaccharides and Phenolic Antioxidants: A Review. *J. Food Sci.* **2010**, *75*, R163–R174. [CrossRef] [PubMed]
7. Tolve, R.; Simonato, B.; Rainero, G.; Bianchi, F.; Rizzi, C.; Cervini, M.; Giuberti, G. Wheat Bread Fortification by Grape Pomace Powder: Nutritional, Technological, Antioxidant, and Sensory Properties. *Foods* **2021**, *10*, 75. [CrossRef] [PubMed]
8. Dikeman, C.L.; Murphy, M.R.; Fahey, G.C. Dietary Fibers Affect Viscosity of Solutions and Simulated Human Gastric and Small Intestinal Digesta. *J. Nutr.* **2006**, *136*, 913–919. [CrossRef] [PubMed]
9. Raigond, P.; Ezekiel, R.; Raigond, B. Resistant Starch in Food: A Review. *J. Sci. Food Agric.* **2015**, *95*, 1968–1978. [CrossRef]

10. Leszczynski, W. Resistant Starch—Classification, Structure, Production. *Pol. J. Food Nutr. Sci.* **2004**, *13*, 37–50.
11. Giuberti, G.; Rocchetti, G.; Lucini, L. Interactions between Phenolic Compounds, Amylolytic Enzymes and Starch: An Updated Overview. *Curr. Opin. Food Sci.* **2020**, *31*, 102–113. [CrossRef]
12. Kamali Rousta, L.; Bodbodak, S.; Nejatian, M.; Ghandehari Yazdi, A.P.; Rafiee, Z.; Xiao, J.; Jafari, S.M. Use of Encapsulation Technology to Enrich and Fortify Bakery, Pasta, and Cereal-Based Products. *Trends Food Sci. Technol.* **2021**, *118*, 688–710. [CrossRef]
13. Bianchi, F.; Tolve, R.; Rainero, G.; Bordiga, M.; Brennan, C.S.; Simonato, B. Technological, Nutritional, and Sensory Properties of Pasta Fortified with Agro-industrial By-products: A Review. *Int. J. Food Sci. Technol.* **2021**, *56*, 4356–4366. [CrossRef]
14. Goñi, I.; Valentín-Gamazo, C. Chickpea Flour Ingredient Slows Glycemic Response to Pasta in Healthy Volunteers. *Food Chem.* **2003**, *81*, 511–515. [CrossRef]
15. Zafar, T.A.; Al-Hassawi, F.; Al-Khulaifi, F.; Al-Rayyes, G.; Waslien, C.; Huffman, F.G. Organoleptic and Glycemic Properties of Chickpea-Wheat Composite Breads. *J. Food Sci. Technol.* **2015**, *52*, 2256–2263. [CrossRef] [PubMed]
16. Granfeldt, Y.; Björck, I.; Drews, A.; Tovar, J. An In Vitro Method Based on Chewing to Predict Metabolic Responses to Starch in Cereal and Legumes Products. *Eur. J. Clin. Nutr.* **1992**, *46*, 649–660. [PubMed]
17. American Association of Cereal Chemists (Ed.) *AACC Approved Method of the AACC*, 10th ed.; American Association of Cereal Chemists: St. Paul, MN, USA, 2000.
18. Atkinson, F.S.; Brand-Miller, J.C.; Foster-Powell, K.; Buyken, A.E.; Goletzke, J. International Tables of Glycemic Index and Glycemic Load Values 2021: A Systematic Review. *Am. J. Clin. Nutr.* **2021**, *114*, 1625–1632. [CrossRef] [PubMed]
19. Johnson, S.K.; Thomas, S.J.; Hall, R.S. Palatability and Glucose, Insulin and Satiety Responses of Chickpea Flour and Extruded Chickpea Flour Bread Eaten as Part of a Breakfast. *Eur. J. Clin. Nutr.* **2005**, *59*, 169–176. [CrossRef] [PubMed]
20. Zięć, G.; Gambuś, H.; Lukasiewicz, M.; Gambuś, F.; Karkoszka, J. Quality and Glycaemic Index of Wheat Bread with Different Shares of Chickpea Flour. *Proceedings* **2021**, *70*, 89. [CrossRef]
21. Fardet, A.; Leenhardt, F.; Lioger, D.; Scalbert, A.; Rémésy, C. Parameters Controlling the Glycaemic Response to Breads. *Nutr. Res. Rev.* **2006**, *19*, 18–25. [CrossRef]
22. Simpson, H.C.R.; Simpson, R.W.; Lousley, S.; Carter, R.D.; Geekie, M.; Hockaday, T.D.R. A high carbohydrate leguminous fibre diet improves all aspects of diabetic control. *Lancet* **1981**, *317*, 1–5. [CrossRef] [PubMed]
23. Stamataki, N.S.; Yanni, A.E.; Karathanos, V.T. Non-Cereal Ingredients for the Attenuation of Glycaemic Response to Bread: A Review of the Clinical Evidence. *Food Funct.* **2016**, *7*, 2926–2936. [CrossRef] [PubMed]
24. Hanhineva, K.; Törrönen, R.; Bondia-Pons, I.; Pekkinen, J.; Kolehmainen, M.; Mykkänen, H.; Poutanen, K. Impact of Dietary Polyphenols on Carbohydrate Metabolism. *Int. J. Mol. Sci.* **2010**, *11*, 1365–1402. [CrossRef] [PubMed]
25. Wang, Y.; Chao, C.; Huang, H.; Wang, S.; Wang, S.; Wang, S.; Copeland, L. Revisiting Mechanisms Underlying Digestion of Starches. *J. Agric. Food Chem.* **2019**, *67*, 8212–8226. [CrossRef] [PubMed]
26. Rocchetti, G.; Rizzi, C.; Cervini, M.; Rainero, G.; Bianchi, F.; Giuberti, G.; Lucini, L.; Simonato, B. Impact of Grape Pomace Powder on the Phenolic Bioaccessibility and on In Vitro Starch Digestibility of Wheat Based Bread. *Foods* **2021**, *10*, 507. [CrossRef] [PubMed]
27. Arp, C.G.; Correa, M.J.; Ferrero, C. Resistant Starches: A Smart Alternative for the Development of Functional Bread and Other Starch-Based Foods. *Food Hydrocoll.* **2021**, *121*, 106949. [CrossRef]
28. Woo, K.S.; Seib, P.A. Cross-Linked Resistant Starch: Preparation and Properties. *Cereal Chem.* **2002**, *79*, 819–825. [CrossRef]
29. Tekin, T.; Fisunoglu, M. A Comprehensive Review Resistant Starch-Containing Bread as a Functional Food: Its Effect on Appetite, Glycemic Index, and Glycemic Response. *Starch/Staerke* **2023**, *75*, 2200291. [CrossRef]
30. Martín Bernabé, A.; Srikaeo, K.; Schlüter, M. Resistant Starch Content, Starch Digestibility and the Fermentation of Some Tropical Starches in Vitro. *Food Dig.* **2011**, *2*, 37–42. [CrossRef]
31. Foschia, M.; Peressini, D.; Sensidoni, A.; Brennan, C.S. The Effects of Dietary Fibre Addition on the Quality of Common Cereal Products. *J. Cereal Sci.* **2013**, *58*, 216–227. [CrossRef]
32. Hemdane, S.; Jacobs, P.J.; Dornez, E.; Verspreet, J.; Delcour, J.A.; Courtin, C.M. Wheat (*Triticum aestivum* L.) Bran in Bread Making: A Critical Review. *Compr. Rev. Food Sci. Food Saf.* **2016**, *15*, 28–42. [CrossRef] [PubMed]
33. Ni, Q.; Ranawana, V.; Hayes, H.E.; Hayward, N.J.; Stead, D.; Raikos, V. Addition of Broad Bean Hull to Wheat Flour for the Development of High-Fiber Bread: Effects on Physical and Nutritional Properties. *Foods* **2020**, *9*, 1192. [CrossRef] [PubMed]
34. Zhou, Y.; Dhital, S.; Zhao, C.; Ye, F.; Chen, J.; Zhao, G. Dietary Fiber-Gluten Protein Interaction in Wheat Flour Dough: Analysis, Consequences and Proposed Mechanisms. *Food Hydrocoll.* **2021**, *111*, 106203. [CrossRef]
35. Girard, A.L.; Awika, J.M. Effects of Edible Plant Polyphenols on Gluten Protein Functionality and Potential Applications of Polyphenol–Gluten Interactions. *Compr. Rev. Food Sci. Food Saf.* **2020**, *19*, 2164–2199. [CrossRef] [PubMed]
36. Saeed, S.M.G.; Ali, S.A.; Naz, J.; Mirza, M.; Elkhadragy, M.F.; Yehia, H.M.; Giuffrè, A.M. Techno-Functional, Antioxidants, Microstructural, and Sensory Characteristics of Biscuits as Affected by Fat Replacer Using Roasted and Germinated Chickpea (*Cicer arietinum* L.). *Int. J. Food Prop.* **2023**, *26*, 2055–2077. [CrossRef]
37. Li, J.; Deng, F.; Han, P.; Ding, Y.; Cao, J. Preparation of Resistant Starch Types III + V with Moderate Amylopullulanase and Its Effects on Bread Properties. *Foods* **2024**, *13*, 1251. [CrossRef] [PubMed]
38. Rosell, C.M.; Rojas, J.A.; Benedito de Barber, C. Influence of Hydrocolloids on Dough Rheology and Bread Quality. *Food Hydrocoll.* **2001**, *15*, 75–81. [CrossRef]

39. Gonzalez, M.; Reyes, I.; Carrera-Tarela, Y.; Vernon-Carter, E.J.; Alvarez-Ramirez, J. Charcoal Bread: Physicochemical and Textural Properties, In Vitro Digestibility, and Dough Rheology. *Int. J. Gastron. Food Sci.* **2020**, *21*, 100227. [CrossRef]
40. Encina-Zelada, C.R.; Cadavez, V.; Monteiro, F.; Teixeira, J.A.; Gonzales-Barron, U. Combined Effect of Xanthan Gum and Water Content on Physicochemical and Textural Properties of Gluten-Free Batter and Bread. *Food Res. Int.* **2018**, *111*, 544–555. [CrossRef]
41. Kowalczewski, P.Ł.; Walkowiak, K.; Masewicz, Ł.; Duda, A.; Poliszko, N.; Różańska, M.B.; Jeżowski, P.; Tomkowiak, A.; Mildner-Szkudlarz, S.; Baranowska, H.M. Wheat Bread Enriched with Raspberry and Strawberry Oilcakes: Effects on Proximate Composition, Texture and Water Properties. *Eur. Food Res. Technol.* **2019**, *245*, 2591–2600. [CrossRef]
42. Tuncel, N.B.; Yilmaz, N.; Kocabiyik, H.; Uygur, A. The Effect of Infrared Stabilized Rice Bran Substitution on Physicochemical and Sensory Properties of Pan Breads: Part I. *J. Cereal Sci.* **2014**, *59*, 155–161. [CrossRef]

MDPI AG
Grosspeteranlage 5
4052 Basel
Switzerland
Tel.: +41 61 683 77 34

Foods Editorial Office
E-mail: foods@mdpi.com
www.mdpi.com/journal/foods

www.ingramcontent.com/pod-product-compliance
Lightning Source LLC
Chambersburg PA
CBHW061626210926
51850CB00009B/2454
9783725868520